D1253426

Electrical Control Systems in Industry

ELECTRICAL CONTROL

Other Books by Charles S. Siskind:

Direct-current Machinery

Electrical Circuits: Direct and Alternating Current

*Electrical Machines: Direct and Alternating Current,
Second Edition*

*Electricity: Direct and Alternating Current,
Second Edition*

Induction Motors: Single Phase and Polyphase

SYSTEMS IN INDUSTRY

By Charles S. Siskind

Associate Professor of Electrical
Engineering, Purdue University

McGraw-Hill Book Company

New York
St. Louis
San Francisco
London
Sydney
Toronto
Mexico
Panama

ELECTRICAL CONTROL SYSTEMS IN INDUSTRY

Preface

Control systems for electric motors have become so vital to the proper performance and protection of modern plant equipment that they are frequently the most essential links in complex industrial applications. These systems may range in extent from the simple practice of merely starting and stopping an electric motor to that of directing the energy flow in a completely automated factory. More often, however, the arrangement may involve one or more of such functions as rapid stopping (braking), reversing, speed changing, travel limits of mechanical equipment (cranes, hoists, machine tools), timing of multimotor drives, and the regulation of current, torque, speed, acceleration, and deceleration. The subject of industrial control, properly regarded as both an art and a science, has become an extremely important and highly developed branch of electrical engineering.

In writing "Electrical Control Systems in Industry," the author has attempted to present the material both logically and progressively, recognizing first, that for classroom instruction, students have, for the most part, limited background and are almost completely unfamiliar with the many kinds of control components and the special "language" of the control engineer. Emphasis has been placed, in the main, on practical circuit analysis and design, although the broad aspect of a subject as extensive as this also involves the design, construction, and operation of the devices that make up the control circuits. In this connection it should be pointed out that the designer generally starts with a given set of specifications which indicate precisely how the system must be made to direct the performance of one or more actuators; he then proceeds to select and interconnect the desired components, mechanically and electrically interlocking and timing certain elements so that, during operation, the system functions to fulfill the necessary requirements.

The control engineer has many "building blocks" at his disposal with which to devise a satisfactory control system. Although most of these are well known and have been standardized, it is frequently necessary to improvise and even develop special models. Those commonly employed are pilot devices, many kinds of switches, relays, contactors, transformers, rectifiers, brakes, protective units, magnetic, electronic, and rotating amplifiers, and many others. Particularly important in most control systems are electromagnetically operated relay-contactor combinations which generally act as direct links between the power source and the motors; moreover, since relays and contactors are, respectively, low-current and high-current devices, considerable amplification often results between the initiating control signal and the power driving the machines. Also, when control systems function automatically, as contrasted with those operated manually, they have the advantages of reliability, flexibility, and convenience in the location and arrangement of the control equipment.

In large automated control systems where many motors are interconnected to drive a long processing or manufacturing line (examples are food handling, linoleum printing, and annealing and tinning of sheet steel), an operator may oversee and supervise the flow of vast quantities of products. In such systems it is generally necessary to coordinate the performance of various parts of the installation by introducing sensing and feedback signals at appropriate places; such signals, properly strengthened by magnetic and rotating amplifiers, regulate such quantities as speed, torque, current, voltage, direction, and position and tend to minimize deviations from desired standards. This interesting aspect of industrial control is treated in Chapter 11 where, after introductory principles are first discussed, an actual practical installation is carefully analyzed.

Except for a few special applications such as *resistance welding, eddy-current brakes,* and *a unique speed-control scheme for d-c motors* (Chapter 10), electronic control has received limited treatment in this book. This is because such systems are not generally suitable for the severe service conditions in industry and involve, moreover, a completely different approach, the use of entirely different kinds of components, and a rather sophisticated kind of circuitry and analysis.

Two completely different types of static-switching circuit have been developed in recent years to provide the control engineer with another powerful tool for the design of complex control systems. Eliminating the need for large numbers of electromagnetic relays with their moving and arcing contacts, one of these utilizes the well-known magnetic amplifier, suitably modified for switching purposes, while the other consists of groups of transistorized components. A number of different schemes are presently available, some employing packaged units, although all make use of one or the other of the devices indicated. Among the advantages possessed by such systems are compact design with considerably reduced space requirements, improved reliability and reduction of service problems, increased speed of operation, extremely low power requirements, and versatility. Since the cost of such systems is generally much higher than comparable relay arrangements, static switching may not replace electromagnetic control for some time to come, except where the advantages listed justify the additional expense. This subject, together with the application of Boolean algebra to static switching systems, is discussed in detail in Chapter 13.

Each of the chapters concludes with a set of questions and, where they help to illustrate control-circuit design and operating practice, a group of problems. An interesting departure from this procedure is the format of Chapter 12 in which a number of circuit-design problems are presented *with their solutions.* For the latter, it is suggested that the student explain their operation and, since there are frequently several ways to satisfy the necessary requirements, propose alternative schemes to fulfill the specifications.

The author is greatly indebted to Dr. J. D. Leitch, vice president, and Mr. A. H. Myles, chief engineer, of the Square D Company for making the excellent facilities of the Cleveland and Milwaukee plants available to him while he was preparing to write this book. He is also grateful to many of his associates and particularly to J. H. Leightty and D. A. Pavek for their specialized assistance while he acted as a consultant for the same company. For carefully reviewing the completed manuscript and suggesting numerous improvements, the author expresses his thanks to H. A. Myles and H. J. Rathbun. Finally, for supplying much useful data and information and many fine photographs, he acknowledges the generosity of such leading control-equipment manufacturers as Allen-Bradley, Cutler-Hammer, and Square D.

CHARLES S. SISKIND

Contents

CHAPTER 1

Electrical Components for Control Circuits

Control-circuit functions. It is significant that electric motors, for the most part, provide the motive power that operates the mechanical machines in modern industrial plants. Moreover, since every motor or group of motors most generally perform several operations, often repeatedly, it is necessary to equip each electrical unit or system with a properly designed controller whose components are sequenced to fulfill, either automatically or manually, the desired control functions.

Although the behavior of many typical installations such as machine tools, pumps, conveyors, air-moving equipment, hoists, cranes, and others is well known and is met by matched characteristics of standard motors, the controllers and circuit arrangements vary rather widely. The reason for the variation is not only that certain results can be accomplished in several ways but also that the number of specified operating combinations is almost unlimited.

Most industrial control systems are either completely automatic or semiautomatic. In the former, an operator may be required merely to initiate one impulse, by the pressing of a button or the movement of a lever, after which the controller assumes and carries out the prearranged functions of the application. In the latter, as in nonrepeating applications, the continued services of an operator are required to initiate each impulse, which, in turn, permits the controller to perform its prescribed functions. In some special installations where automatic equipment may be unsuitable, unnecessary, or unwarranted because of cost, manually operated controllers are employed.

The first requirement of a controller is, of course, to *accelerate the motor* and the application it drives. This must be done as smoothly as possible and without permitting the current inrush to exceed permissible maximum values. Moreover, the period of acceleration, determined by the developed starting torque, must be short enough to prevent high motor temperature and sufficiently long to avoid mechanical shock to the moving parts of the driven machine. The controller and its circuit must also incorporate one or more safety features; these will always include some type of *overload* protection and, where required, must ensure the motor against the possibility of *overspeeding* (open field protection of shunt and compound motors, for example), *phase-reversal* and *open-phase* protection of polyphase *motors*, and such others as *over-travel* and *reversed-current* protection. *Stopping a motor* is obviously important, and this can be accomplished by opening the main switch or circuit breaker, pressing a STOP button to deenergize the control circuit, or permitting a limit switch or other automatic device such as a temperature-sensitive or pressure-sensitive element to disconnect the motor circuit. In some installations it is essential to bring a motor to a quick stop by *braking*, and this may involve an electrical scheme such as *dynamic braking* or *plugging* and/or a brake that is applied mechanically. When the latter is used in conjunction with electrical braking, it is often done to hold a stationary motor (hoist or crane) or to provide an emergency stop. In addition, controllers frequently make provision to *reverse the direction of rotation* of a motor, *control its speed, sequence speed changes* in a multispeed motor and/or motor operation of a multimotor drive, and many others.

It should be clear from what has already been said that control circuits may be comparatively simple or extremely complex, depending upon the way in which the system must perform. In any event the design of the circuit and its components must fulfill the following requirements: (1) Operation must proceed in accordance with prescribed specifications; (2) the circuit must be completely safe and must fail safe; (3) there must be no "sneak" circuits to make the system malfunction; (4) the control system must have reasonably good reliability and continuity under abnormal conditions of operation; (5) the controller must have a minimum of components; (6) all parts and components should be easily serviced, adjusted, and replaced.

Symbols and circuit diagrams. Industrial control-circuit diagrams are generally drawn with symbols that have been standardized by the American Standards Association (ASA). Moreover, since such symbolic representation is rather simple, in contrast to former nonuniform practices of manufacturers who attempted to illustrate devices and machines as they appeared physically, much confusion to users and service personnel is avoided. A further advantage of standardization is that multielement devices can be shown with individual parts properly labeled and conveniently located on the drawing where they can be traced easily. This aspect of complex diagrams will be recognized later when it will be necessary to interpret a control system by "jumping" from one part of a circuit to another as the sequence of operations progresses.

Switches. One group of devices in control systems acts to make, break, or change the connections in electric circuits. Called *switches*, there are two general types of them; those generally operated manually are knife, cam, and drum switches, while those actuated automatically are limit, float, pressure, temperature-operated (bimetallic), plugging, and electron switches. Figure 1·1 illustrates the standard symbols for some of the

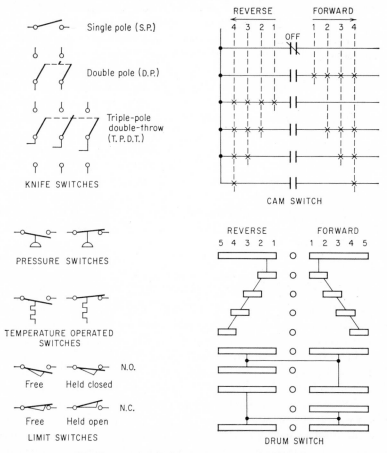

Fig. 1·1 Symbols for several types of switches.

more common of them. In practice, a cam or drum switch usually performs the function of a *master*, under which condition it is designated a *master* switch and dominates the operation of contactors, relays, or other magnetically operated devices. As the sketch in Fig. 1·1 shows, a group of cams of a cam switch are rotated when the master handle is turned from one position to another, and these in turn open and close designated contacts. A drum switch, on the other hand, is one in which electrical contacts are made on segments or surfaces on the periphery of a rotating cylinder or sector or by the operation of a rotating cam.

Protective Devices. As was previously indicated, all electric circuits must be provided with some type of overcurrent protection which, under conditions of extremely high values of current, act almost instantly to disconnect electrical equipment from the power source while for moderate overloads performs its function with some time delay. Several kinds of protective devices are fuses, thermal elements, and air or oil circuit breakers. Many unique schemes have been developed to make such units operate, among which are solder-pot heaters, heated bimetallic strips, electromagnetic constructions, induction heaters, assemblies using special alloys such as *Invar* to produce a lockout action, and others. The most common time-delay (inverse-time) device is, of course, the fuse, which, containing a circuit opening fusible member, is directly heated and destroyed by the passage of overcurrent through it. Figure 1·2 illustrates the standard symbols for several types of protective devices.

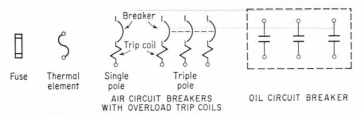

Fuse   Thermal element   Single pole   Triple pole

AIR CIRCUIT BREAKERS
WITH OVERLOAD TRIP COILS

OIL CIRCUIT BREAKER

Fig. 1·2   Symbols for several types of protective devices.

Push Buttons. The operation of automatically controlled electrical systems are generally initiated by pilot devices, the most common of which are *push buttons*. The latter are always actuated mechanically which, in turn, close and/or open auxiliary circuits that eventually operate contactors or other devices in the main power circuits. Two classes of push buttons are available, namely, (1) the momentary contact type which remains in a depressed position when pressure is applied to the button and is released by a spring when pressure is removed and (2) the maintained contact type which, designed with two buttons, has a snap-action mechanism that maintains certain contacts in closed or open positions when one button is pressed and resets the contacts to the original maintained position when the second button is pressed.

Many contact combinations are available in standard push buttons, and when two or more units are "ganged," rather complex operations can be performed simultaneously. Symbols for several of the more common contact arrangements are illustrated in Fig. 1·3. The single bar

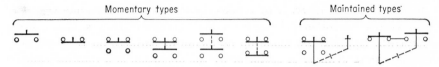

Fig. 1·3 Symbols for several types of push buttons.

with normally open (N.O.) contacts is generally used as START button, while the single bar with normally closed (N.C.) contacts is the customary STOP button. Most so-called heavy-duty buttons for comparatively high current-carrying and current-interrupting service have two sets of contacts, often with one set of N.C. contacts and one set of N.O. contacts. When the button is pressed in such units, the N.C. *back* contacts open *before* the N.O. *front* contacts close. (This feature is particularly important in reversing controllers because it provides safe electrical interlocking.) Nameplates are often fastened to push buttons to indicate their functions (Fig. 1·4), while others have the designated purposes

Fig. 1·4 Cabinet containing push buttons and indicating lights (top row) for a machine-tool application.

molded directly on the face of the buttons. Moreover, several colors are used for identification, with *black* commonly employed for START buttons and *red* for STOP buttons.

Relay and Contactors. When pilot devices are operated in *magnetic* control systems in industry, two types of electromagnets, relays and contactors, are energized or deenergized, and these, in turn, function to apply, remove, or change the power demands of the main machines. The control circuit, consisting of many kinds of carefully arranged and properly sequenced elements is, in this respect, the link between the electric power source and the performing motors, generators, or other heavy equipment. It should be clear, therefore, that, as indicated, relays and contractors are *electromagnetic* devices in the sense that magnetic forces are produced when electric currents are passed through coils of wire; in response to such forces contacts are closed or opened by the motion of plungers or pivoted armatures.

As defined by the National Electrical Manufacturers Association (NEMA), a *relay* is a device that is operative by a variation in the conditions of one electric circuit to effect the operation of other devices in the same or another electric circuit; a *contactor*, on the other hand, is a device for repeatedly establishing and interrupting an *electric power circuit.* It is important to recognize the difference between the two, noting particularly that the relay, serving a *secondary* role, causes other devices to function whereas the contactor is the *primary* unit, doing its work in the main power circuit. Although the distinction, from a manufacturing point of view, is well defined, it should, however, be pointed out that relays are sometimes used as contactors where the power requirements are low enough to be within the capabilities of the small devices whereas contactors have been used on occasion in heavy-current relay circuits. The symbol of the operating coil of relays or contactors is a *circle* with the proper identifying legend placed within or adjacent to it. Figure 1·5 illustrates several such devices with *CR* for

Fig. 1·5   Symbols for several relay and contactor components.

control relay, *UV* for undervoltage relay, *FA* for field-acceleration relay, *M* for main contactor, *DB* for dynamic-braking contactor, and *F* and *R* for forward and reverse contactors respectively.

Timing Relays. One of the more important advantages of automatically controlled circuits is that a sequence of operations can be *timed* with great accuracy. This is accomplished by the use of timing relays, many types of which are available and can be adjusted to give time delays of as little as a fraction of a second to as much as several minutes. More-over, extremely long time delays, up to several hours, are possible with timing relays that are motor-driven. Since most industrial control systems do not run through unattended repetitive cycles, the timing relays for such installations are generally noncyclic. Common designs are pneumatic timers, inductive timers based on the principle of flux decay, capacitive timers that employ unique resistor-capacitor (RC) circuit elements for delayed-action, relays that are timed by dashpots, and many others. For operations that repeat themselves without atten-tion, such as traffic-signal and air-conditioning systems, cyclic timers must be employed; the latter are generally operated by small synchronous motors or are connected in ON-OFF circuits that are thermostatically controlled.

A timing relay can be arranged to _start its timing action_ in one of two ways, namely (1) at the instant the coil of the device is _energized_, i.e., after energization, in which case it is called an ON-DELAY relay, and (2) at the instant the coil of the device is _deenergized_, i.e., after deenergization, under which condition it is designated an OFF-DELAY relay. Associated with either of them may be one or more sets of N.O. contacts, N.C. contacts, combinations of both N.O. and N.C. contacts, and, in addition to these, sets of instantaneously operated contacts. Symbolically, timing relays are represented by the legend TR placed within a circle and, where there may be some doubt as to its function, the designation ON-DELAY or OFF-DELAY outside the circle. The associated contacts must, however, be labeled TR-T.O. for _timing relay–time opening_ or TR-T.C. for _timing relay–time closing_. For example, N.O. contacts that are timed by an ON-DELAY relay would be labeled TR-T.C. while N.C. contacts that are timed by an ON-DELAY relay are designated TR-T.O.; also, N.O. contacts associated with an OFF-DELAY relay are marked TR-T.O. whereas N.C. contacts that are timed by an OFF-DELAY relay are indi-cated by TR-T.C. Figure 1·6 illustrates the symbolic notations described as well as alternate contact designations.

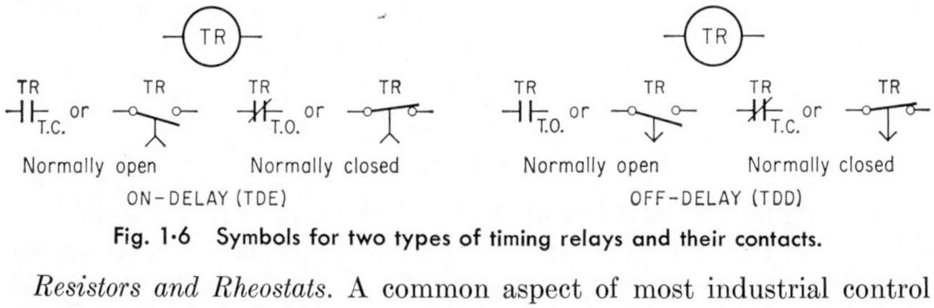

Fig. 1·6 Symbols for two types of timing relays and their contacts.

Resistors and Rheostats. A common aspect of most industrial control circuits is the use of resistors and rheostats for the purpose of limiting current flow during certain operating periods and/or to provide a means of making voltage adjustments. However, since they are incapable of storing energy, as can inductors and capacitors, a current flow through a given resistor always involves an instantaneous conversion of electrical energy into heat or radiant energy which has a value such that the product of the resistance and the square of the current gives the rate of energy conversion.

Metallic resistors, usually ferro-chrome-nickel alloys, are extremely popular because they are mechanically strong and hard, have comparatively high resistivities, resist oxidation and alteration at high temperatures, and have negligibly small temperature-resistance coefficients. Since banks of resistors for most industrial control circuits need not have a particularly high degree of accuracy, with acceptable deviations from calculated ohmic values of as much as ± 10 per cent, these units are not usually manufactured to very close tolerances. For many installations resistor banks are built up of standard grids of varying cross-sectional areas and lengths and are tapped at convenient points. The current-varying capacity of resistors must, of course, be high enough to prevent excessive temperature rise, no greater than 375°C for bare materials, and this generally implies that cross-sectional areas must be selected with regard to heat-dissipating ability. The latter is based not only on actual values of current but also upon the duty cycle of the application.

A rheostat is an adjustable resistor so designed that its resistance can be changed without opening the circuit in which it is connected. Constructed in many ways, it generally has a sliding contact that can be moved by a rotating arm over the bare coiled resistance wire or over a succession of closely spaced studs to which the resistor sections are connected. The principal function of rheostats in industrial control systems is for current adjustment in the field circuits of generators and motors or in the various circuits where they introduce needed artificial voltage drops.

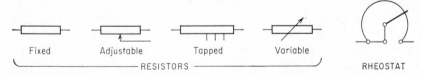

Fixed   Adjustable   Tapped   Variable

RESISTORS   RHEOSTAT

Fig. 1·7   Symbols for resistors and rheostats.

Figure 1·7 illustrates the symbols that have been adopted to represent resistors and rheostats, and Fig. 1·8 shows several types and construction of resistors mounted on a display board.

Rotating Electric Machines. The main power equipments in most electrical systems are the rotating machines, i.e., generators and motors,

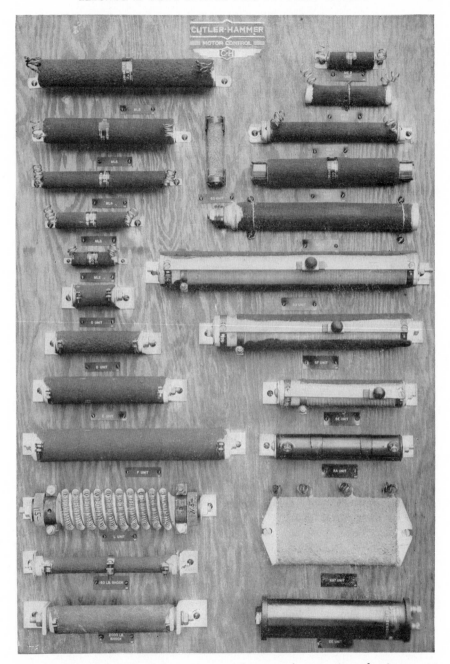

Fig. 1·8 Display board illustrating several types and constructions of resistors.

although small units are used occasionally in control circuits. There are many types of such energy-conversion machines, each of which develops definite operating characteristics and performs either as a motor or as a generator. However, under certain conditions, as when installed on a hoist, machines may have the dual function of acting as a motor when driving the load (during the hoisting period) and as a generator when the load drives the electrical machine (during the lowering or overhauling period). Moreover, d-c motors must not be connected to an a-c source and vice versa, except in the case of d-c/a-c machines (universal type), when they will generally operate satisfactorily on direct current or on alternating current up to about 60 cps.

Schematic diagrams with identifying legends of the most common types of machines are shown in Fig. 1·9. Note particularly that those

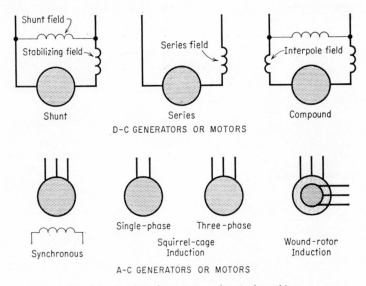

Fig. 1·9   Symbols for rotating electrical machines.

parts carrying the power currents (armature, series, and interpole fields of d-c machines and stator and rotor windings of induction machines) are drawn with heavy lines while the low-current shunt-field windings of d-c and synchronous machines, respectively, are represented by fine lines.

Other Components for Control Circuits. In addition to the foregoing devices and machines, found most frequently in electrical control systems, there are a number of other types of equipment that appear often. These include such components as transformers, rectifiers, capacitors, inductors and reactors, magnetic amplifiers, rotating amplifiers, brakes, synchros, and many devices that perform special operating and protective functions.

Depending upon the service which must be rendered, transformers are constructed with secondaries insulated from primaries (two-winding types) or as autotransformers where a single continuous winding is tapped at convenient points. Important applications of the foregoing are *control transformers*, specially designed to provide a safe, reasonably constant, low-voltage source for relays, contactors, and other electromagnetic devices; *compensators* for polyphase motor starters; transformers that serve with rectifiers to supply direct current from a single-phase or polyphase a-c source; *potential* and *current transformers* used in connection with low-reading instruments to measure comparatively high voltages and currents; and others. Figure 1·10 illustrates how transformers and transformer-rectifier combinations are represented schematically.

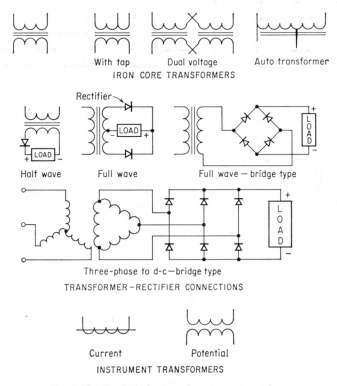

IRON CORE TRANSFORMERS

With tap   Dual voltage   Auto transformer

Half wave   Full wave   Full wave — bridge type

Three-phase to d-c — bridge type

TRANSFORMER – RECTIFIER CONNECTIONS

Current   Potential

INSTRUMENT TRANSFORMERS

Fig. 1·10   Symbols for transformers and rectifiers.

Capacitors perform many important functions in control circuits, several of which are (1) to suppress arcing across relay contacts when the latter are opened in d-c circuits; (2) to minimize voltage ripple, i.e., to "smooth out" the voltage wave in rectified d-c circuits; and (3) to provide time delay for the operation of relays when connected in series with properly selected relays. The symbols for a capacitor as for a reactor (also employed in control circuits for such functions as starting and speed adjustment of motors and in connection with multipurpose saturable reactors), are shown in Fig. 1·11.

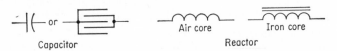

Capacitor   Reactor

Fig. 1·11   Symbols for capacitors and reactors.

Circuit diagrams. The wiring connections showing how the various components of a control system are joined together and to the equipment that is to be operated can be illustrated in several ways. For shop-wiring purposes and for servicing in the field complete connections of the panel must be used. These not only indicate the relative location of the individual units such as transformers, relays, contactors, switches, resistors, reactors, capacitors, overload devices, etc., but have identifying letters and numbers on all leads so that wire tracing and testing can be simplified. Generally, wires that feed the main power circuits are drawn with heavy lines, whereas those leads that connect control-circuit devices and carry comparatively low values of current are shown with fine lines. Another aspect of such diagrams is that a bundle of numbered and lettered control wires is usually shown as a single line that runs from a conveniently located terminal block to emerging points and to the various components through wire channels, conduit, or other means. Figure 1·12, representing a sketch of this type for a high-voltage reactance starter connected to a 3-phase squirrel-cage motor, should be studied with reference to the foregoing remarks and particularly to identify many of the symbols previously given. Explanations of circuits similar to this will be considered subsequently, but for the present its general form should be noted.

Since the complete wiring diagram of the actual installation (Fig. 1·12) is rather complex and somewhat difficult to trace and is used primarily by shop and service men, it is generally supplemented by a schematic sketch or *elementary diagram* that greatly clarifies the operation of the control system. A diagram such as the latter not only is traced much more readily but provides a simple "picture" of the various interconnected components and how they function progressively. It should be pointed out in this connection that, once the pilot device at the START-STOP station (or control panel) initiates a signal, operation of the equipment proceeds automatically in accordance with some prearranged plan.

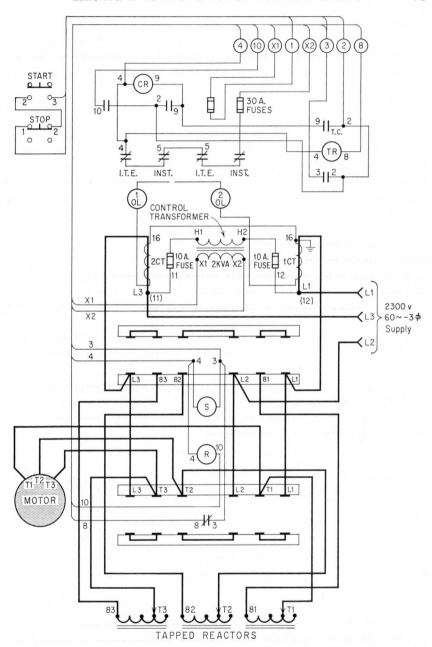

Fig. 1·12 Complete wiring diagram of a reactor starter connected to a 3-phase induction motor.

Moreover, the individual units of the control system are disposed on the diagram to simplify and facilitate the drawing arrangements and are not necessarily placed in the same relative positions as they are actually located on the panel or, for that matter, in individual units. This means that coils and contacts, for example, normally parts of complete assemblies, are often shown to occupy widely separated portions of the schematic and must frequently be traced by "jumping" from one part of the sketch to another as the sequence of actions proceed. The latter is not particularly difficult as, with experience, the student acquires a unique skill in analyzing this type of diagram, one that will hereafter become a standard form in this book. Figure 1·13 illustrates the elementary diagram that corresponds to the complete wiring of Fig. 1·12.

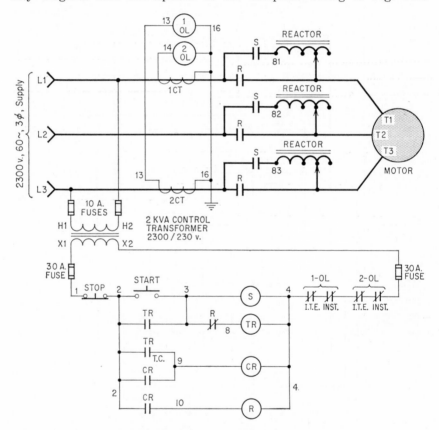

Fig. 1·13 Elementary wiring diagram of a reactor starter connected to a 3-phase induction motor (see Fig. 1·12 for complete wiring diagram).

Although both are exactly similar in so far as actual operation is concerned, note particularly the greatly simplified version of the schematic and the complete absence of detailed complex wiring connections.

There are, of course, other kinds of diagrams that represent special types of equipment and make use of certain standardized symbols and wiring connections peculiar to such components. The latter include grid resistors used in motor circuits for accelerating and speed-control purposes, static switching devices, magnetic and rotating amplifiers, and others. These will be considered in some detail in appropriate portions of this book.

Another aspect of control-circuit diagrams, and Figs. 1·12 and 1·13 illustrate this in part, is the labeling of the many kinds of devices and equipment. Symbols for a great many of the latter have been standardized by NEMA and, with others, will be used in the diagrams that follow. Although most of these are usually designated by the first letters of the various components (*F* for *forward*, *R* for *reverse*, *H* for *hoist*, *L* for *lower*, *TR* for *timing relay*, *S* for *start*, *R* for *run*, *J* for *jog*, *CR* for *control relay*, *P* for *plugging*, and many others) and are obvious to those familiar with control equipment and operations, some abbreviations may not readily identify the proper functions. All symbols will, however, be properly defined and, with use, will soon become a necessary part of the control engineer's language.

Wireless diagrams are occasionally used for complex circuits, primarily to reduce drafting time and, as is sometimes claimed, to save wiring time. The devices are generally outlined and shown in their proper relative positions on the drawing and are labeled and given terminal markings similar to those used on conventional diagrams. With all connecting lines omitted, points that are normally connected are given like numbers and a table is made to indicate how the several components are interconnected. Thus, for example, if a number 4 in one column is shown opposite a *2CR*, *3CR*, *S*, *R*, *TR*, and *UV*, the wireman will know that the six devices indicated are to be connected.

Single-line diagrams are frequently employed to represent complex installations in which a great many machines, regulating equipments, static devices, and numerous components are arranged to function as a system. The various parts are generally shown connected together by single lines that indicate how the potentials are applied and the currents flow to provide for system stability. Figure 1·14, for example, illustrates how motors, generators, static equipment (rectifiers), and voltage and current regulators are interconnected for an installation in which 52-in. hot-rolled sheet steel is slitted as it passes continuously and at high speed from a mandrel at one end of a long line to a windup reel at the other. Although no attempt is made to show actual wiring connections, the diagram does indicate rather clearly how the material is processed and the way in which properly identified parts are related. Note particularly that the various motors that drive the line are *electrically* interlocked by voltage and current regulators that act on the shunt fields of the

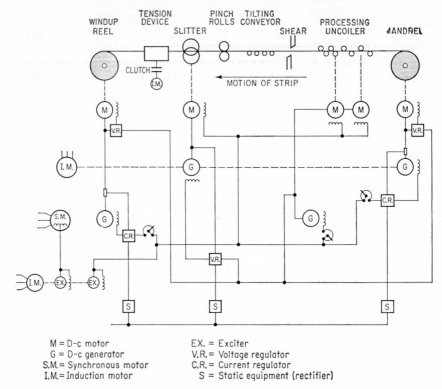

M = D-c motor
G = D-c generator
S.M.= Synchronous motor
I.M.= Induction motor

EX. = Exciter
V.R.= Voltage regulator
C.R.= Current regulator
S = Static equipment (rectifier)

Fig. 1·14 One-line diagram of a slitter steel-strip processing line.

generators and motors; the steel strip is thus maintained at proper tensions as it passes at constant speed along the entire length of the line.

QUESTIONS

1. Specify several important functions that are frequently performed by control circuits, and list a number of requirements such circuits must possess if satisfactory operation is to exist.
2. Name several kinds of *pilot device* normally employed in electric circuits to initiate control functions.
3. Distinguish between *cam-* and *drum*-type master switches with regard to mode of operation.
4. What purpose is served by an overcurrent protective device in a control circuit? What damage is likely to result if proper overcurrent protection is omitted in a control circuit?
5. Distinguish between the control functions of relays and contactors.
6. Why are timing relays often indispensable in industrial control systems that are automatically controlled? List several types of timing relays.
7. Distinguish between ON-DELAY and OFF-DELAY timing relays and indicate how these operate with normally open (N.O.) and normally closed (N.C.) contacts.

8. List a number of functions that are frequently performed by *resistors* and *rheostats* in control circuits. What important properties must such units possess if they are to provide satisfactory service?

9. Make a list of the various types of d-c and a-c energy convertor, indicating particularly the several kinds of d-c motor and polyphase and single-phase a-c motor.

10. Distinguish between capacitors and inductors, current and potential trans-formers, transformers used in power circuits and control circuits, rotating and magnetic amplifiers, mechanical and electrical brakes.

CHAPTER 2

Manual Starters for

D-C and A-C Motors

Types of d-c motors and their fields. There are three general types of d-c motor, and these differ from one another only by the way in which the machine is excited. In one of these, the *shunt motor*, excitation is provided by a winding, made up of as many coils as poles, that is connected in parallel with the armature. The *series motor*, on the other hand, receives its excitation from a winding that is in series with the armature and carries load current. Finally, when excitation results from the combined action of both a shunted field winding and a series field winding, the machine is called a *compound motor*.

Because d-c motors are generally (though not always) operated from a constant-potential source, the shunt field, normally connected across the armature that receives full line voltage, creates a magnetic field that is essentially constant in magnitude. Moreover, since the shunt-field winding receives a current that is completely independent of the armature or load current, it is usually designed to develop its required ampere-turns by having a comparatively large number of turns of fine wire. Its resistance is, therefore, made high enough to limit the shunt-field current to about 1 to 4 per cent of rated motor line current. In contrast, the current in the series field of a series or compound motor is the rather high load current. The series field is, for this reason, constructed with few turns of rather heavy wire so that it not only will be capable of passing large currents but will, because the resistance is low, incur a low voltage drop. Figure 1·9 illustrates the schematic wiring connections of the three types of motor, with the series field of the compound motor shown connected to carry the armature current (the long-shunt connection). In the short-shunt connection sometimes used, the shunt field is directly in parallel with the armature, in which case the series-field current is the line current.

In addition to the foregoing three kinds of fields, properly called *working fields*, modern motors are usually equipped with one or more types of field windings that exercise a corrective influence upon the operation of the motor under load. Called *corrective fields*, they tend to offset such detrimental effects of armature reaction as poor commutation, motor instability at high speeds, and commutator flashover under conditions of suddenly applied overloads. The most widely used corrective field is the *interpole* and its winding that is permanently connected in series in the armature circuit; this field maintains the magnetic neutrals in the same positions under all load conditions and thereby permits the motor to commutate well, i.e., without sparking at the brushes. The *stabilizing field*, applied only to shunt motors that must be made to operate at high speeds by shunt-field weakening, is a light series field placed directly over the shunt winding whose moderate flux tends to prevent runaway operation or instability that may result from the demagnetizing effect of armature reaction. A third type of corrective field is the *compensating winding* placed in slots or holes in the main pole-face cores. Also connected in series in the armature circuit, this winding creates a magnetic field that tends to offset that effect of armature reaction which acts to *distort* the flux-density distribution under the pole faces; left uncorrected, such flux distortion tends to encourage flashover between brushes under conditions of suddenly applied overloads.

It should be understood, although this matter will be considered in some detail subsequently, that the three general types of d-c motor develop completely different and desirable speed-load and torque-load characteristics. The shunt motor, for example, operates at nearly constant speed over its normal-load range, has a definite stable no-load speed, and is particularly adapted to wide speed adjustments; it suffers from the disadvantage, however, that it has limited starting and overload-torque capability. The series motor, on the other hand, has excellent starting and overload-torque characteristics, although speed regulation is poor and high runaway speeds at light loads are possible; moreover, speed adjustment of the motor is a little difficult. The compound motor, being a sort of combination shunt-series machine, behaves somewhat better than the shunt type from the standpoint of starting and overload torque, has a definite stable no-load speed, and tends to change speed by as much as 25 per cent between full load and no load; the speed of this motor is as readily adjusted as is the shunt type of machine.

Principles of d-c motor acceleration. All d-c motors (and this applies equally to a-c motors) are simultaneously generators while they are motoring. This implies that two fundamental actions—*motor action* and *generator action*—take place at the same time *while the armature of a motor is rotating* as it develops torque. Obviously, motor action predominates under this condition because electrical energy, in contrast to mechanical energy, drives the armature, and this means, of course, that, in the presence of magnetic flux created by the field, armature current is delivered by the source. Thus, fulfilling the principle of motor action, which states that rotational torque is developed when current-carrying conductors are properly disposed in an existing magnetic field, the armature revolves. But in doing so, the very conductors that experience force actions cut magnetic flux and therefore produce generator action. Now then, since the *generated emf* in the armature, always less than the *impressed voltage, opposes* the latter, it should be clear that the magnitude of the armature current will depend upon the *difference* between the two voltages indicated. In fact, it is the generated voltage, properly called a *counter emf*, which exercises a limiting effect upon the armature current and causes the motor to adjust its speed and torque automatically to the varying demands of the load. Summarizing the foregoing it should therefore be understood that, by Ohm's law, the armature current I_A is directly proportional to the *difference* between the impressed armature voltage V_A and the cemf E_c and inversely proportional to the armature-circuit resistance R_A; thus, $I_A = (V_A - E_c)/R_A$. Moreover, since the cemf E_c—a generated voltage—depends upon the rate at which flux is cut, or specifically, the product of flux ϕ and speed S, the following basic equation can be written:

$$I_A = \frac{V_A - k\phi S}{R_A} \tag{2·1}$$

where k is a constant of proportionality.

At the instant a motor is started, the cemf E_c is, of course, zero because the armature is not revolving. This means that some external resistance must be inserted in series with the low armature-winding resistance to offset the lack of cemf if excessive values of armature current are to be avoided. As the motor accelerates, the so-called *accelerating resistance* may be cut out (or short-circuited) in steps, and when the resistance is entirely removed, the armature of the motor is connected directly across the line and is running at full speed.

In practice the accelerating resistance is generally cut out (or short-circuited) in several steps, each of which is chosen so that maximum current or torque peaks during acceleration are adjusted on the basis of good commutation and torque requirements of the driven mechanical load. Usually, however, the value of the total accelerating resistance is selected to permit the armature to take about 125 to 175 per cent of rated current so that the motor will be capable of starting under load. Moreover, when current peaks are properly limited as successive sections of the accelerating resistance are cut out, line-voltage disturbances, i.e., voltage dips, are avoided. Also, in the case of certain applications such as hoists, elevators, and trolley or bridge motions on cranes, jerky acceleration, i.e., "jack-rabbiting," must be avoided, and this means that acceleration resistances must be carefully designed to provide a minimum of torque variation as the load is brought up to speed. Figure 2·1 illustrates

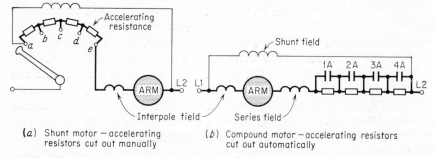

(a) Shunt motor — accelerating   (b) Compound motor — accelerating resistors
      resistors cut out manually   cut out automatically

Fig. 2·1   Sketches illustrating how accelerating resistors are connected in armature circuits of shunt and compound motors.

how accelerating resistances are connected in the armature circuits of shunt and compound (long-shunt) motors; a sketch similar to that of Fig. 2·1b without the shunt field would apply to a series motor.

Example 1. The armature of a 75-hp 230-volt 275-amp shunt motor has a resistance of 8 per cent of V/I. If the inrush current to the armature is to be limited to 150 per cent of rated line value, (**a**) calculate the total resistance of the accelerating resistor. (**b**) What approximate inrush current can be expected if the motor is started directly from the 230-volt source without accelerating resistance?

Solution

a. $R_{\text{accel}} = \dfrac{230}{1.5 \times 275} - 0.08 \left(\dfrac{230}{275}\right) = 0.557 - 0.067 = 0.49$ ohm

b. $I_{\text{inrush}} = \dfrac{230}{0.067} = 3{,}440$ amp

In the foregoing example, it should be clear that the inrush current would be extremely high, about 12.5 times full-load value, if the motor were started without a limiting resistance in the armature circuit.

Actually, however, the inrush would be somewhat less than the calculated value because the armature inductance would delay current buildup slightly, by which time the armature would be rotating and developing some cemf.

To extend further this discussion of the behavior of a motor during the accelerating period, it will be desirable to consider graphically the operating characteristics of a shunt motor. Referring to Fig. 2·1a, which schematically represents a five-step manual starter (four accelerating resistors), assume that the inrush current peaks, five of them, will be limited to $1.5I_A$, where I_A is the full-load armature current, and that transition, i.e., cutting out resistance steps, always occurs when the current drops to I_A as the motor accelerates. At the instant the starter arm is moved to position a, there is an initial current inrush of $1.5I_A$. The motor then accelerates along line aa' of Fig. 2·2a and would theoretically continue to 100 per cent speed if the mechanical load were zero. However, when the current drops to I_A, the arm is moved to point b on the starter (Fig. 2·1a); the armature current again rises instantly to $1.5I_A$, and the motor accelerates along curve bb'. At point b' (Fig. 2·2a) when the current again reaches a value of I_A, the arm is moved to point c (Fig. 2·1a) with a repetition of the acceleration along curve cc'. Finally, when the arm has reached the last position, i.e., point e, and all the accelerating resistance is cut out, the motor is operating normally with the armature connected directly across the power source. The same sequence of current change and motor acceleration would occur in an automatically operated starter of the proper design, where, in Fig. 2·1b, contacts $1A$, $2A$, $3A$, and $4A$ would close in sequence and short-circuit successive sections of the accelerating resistor.

While the current-speed changes are taking place as explained, the current-time variations follow the step pattern of Fig. 2·2b. It will be noted in this diagram that the current tends to level off to a value of I_A after each of the five inrush peaks of $1.5I_A$, assuming, of course, that the motor must develop rated torque after each accelerating period is completed; the motor must obviously develop *more* than rated torque while it is accelerating between a current peak and a transition point.

The speed-time graph of Fig. 2·2c illustrates how the motor accelerates gradually to full speed, i.e., without sudden and violent changes. The latter point is particularly important in certain applications where smooth acceleration is accomplished by using starters with many steps. In other installations where extremely rapid speedup is either desirable or permissible and where high current inrushes are not particularly objectionable, comparatively few acceleration steps may be used. In fact, small d-c motors having ratings up to 2 or 3 hp, and in special installations as high as 7.5 hp, are sometimes started by being connected directly to the line; in such cases the high initial currents are of short duration and the motors and coupled loads are capable of withstanding the shock of high torques suddenly applied.

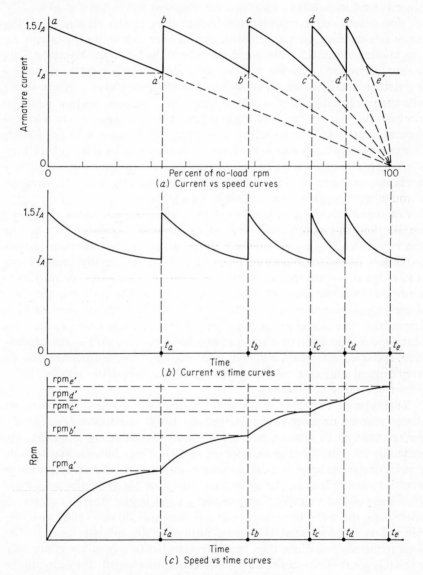

Fig. 2·2   Shunt-motor acceleration.

Types of a-c motors. Unlike d-c motors, whose armatures must be equipped with commutators and brushes to change the impressed direct to alternating current, a-c motors (except for some very special constructions) have no commutators and armature windings in slotted armature cores that are fed directly from the power source. Moreover, since the a-c impedance of a stator winding that surrounds a good quality of magnetic steel is considerably higher than is the resistance of the armature winding of a d-c motor, inrush currents to the former, with the application of full voltage, is very much less than in the latter. This means, therefore, that a-c motors need not, for the most part, be restricted to the usual current-limiting starting procedures of d-c motors, although reduced-voltage methods are frequently used, not so much to protect the motors against high inrush currents and severe electromagnetic stresses as to minimize line-voltage disturbances. In particular, it can be said that initial surge currents to a-c motors with full-voltage starting rarely exceeds six to seven times rated value, and such high currents not only are of short duration but pass directly into the windings through wired connections, not sliding contacts.

Single-phase Motors. Because of the varying requirements of numerous applications in industry, on the farm, in business establishments, and in the home, many types of single-phase motors have been developed. Such machines, manufactured in very large numbers, are usually small in size and range in output from a fraction of one horsepower to several horsepower. Depending upon how and where they are to be used, they are designed and constructed in many ways to develop one or more of such important characteristics as high starting torque, quiet operation, constant speed, operation on alternating or direct current with equal satisfaction, speed adjustability, and others. Moreover, these motors are usually very rugged and can be expected to deliver dependable service under severe operating conditions.

The type of motor that performs with about equal satisfaction on direct current or alternating current up to 60 cps is the familiar d-c *series motor.* Such motors, with or without compensating windings, are generally built in small sizes, operate at rather high speeds, and include special design features so that commutation and armature-reactance difficulties are minimized. An important aspect of these reliable motors is that the ratio of weight to horsepower output is low (this is because of their high speeds of operation, often as much as 30,000 rpm); also significant is the fact that the speed automatically adjusts itself to the load requirements. Since they can be connected to any of the commonly available sources of supply, direct or alternating current, they are appropriately called *universal motors.*

The induction principle is responsible for the operation of several types of single-phase motor. The principle involves the creation of a revolving magnetic field, several methods having been developed for doing this in single-phase motors. An extremely popular small motor, making use of one revolving-field scheme, is the shaded-pole type, used for such applications as the operation of fans, small valves, turntables, and other similar devices. Although its efficiency, power factor, starting torque, and overload capacity are low, it does possess such advantages as low cost, reliability, and ease of speed adjustment.

One of the most widely used of small single-phase motors is the so-called split-phase type, several designs of which are available. The *standard* split-phase motor, employed in such applications as refrigerators, washing machines, and small machine tools, develops 150 per cent (or more) starting torque (i.e., 1.5 times full-load torque), has good overload capacity, and is reasonably quiet running and trouble-free. When starting torques in excess of 150 per cent are needed for pumps, certain farm equipment such as hammer mills, and the like, the *capacitor-start* split-phase motor is available. Also, for extremely quiet operation the *capacitor* split-phase motor is employed, although in this design the starting torque is somewhat limited. An added advantage of this motor, and one that will be considered in some detail in a later chapter, is that it lends itself readily to speed adjustment. For applications requiring both high starting torque and quiet operation the *two-value capacitor* split-phase motor is a first choice; it develops characteristics that are rather similar to the two-phase induction motor.

Repulsion, Repulsion-induction, and Repulsion-start Motors. These motors are other types of single-phase machines that operate on the induction principle. Although they develop high values of starting torque, they are somewhat more expensive and less trouble-free than the capacitor-start or two-value capacitor split-phase designs. For these reasons they have been largely replaced by the latter machines except in applications that require *sustained* high starting torque or a high degree of speed adjustment; the latter characteristics are supplied, respectively, by the repulsion-start and repulsion motors.

Synchronous Motors. As the name implies, these operate at synchronous speed under all conditions of load. There are several constructions of such machines, although they are usually manufactured in very small ratings. Depending upon the way in which they are made or their principle of operation, they have special names such as *reluctance* motors, *sub-synchronous-reluctance* motors, and *hysteresis* motors. They are used occasionally in control systems to provide accurately timed impulses and particularly when the time delays are extremely long or repetitive.

Polyphase Motors. Two general classifications of polyphase (3- and 2-phase) motor that are widely employed are the *induction* and *synchronous* types. The former are manufactured in sizes ranging from the fractionals (less than 1 hp) to those developing outputs of many thousand horsepower. Synchronous motors, on the other hand, are generally used in applications requiring ratings of 50 hp or more, although a special *synchronous-induction* motor has recently been developed that is built in sizes to 75 hp, operating at synchronous speed up to loads somewhat beyond normal rating, after which, on heavier loads, it runs as an induction motor at below-synchronous speed.

Except for certain special designs and constructions, all polyphase motors have stator cores and windings that are similar and, when excited from 3- or 2-phase power sources, create the synchronously revolving fields necessary for motor action. The distinguishing difference between synchronous and induction motors is, however, the rotor construction and the fact that the former is a doubly fed machine in the sense that a winding on the rotor is excited from a d-c source. Concerning rotor constructions it can be stated briefly that a squirrel-cage or core-winding arrangement similar to that of the stator is used in the induction motor whereas a salient-pole and winding configuration is employed on the synchronous machine. Operationally, the synchronous motor runs at an average *constant* speed and can have its power factor adjusted by simple d-c field control; the induction motor, on the other hand, generally operates below synchronous speed, slipping more behind synchronism with increasing loads, and performs at lagging power factors that depend upon design features.

Principles of polyphase induction-motor acceleration. As previously indicated, the stators of the two general kinds of polyphase induction motor are similar, but as the names imply, the squirrel-cage and the wound-rotor constructions of the revolving sections differ widely. The squirrel cage is, of course, much simpler and less costly to manufacture than is the wound rotor but, in operation, suffers from the disadvantage that speed adjustments are somewhat more difficult than in its counterpart. Moreover, wound-rotor motors lend themselves to starting-torque control, a procedure not possible with the squirrel-cage design, the resistance of which is fixed. These subjects will be considered subsequently when control methods and problems are discussed in some detail.

Squirrel-cage Motors. An important and interesting aspect of induction motors with squirrel-cage rotors is that their construction, involving such features as resistance, shape of rotor bars, kind of rotor-bar material, whether one or two squirrel-cages are employed, cross section of end rings, and others, largely affects the starting torque and operating characteristics. Also, the type of squirrel-cage construction, designated by NEMA classes A, B, C, D, and F will generally indicate whether line starting or reduced-voltage starting is permissible or satisfactory. These matters will also be treated fully in a later chapter, but for the present it will be desirable to introduce the subject in a rather general way.

Since the squirrel cage of an induction motor acts like the secondary of a short-circuited transformer at the instant of starting, the *equivalent motor impedance*, i.e., stator and rotor, is rather low. Comparatively, it is, however, somewhat larger than is the armature resistance of a d-c motor. In fact, if this kind of motor is started by connecting it directly to the polyphase source, i.e., without some current-limiting means, it will take an inrush current of approximately 4.5 to 6.5 times rated value. This is usually not particularly excessive, assuming, of course, that the motor accelerates quickly enough to limit heating. Then as the rotor gains speed, a cemf is developed in the stator which, as in the d-c motor, causes the stator and rotor currents to drop to values capable of sustaining sufficient load torque. *Line start* motors of this type are very widely employed, requiring either a manually operated switch or a push-button START-STOP station with an automatic starter for this purpose. Figure 2·3 shows a schematic diagram of such a line-start arrangement

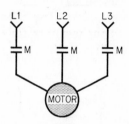

Fig. 2·3 Three-phase motor connected for line starting.

(with control circuit omitted) and indicates merely that the motor will start and accelerate to full speed when the *M* (*main*) contacts close.

If it is desired to limit the starting currents to lower values than those indicated above, one of four schemes may be employed. In each case less than rated voltage is applied to the motor terminals during a major part of the accelerating period, under which condition the inrush current is reduced in proportion to the voltage reduction. However, since the starting torque is proportional to the *square* of the initially applied voltage, the motor is likely to accelerate slowly at the reduced voltage, particularly when heavy or high-inertia loads must be started. It is imperative, therefore, that starters be selected with due regard to loading requirements, since <u>long accelerating periods tend to cause excessive motor heating</u>, especially when these are repeated frequently.

Four types of *reduced-voltage starters* commonly employed for polyphase squirrel-cage motors involve the following principles: (1) the insertion of line resistors to introduce artificial voltage drops, (2) the insertion of line reactance to introduce artificial voltage drops, (3) the use of special stepdown autotransformers (called compensators), (4) connecting the stator winding in *star* (**Y**) during the accelerating period and reconnecting the winding in *delta* after the motor comes up to nearly full speed.

Simplified schematic diagrams of the foregoing four reduced-voltage starting methods for 3-phase squirrel-cage motors (without control circuits) are illustrated by Fig. 2·4. The sketches of Fig. 2·4a and b merely

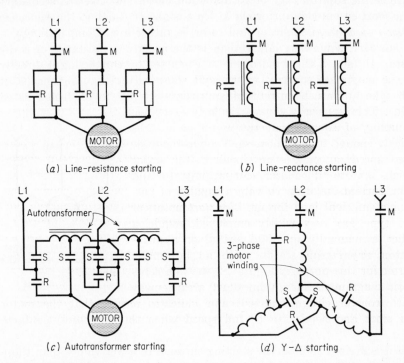

Fig. 2·4   Sketches illustrating four reduced-voltage starting methods for 3-phase squirrel-cage induction motors.

indicate, respectively, that resistors and reactors incur voltage drops during the accelerating period, when the M (*main*) contacts only are closed, and thus permit the motor to start under reduced-voltage conditions. After the motor reaches full or nearly full speed, the R (*run*) contacts close to short-circuit the resistors or reactors when a contactor is energized, under which condition normal operation continues. Figure 2·4c represents a typical autotransformer (compensator) starting arrangement in which the customary open-delta (V) connection is employed. In the actual compensator the two windings shown are placed on the two outside legs of a three-legged core and reduced-voltage taps are brought out at the 50, 65, and 85 per cent points. As indicated, when the S (*start*) contacts close, the motor starts on reduced voltage, determined by the tap selected, and after the motor accelerates to full or nearly full speed, the S contacts open and the R contacts close. When this happens, the compensator is completely disconnected from the circuit and the

motor runs properly from the full-voltage source. The Y-delta starting
method is interesting and economical in that no external equipment such
as resistors, reactors, or transformers are needed to provide each phase
with reduced voltage during the accelerating period. However, when this
scheme is employed, it is obviously necessary that the motor be designed
to run normally when the three winding phases are connected in *delta*.
Thus, in Fig. 2·4d, when the *M* and *S* contacts close, the motor starts
star-connected with about 58 per cent voltage $(V/\sqrt{3} = 0.58V)$ across
each phase; then, after reaching full speed, the *S* contacts are opened
first, after which the *R* contacts close to connect the three winding phases
in delta for full-voltage operation.

Wound-rotor Motors. The ohmic value of the rotor resistance of a polyphase induction motor greatly affects three important operating characteristics; these are (1) the starting torque, (2) the starting current, and
(3) the speed under varying load conditions. This means, of course, that,
for a given impressed voltage, the starting torque, starting current, and
full-load speed of a squirrel-cage motor *cannot* be adjusted because the
rotor resistance is fixed. For most applications these limitations are not
particularly serious, although there are others whose motors must lend
themselves to adjustment of one or more of the characteristics indicated.
Induction motors with phase-wound rotors fulfill these requirements.

A simplified diagram showing how external resistors are connected in
the rotor circuit of a wound-rotor motor is illustrated by Fig. 2·5. In

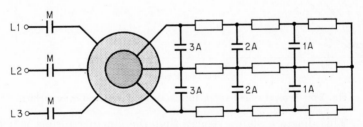

Fig. 2·5   Sketch illustrating wiring connections for a wound-rotor induction motor.

practice the total external resistance is selected on the basis of the desired
starting torque and the maximum permissible starting current, while
the number of sections or steps, beyond the minimum specified by NEMA,
is selected on the basis of the number of speed points (when speed adjustment is necessary) and an acceptable smoothness of acceleration. It
should further be pointed out that, the higher the external resistance,
the lower will be the inrush current. This should be clear because the
rotor circuit is, in reality, the secondary of a transformer whose total
impedance affects the magnitude of not only the current in that circuit

but, by transformer action, the current taken by the stator, i.e., the primary.

To illustrate graphically how the external resistance in the rotor circuit affects the starting torque and full-load speed, Fig. 2·6 has been

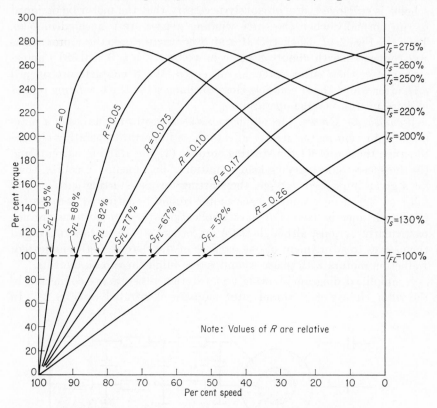

Fig. 2·6 Typical speed-versus-torque curves for wound-rotor motors.

drawn. This diagram shows clearly that the starting torque is a minimum (about 130 per cent) when the external resistance is zero, rises to a maximum (about 275 per cent) for a comparative external resistance of 0.10, and then drops to about 250 and 200 per cent, respectively, as the relative resistances increase further to 0.17 and 0.26. Furthermore, note particularly that the speeds for full-load torques diminish progressively from about 95 to 52 per cent of synchronous speed as the external rotor resistance is increased from zero to 0.26; this definitely indicates that the speed of a phase-wound motor can be readily adjusted by external rotor-resistance control.

Part-winding Starting. The windings of 3-phase squirrel-cage motors are frequently constructed to operate at either of two standard voltage sources, e.g., 220 and 440 volts. This is accomplished by dividing the winding into two identical sections, each one of which is capable of creating the proper number of pole-flux patterns when energized by the lower voltage supply. In such dual-voltage machines nine leads are brought out and are interconnected, so that the winding is in the form of a series star for 440 volts and a two-parallel star for 220 volts. A schematic diagram showing how the terminal leads are marked in accordance with NEMA standards is given in Fig. 2·7.

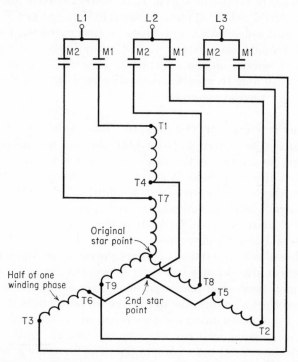

Fig. 2·7 Standard 220/440-volt squirrel-cage induction motor connected for *part-winding* starting.

If the motor is to be connected for *part-winding* starting, terminals $T4$, $T5$, and $T6$ are first joined to make a second star point (the first permanent star point is made originally, as shown, by connecting the far ends corresponding to terminals $T7$, $T8$, and $T9$), and normal operation results on 220 volts when the two sections of the winding are in parallel, i.e., two-parallel star. The starter is then arranged to connect $T1$, $T2$, and $T3$, representing one section, to the line through the $M1$ contacts when a START button at the START-STOP station is pressed; then after a short interval provided by an ON-DELAY timing relay, a second set of contacts $M2$ connect $T7$, $T8$, and $T9$ to the line so that both sections of the winding are joined in parallel. An accelerating procedure such as this usually limits the inrush currents to about 60 to 70 per cent of the locked-rotor value with both sections paralleled and causes the motor to develop a starting torque that is somewhat less than 50 per cent. A serious objection to this method of starting standard 220/440-volt motors is that torque dips appear in the speed-torque curve, irregularities that may, under some loading conditions, prevent a machine from accelerating sufficiently on the first step. It is, therefore, recommended that this scheme should not be applied to stock motors unless the latter are expected to start unloaded or, with limited inrush currents, the machines are actually brought up to speed on the second step. For motors that have received special design treatment, however, the part-winding method has been able to provide reasonably good starting service without the need for accessory equipment.

Faceplate starters for d-c motors. Two widely used manually operated starters and controllers for both d-c and a-c motors are the faceplate and drum types; the distinction between starters and controllers is that the former merely permits the operator to accelerate the motor while the latter incorporates, in addition, such other functions as speed control, electrical braking, and reversing.

Two general constructions of faceplate starter, *three-point* and *four-point*, are employed for shunt and compound motors; the terms indicated refer to the number of wiring terminals that are provided for connection to the motor and the source. In such starters a steel operating arm is fitted with a spring-loaded brass shoe that wipes over a set of contact studs as the arm is moved clockwise from the OFF to the ON position; in doing so (see Fig. 2·1a) a set of accelerating resistors is cut out as the motor comes up to speed. To minimize arcing at the contacts and to avoid a too rapid motion of the lever arm, more contact points or steps are generally used on manual starters than with carefully adjusted and regulated automatic devices; also, the last contact stud is made larger than the others because it carries current continuously. For the smaller sizes, the resistance material is usually iron wire or some ferro-chrome-nickel alloy that resists oxidation at high temperatures, while heavy cast-iron or special alloy stampings, arranged in built-up sections with properly spaced taps, are common to the larger starters. Moreover, as previously explained (see Fig. 2·1), the magnitudes of the various acceleration resistors are determined by the permissible inrush currents—between 125 and 200 per cent of rated value—and the sizes of the fuses that must protect the motor.

Figure 2·8 shows a wiring diagram of a three-point starter connected to a compound motor, with Fig. 2·8*b* illustrating schematically how the

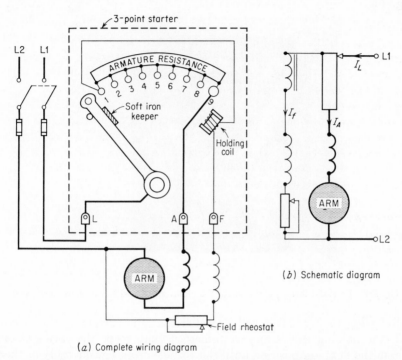

(*b*) Schematic diagram

(*a*) Complete wiring diagram

Fig. 2·8   Diagrams showing a three-point manual starter connected to a compound motor.

line current divides into *two* parallel paths, the low-current shunt-field circuit and the armature or *load* circuit. Note particularly that full voltage is applied to the shunt field on the first point—this is, therefore, a *full-field point* as well as an armature-current limiting point—and that the very resistance that is cut out of the armature circuit (comparatively low) is cut into the shunt-field circuit. This is because maximum field strength is desired, for highest developed torque, at the instant the motor is started. Since the resistors are used only during the accelerating period, they are designed for short-duty service and usually for inrush or maximum currents of about 150 per cent of full-load value. For *normal-starting* service such resistors would be NEMA class 115, and for *heavy-starting* service NEMA class 135. [The last figure in the class designation, i.e., number 5, implies that the inrush current would be 150 per cent F.L.C. (full-load current), while the middle number indicates the duty cycle; the middle number 1 means 5 sec ON out of 80 sec, and the middle number 3 stands for 10 sec ON out of 80 sec.] A three-point starter mounted in a steel case with the front protecting cover removed is depicted in Fig. 2·9.

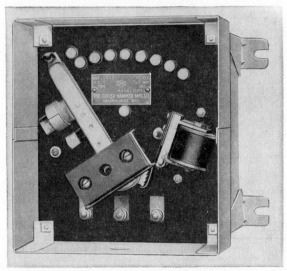

Fig. 2·9   Enclosed three-point manual d-c motor starter with front cover removed.

Referring to Fig. 2·8 it will be noted that the current in the *holding coil* (the holding coil is an electromagnet that keeps the spring-loaded arm in the FULL-ON position until the voltage drops considerably or there is a power failure, in which case the force of the spring returns the arm to the OFF position) is in series with the shunt field and the current in the two are the same. This means that, if a field-rheostat resistance of sufficient ohmic value is cut into the field to raise the speed (this will be discussed later under the subject of speed adjustment), the magnetomotive force of the holding coil may be reduced to a value too low to hold the arm in the FULL-ON position; the arm would then return to the OFF position automatically and the motor would stop. To avoid such faulty operation for high-speed shunt-field-controlled motors, the holding coil is removed from the shunt-field circuit and is energized directly, through a limiting resistor, by the main power source; this change results in a four-point starter.

A wiring diagram of a four-point starter connected to a shunt motor is shown in Fig. 2·10, with Fig. 2·10*b* illustrating schematically how the line current divides into *three* parallel paths. As mentioned above, the holding-coil and shunt-field currents are independent of each other, so that speed changes by field rheostat adjustment will not affect the electromagnetic pull exerted by the holding coil.

One of the objections to the use of the simple three- and four-point starters (Figs. 2·8 and 2·10) is that a motor may be started inadvertently with considerable field-rheostat resistance in the shunt-field circuit; this

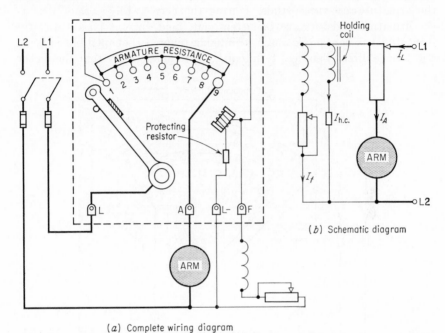

(*b*) Schematic diagram

(*a*) Complete wiring diagram

Fig. 2·10 Diagrams showing four-point manual starter connected to a shunt motor.

may occur when the motor is stopped after having been operated at high speed and the operator neglects to set the rheostat in the *all-out* position before a new start is made. When this happens, the inrush armature current may rise well above its intended value because starting torque is proportional to both armature current and flux, and any deficit in the latter must be made up by an increase in the former if the load is to be accelerated. In fact, the armature current may rise to an abnormally high value and cause fuses to blow or the circuit breaker to open if the motor is started with a weak field. The objection indicated can be avoided if a compound faceplate starter is employed, a construction that combines both accelerating resistors for starting and field rheostat for speed adjustment.

Figure 2·11 shows a wiring diagram of a compound motor connected to a compound starter, an arrangement that not only makes certain that the motor starts with full field but provides a neat unit construction of armature and field-rheostat resistors. The arrangement requires a double arm, one longer than the other, and mounted and pivoted at the same hub post. An operating handle, fixed to the end of the free brush-carrying *long arm*, is moved over a set of field-rheostat studs in a clockwise direction and *pushes* the *shorter spring-loaded arm* whose brush wipes over the armature-resistance studs. During the accelerating period, i.e., as the armature resistors are being cut out, field current passes directly from stud 1 to contacts *a* and *b* (through a pivoted U-shaped brass piece) and thence directly to field terminal *F*. Note particularly that the motor

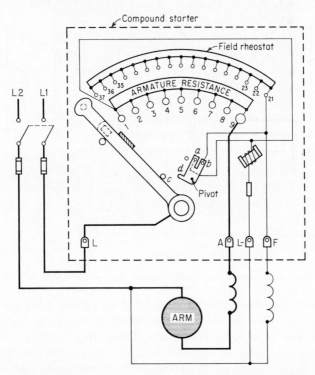

Fig. 2·11   Diagram showing a compound starter connected to a compound motor.

starts on full field but, as in the plain starters, the armature resistors are inserted in the field circuit as they are removed from the armature circuit. When the short arm reaches its final position, it is held there by the low-voltage release (holding) coil. At the same time, a small insulated piece *c* that is fastened to the short arm engages the pivoted U-shaped brass segment at *d* to rotate it counterclockwise; this motion disconnects the brass segment from the armature resistor at point 1 and connects the line terminal directly to stud 21 and field terminal *F*. The motor will thus be running on full field with all accelerating resistors cut out. Now then, if the long arm, *which is free*, is moved counterclockwise, it can be made to rest by friction at any higher numbered stud and cause speedup; this motion is equivalent to cutting in resistance with a field rheostat.

A typical four-point compound motor starter is shown in Fig. 2·12.

Fig. 2·12   Enclosed four-point compound manual d-c motor starter with front cover removed.

Faceplate starters for a-c motors. Although induction motors with wound rotors are used much less frequently than those with squirrel-cage rotors, they do possess characteristics that make them particularly suitable for (1) starting high-inertia loads and (2) applications that require wide-range speed adjustments. The first of these is fulfilled by inserting the correct values of external rotor resistance when the motor is started, and the second is accomplished by varying the external resistance for different load torques after the motor is accelerated.

For comparatively small 3-phase wound-rotor motors, manually operated faceplate starters are frequently used for acceleration as well as speed-adjustment purposes. Such a device consists of three banks of star-connected resistors that are properly tapped and joined to three sets of copper segments mounted on a slate or other insulating base. A three-spoke spiderlike contact arm that forms a wye point at its center is arranged to wipe over the segments as the arm is rotated. In one construction the starting arm is spring loaded so that it will return to the *all-in* position when it is released or when a low-voltage-holding coil, connected to one pair of line wires, releases the arm; the resistors in this type of starter are designed for short-duty service. In another construction the arm is not spring loaded and can be left at any segment position for purposes of speed adjustment; continuous-duty resistors must, of course, be used with this type of control device.

Since the faceplate controller functions only in the rotor circuit and it is essential that the motor be started with the arm in the *all-in* position, a separate stator starter, consisting of a push-button START-STOP station, a contactor, and a normally closed interlock *on the starting arm*, is often provided. As illustrated in Fig. 2·13, the auxiliary contact is connected in series with the START push button so that the stator line contactor can be energized only when the rheostat in the rotor circuit is in the starting position.

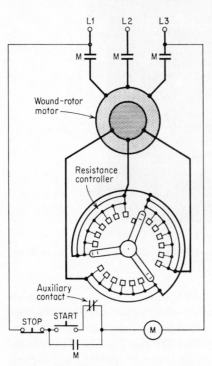

Referring to Fig. 2·13, note that the arm is in the *all-in* position, i.e., with maximum resistance in the rotor circuit, and that the stator can be energized and the motor started by pressing the START button. When this is done, the *mechanically actuated auxiliary contact* and the START button are short-circuited by the electrical interlock M as the M contactor picks up. The motor then starts with full external rotor resistance and develops high starting torque with minimum inrush current. When the arm is rotated clockwise, the auxiliary contact is opened *mechanically* (with no effect)

Fig. 2·13 Diagram showing a wound-rotor motor connected to a faceplate manually operated resistance controller.

as the former leaves the first point, and depending upon the segment to which the arm is moved where it is held by friction between contacts, the motor will accelerate toward full speed. Should there be a power failure or a sufficiently large voltage dip, the M contactor will drop out and the motor will stop. However, upon return of power the motor will not start because the electrical interlock M and the auxiliary contact are open. To restart the motor it is obviously necessary to return the arm to the first position where the auxiliary contact is closed mechanically, under which condition the motor can be started in the usual way.

Figure 2·14 illustrates a rotor resistance controller with the front cover removed. It clearly shows the construction of the wye arm with the auxiliary contact arrangement and terminals. The resistors are mounted in the rear and are not visible.

Fig. 2·14   Resistance controller for a wound-rotor motor.

Drum controllers for d-c motors. Although used frequently for motors in the low-horsepower ranges, i.e., up to about 10 hp, and occasionally for those rated from 15 to 30 hp, faceplate starters have several mechanical and electrical disadvantages and limitations. They cannot, for example, be constructed to render service that is severe or rugged, because arcing at the sliding contacts tends to roughen the surfaces and contact roughness makes it difficult to maintain proper contact pressures. Thus, if contact pressure is increased sufficiently to improve electrical conductivity, a bending strain is often imposed upon the bearing support where the slate or other insulating material of the panel may crack. Other objections are that they do not lend themselves readily to operations such as reversing the direction of rotation of a motor and electrical braking or to control circuits other than those generally standardized.

The sturdier, more compact drum controller overcomes the foregoing and other disadvantages, since it not only can be constructed to withstand much abuse but can be equipped with arcing barriers between contacts and with electromagnetic blowout protection. Also, it is easily arranged for most circuit combinations, however complex, and can, moreover, be readily modified in the field when this is necessary.

As Fig. 2·15 illustrates, a drum controller has two vertical rows of stationary contact fingers that are attached to, but insulated from, the

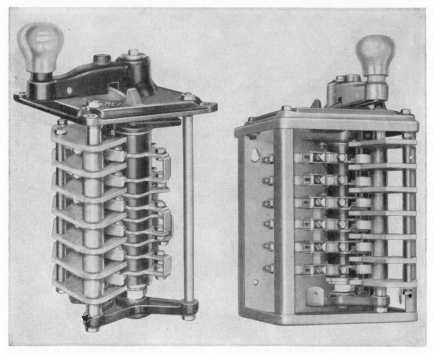

Fig. 2·15   Two views of drum controller.

framework. Between the rows of fingers is a cast-iron spider or drum that is mounted on a square steel shaft from which it is insulated by a mica sleeve; a handle at the top engages the extended steel shaft so that the drum can be rotated. Also, attached to the drum are a set of insulated copper segments of different lengths which touch the fingers, with the latter backed up by follow-up springs that maintain strong uniform pressure between contacts. As the controller handle is rotated, a roller that rides over a notched wheel near the top of the drum is forced into the notches by a spring at each operating position. Since the length and location of all segments can be changed easily, the duration of the closing and opening of all sets of contacts can be readily adjusted to fulfill practically any control sequence of a motor.

A wiring diagram of a nonreversing drum controller connected to a compound motor is shown in Fig. 2·16. Note particularly that the vertical

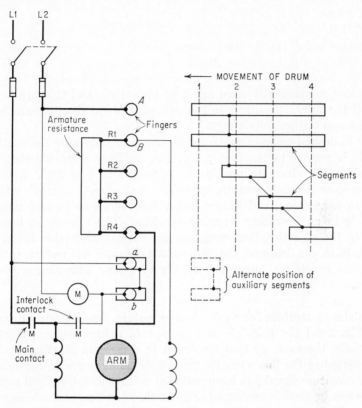

Fig. 2·16   Nonreversing drum controller connected to a compound motor.

row of circles stands for the stationary contact fingers while the horizontal steps represent the segments of the drum as they would appear if rolled out flat; also the vertical broken lines indicate the positions of the controller fingers in relation to the segments at successive positions of the drum. Unlike the faceplate starter, where undervoltage (or no-voltage release) protection is provided by a spring-loaded arm and an electromagnetic holding coil, the drum controller, whose arm can be moved to and held in any position, requires the use of a contactor for this function. Observe that, before the motor is started, contactor M is energized through a pair of auxiliary fingers a and b when the drum is in the OFF position only; thus the main contact M is closed so that operation can begin, and interlock contact M is closed to provide a holding circuit for the contactor after the drum is rotated to any running position, even though contacts a and b are disconnected. When the drum is moved to the first position, the shunt field is energized through fingers A and B and the armature is connected to the source through resistances $R1$ to $R4$. Furthermore, rotation of the drum to positions 2, 3, and 4, respectively, cut out resistances $R1$-$R2$, $R2$-$R3$, and $R3$-$R4$ as the motor accelerates to full speed. Should there be a power failure or a sufficiently large voltage dip, contactor M will drop out and cause the main and interlock contacts M to open to stop the motor. Obviously, to restart the motor, contactor M must again be picked up, and this can happen only if the arm is returned to the OFF position to short-circuit fingers a and b and restore full voltage.

If it is desired to have contactor M energized only *after* the controller handle is moved to the first point, the auxiliary segments should be moved to the alternate position as indicated in Fig. 2·16. With this arrangement the main contacts open the circuit when the drum is returned to the OFF position, and the contactor coil carries no current until the drum is in an operating (running) position. The scheme is, however, open to the objection that complete low-voltage protection is not provided. With the drum on the first point the motor will restart, with full armature resistance, upon the return of power following a voltage failure.

Resistance starters for squirrel-cage motors. As previously discussed (see Fig. 2·4a) the principle of this type of reduced-voltage starter is essentially the same as that employed in d-c motor starting; resistors are inserted in the line wires (in the d-c motor they are placed in series in the armature circuit) to incur artificial voltage drops. Several arrangements can be used involving equipment that can be operated manually, automatically, or semiautomatically. In one scheme the resistors are made up of stacks of specially treated graphite discs placed in steel tubes whose inside surfaces are lined with an insulating refractory material. When a mechanical force is applied to the columns of graphite discs, they are compressed, with the result that resistance diminishes steplessly with increasing pressure. Since such units provide considerable ranges of resistance and have, in addition, good heat-dissipating capacity, they are ideal for induction-motor starters.

A diagram showing the wiring connections for a semiautomatic type of resistance starter, used for 60- to 200-hp squirrel-cage motors, is given in Fig. 2·17. A steel cabinet houses the resistor units, all necessary contacts, contactors, overload devices, and other components; also, a lever mounted on the outside of the enclosure is operated to apply varying equalized pressures to the stacks of graphite discs.

With the lever in the DOWN or OFF-RUN position, the motor circuit is open. To start the machine, the lever is raised quickly to the START position where a set of pilot contacts closes at s and s', which, in turn, energize the start contactor S; this action causes the three start contacts S to close in the resistor circuits and closes the interlock contact S. The motor is now in series with the three line resistors and, therefore, starts at reduced voltage. As the rotor accelerates, the handle is raised further

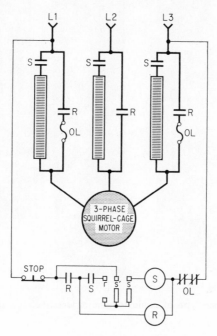

Fig. 2·17   Diagram of a reduced-voltage *resistance* starter for a 3-phase squirrel-cage motor.

from the midposition. This applies increasing equalized pressures to the graphite discs in the resistor tubes and reduces the resistances uniformly and steplessly. After the lever reaches the top point of its travel, the COMPLETE START position, contact r closes to establish a path through s', r, and interlock contact S; this causes the R contactor to pick up and close the three run contacts R (to short-circuit the line resistors) as well as the interlock contact R. The motor is thus connected directly across the line, and since the R contactor is now energized through its own interlock contact R, the lever can be returned to the OFF-RUN position. When the latter is done, the S contactor drops out and all its associated contacts open. Pressing the STOP button will, of course, deenergize the R contactor to stop the motor, as will the opening of either overload OL or a voltage failure.

Figure 2·18 shows a reduced-voltage line-resistor starter of the type described above, mounted in a steel cabinet.

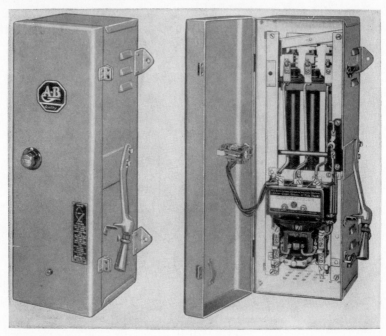

Fig. 2·18   Line-resistance type of semiautomatic reduced-voltage starter. (*Allen-Bradley Co.*)

If fixed resistors like those illustrated in Fig. 2·4a for a 3-phase motor are to be used to limit the inrush current, calculations can be made for their values, assuming that, in addition to available nameplate data, the following information is given: blocked-rotor motor power factor, inrush current with rated applied voltage, inrush current with inserted line resistors. Referring to Fig. 2·19, which represents the motor and the phasor relationships on a *per-phase* basis,

$$I = \frac{E_R}{R} = \frac{E_M}{Z_M}$$

from which

$$R = Z_M \frac{E_R}{E_M}$$

ɤ here Z_M is the motor impedance with the rotor blocked. Also

$$E_{R_M} = E_M \cos \theta_{BR} \quad \text{and} \quad E_{X_M} = E_M \sin \theta_{BR}$$

where E_{R_M} and E_{X_M} are, respectively, the motor resistance and reactance drops at the required inrush current.

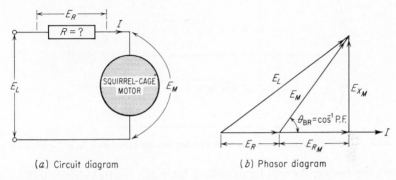

(a) Circuit diagram   (b) Phasor diagram

Fig. 2·19 *Per-phase diagrams* illustrating method to determine line resistance in a resistance starter for a polyphase motor.

Example 2. A 50-hp 440-volt 3-phase 54-amp 60-cycle squirrel-cage induction motor takes an inrush current of six times rated value when connected to a 440-volt source. If it is desired to limit the starting current to $3.5 I_{FL}$, what value of resistance should be used in each line wire, assuming a blocked-rotor power factor for the motor of 0.5?

Solution

$$I_{\text{inrush}} = 3.5 \times 54 = 189 \text{ amp}$$

$$E_L = 440/\sqrt{3} = 254 \text{ volts} \qquad E_M = \frac{3.5}{6.0} \times 254 = 148 \text{ volts}$$

$$E_{R_M} = 148 \times 0.5 = 74 \text{ volts} \qquad E_{X_M} = 148 \times 0.866 = 128 \text{ volts}$$
$$(E_R + E_{R_M}) = \sqrt{(254)^2 - (128)^2} = 219 \text{ volts}$$
$$E_R = 219 - 74 = 145 \text{ volts}$$
$$R = {}^{145}\!/_{189} = 0.77 \text{ ohm}$$

Reactance starters for squirrel-cage motors. For comparatively large squirrel-cage induction motors operating from upper-voltage circuits, i.e., 440, 550, 1,100 and 2,300 volts, line reactors are often used to advantage for starting purposes. These generally occupy less space and are more economical than resistors and can, in addition, be provided with taps to permit a choice of several motor-starting voltages. The simplified diagram of Fig. 2·20, without the control circuit, illustrates the application of this device with taps to yield 50, 65, and 80 per cent starting voltage

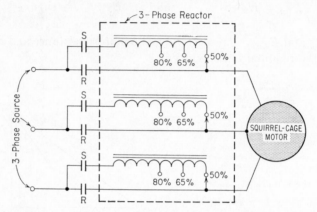

Fig. 2·20   Simplified circuit diagram of a reactor starter connected to a 3-phase motor.

across the motor. As will be described later the control circuit will function to close the *S* (start) contacts first to permit the motor to start on reduced voltage, close the *R* (run) contacts after the motor has accelerated to almost full speed, and then open the *S* contacts to disconnect the reactor.

The value of each of the line reactors can be determined in a manner similar to that given in the foregoing article for the resistance starter, with the exception that the reactance-voltage drop must be represented to *lead* the starting current by 90° whereas the line-resistance voltage drop is in phase with the starting current. Proceeding as before and referring to Fig. 2·21, the following equations represent the motor and

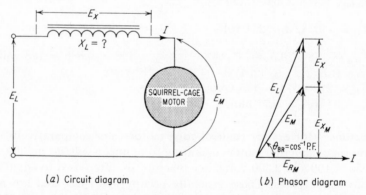

(*a*) Circuit diagram   (*b*) Phasor diagram

Fig. 2·21   *Per-phase diagrams* illustrating method to determine line reactance in a reactance starter for a polyphase motor.

phasor relationships on a *per-phase* basis:

$$I = \frac{E_X}{X} = \frac{E_M}{Z_M}$$

from which

$$X = Z_M \frac{E_X}{E_M}$$

where Z_M is the motor impedance with the rotor blocked.

Example 3. A 600-hp 2,300-volt 3-phase 140-amp 60-cycle squirrel-cage induction motor takes an inrush current that is 6.5 times rated value when connected to a 2,300-volt source. Calculate the reactance in each of the line wires

Fig. 2·22 Three-phase reactor used in starting a polyphase induction motor.

for a starting *motor* voltage of 1,150, under which condition the inrush current will be limited to $3.25 I_{FL}$. Assume a motor blocked-rotor power factor of 0.4. Also determine the reactances for the 80 per cent tap.

Solution

$$I_{\text{inrush}} = 0.5 \times 6.5 \times 140 = 455 \text{ amp}$$
$$E_L = 2{,}300/\sqrt{3} = 1{,}330 \text{ volts} \qquad E_M = 0.5 \times 1{,}330 = 665 \text{ volts}$$
$$E_{R_M} = 665 \times 0.4 = 266 \text{ volts} \qquad E_{X_M} = 665 \times 0.917 = 610 \text{ volts}$$
$$(\cos^{-1} 0.4 = 66.4°; \sin 66.4° = 0.917)$$
$$(E_X + E_{X_M}) = \sqrt{(1{,}330)^2 - (266)^2} = 1{,}300 \text{ volts}$$
$$E_X = 1{,}300 - 610 = 690 \text{ volts}$$
$$X = {}^{690}\!/_{455} = 1.52 \text{ ohms}$$

For the 80 per cent tap

$$I_{\text{inrush}} = 0.8 \times 6.5 \times 140 = 728 \text{ amp}$$
$$E_{R_M} = 1{,}064 \times 0.4 = 427 \text{ volts}$$
$$E_M = 0.8 \times 1{,}330 = 1{,}064 \text{ volts}$$
$$E_{X_M} = 1{,}064 \times 0.917 = 977 \text{ volts}$$
$$(E_X + E_{X_M}) = \sqrt{(1{,}330)^2 - (427)^2} = 1{,}260 \text{ volts}$$
$$E_X = 1{,}260 - 977 = 283 \text{ volts}$$
$$X = {}^{283}\!/_{728} = 0.39 \text{ ohm}$$

A photograph of a 3-phase reactor used in starting a large induction motor is shown in Fig. 2·22.

Autotransformer starters for squirrel-cage motors. A popular type of manual starter for polyphase squirrel-cage motors makes use of the autotransformer for reduced-voltage starting. As generally constructed it consists of a three-legged laminated core around the outer two legs of which are placed coils, each one tapped at points that are approximately 50, 65, and 80 per cent from one end. Referring to the schematic diagram of Fig. 2·4c, note that connections are made to the line and motor terminals through two sets of contacts, START S and RUN R, often immersed in oil. Since the starting torque of a motor is proportional to the *square* of the applied voltage, connections to the above-indicated tappings will yield values that are 25, 42, and 64 per cent, respectively, of the full-voltage starting torques. Therefore, for motors that develop starting torques that are $1.5T_{FL}$ when rated voltage is applied, the starting torques can be adjusted to values of about $0.38T_{FL}$, $0.63T_{FL}$, and $0.96T_{FL}$; these torques are usually sufficient to start most loads. Moreover, the inrush currents on the *primary* (line) sides of the transformers are reduced by approximately the *square* of the secondary-to-primary voltage ratio, because (1) the actual *motor* starting current depends upon *its* impressed voltage and this is reduced by the step-down transformers and (2) the primary or *line* current is the same percentage of the secondary or *motor* current as is the secondary-to-primary voltage ratio.

A complete wiring diagram of a manually operated autotransformer starter connected to a squirrel-cage motor is shown in Fig. 2·23. All

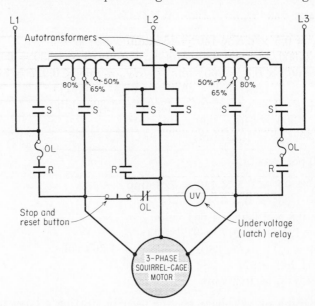

Fig. 2·23   Complete diagram of a reduced-voltage (autotransformer) starter connected to a squirrel-cage motor.

components, i.e., transformers, contacts, overload protection, undervoltage relay, STOP button, and mechanical latching device, are placed in a steel cabinet. A spring-loaded lever, mounted on the outside of the cabinet, stands normally in a vertical position and is pulled back to start the motor; this action closes six START contacts S, energizes the two autotransformers that are connected in open delta, and impresses reduced voltage (65 per cent in Fig. 2·23) across the motor terminals. After the rotor has accelerated to about full speed, the lever is quickly pushed forward; this instantly opens the START contacts and *then*, a split second later, closes the three RUN contacts R. The motor is now connected across the full-voltage source, while the autotransformers, disconnected from the power supply, are deenergized and take no magnetizing current. A latch, fixed to the lever mechanism, is made to drop into a notch so that the operator is prevented from moving the handle accidentally to the RUN position first. When the handle is quickly pushed from the START to the RUN position, however, the latch is kicked free to permit the lever to move forward. An electromagnet, shown as an undervoltage (latch) relay in the diagram, continues to hold the lever in the RUN position until the STOP button (passing through the front panel) is pressed; moreover, if there should be a voltage failure or a voltage dip of sufficient magnitude, the UV relay will release and trip the holding mechanism. Another important feature is the thermal overload protection that is provided to trip the lever under conditions of sustained high currents; also the STOP button is so constructed that it actuates a reset mechanism for the overload relay.

Fig. 2·24   Autotransformer type of manual reduced-voltage starter. (*Allen-Bradley Co.*)

Figure 2·24 illustrates a complete autotransformer reduced-voltage starter mounted in a steel cabinet and with the door open, and Fig. 2·25 shows the general construction of the compensator itself.

Example 4. The following information is given in connection with a 25-hp 3-phase 220-volt 60-cycle 1,750-rpm squirrel-cage motor:

$$I_{FL} = 62 \text{ amp} \qquad T_{FL} = 75 \text{ lb-ft} \qquad I_{ST} \text{ (at 220 volts)} = 5.5 I_{FL}$$
$$T_{ST} \text{ (at 220 volts)} = 1.5 T_{FL}$$

If the 65 per cent taps on an autotransformer type of reduced-voltage starter are used to accelerate the motor, calculate (a) the starting torque, (b) the motor and line side starting currents, neglecting the exciting current of the transformers.

Fig. 2·25   Compensator showing three-legged core and tapped coils.

Solution

a. $T_{ST} = 1.5 \times 75 \times (0.65)^2 = 47.5$ lb-ft

b. I_{ST} (motor side) $= 5.5 \times 62 \times 0.65 = 222$ amp

I_{ST} (line side) $= 5.5 \times 62 \times (0.65)^2 = 144$ amp

In contrast to the line-resistance type of starter, where the power circuit is *not* opened during the entire accelerating period, the autotransformer method as described above makes it necessary to disconnect the motor from the source during the short switching interval, i.e., when the motor is switched from reduced voltage to full voltage. This practice may result in a transient current inrush that is extremely high, since it depends upon both the exact instant line voltage is reapplied and the length of the transition period. At the instant the motor is disconnected from the reduced voltage source, the current in the *stator winding*, of course, drops to zero. However, in the closed circuit of the squirrel-cage rotor the current will *not* drop to zero but will change to whatever value is necessary to maintain the motor flux at that time. At disconnection, therefore, the rotor flux is "trapped" and no longer moves along the rotor surface as was the case when the stator winding served to create a synchronously revolving field and the rotor was rotating at less than synchronous speed. The rotor flux is therefore "tied" to the rotor, not the stator, and as it decays, it induces a voltage in the rotor which acts to sustain its slowly decaying current. Moreover, as the d-c rotor field revolves, a-c voltages at less than line frequency are generated in the stator winding, the phase relation of which is continually changing with respect to the line voltage. Thus, if full voltage is applied at the instant when the two voltages—line and generated—are in phase, and particularly when the peaks are additive, a high inrush current of short duration may flow. This current may, in fact, be higher than that taken by the motor if initially connected to full voltage, a condition that not only defeats the purpose of the autotransformer but may open the circuit breakers.

Another objection to the compensator type of starter is that the exciting current taken by each of the transformer windings is often a rather large component of the total line current. This is because such devices are generally manufactured in standard sizes, each of which is expected to serve a range of motor ratings, say 5 to 15 hp, and, being used for short periods only, are designed to be constructed as economically as possible. Thus, for example, if a 15-hp autostarter is used with a 5-hp motor, its excessive magnetizing current becomes an appreciable part of the total inrush current, with the result that the latter may be as much or even more than if the motor were started directly from the full-voltage supply. This, too, reduces the effectiveness of the autotransformer, since the starting currents are larger than desired and predicted values. An example should make this discussion clear.

Example 5. A 5- to 15-hp 220-volt 3-phase autotransformer type of starter takes a magnetizing current of 11 amp. If the 65 per cent taps are used, calculate (**a**) the inrush current on the line side of a 15-hp motor if $I_{FL} = 39$ amp and $I_{ST} = 5.5I_{FL}$ when started on full voltage, (**b**) the inrush current on the line side of a 5-hp motor if $I_{FL} = 13.6$ amp and $I_{ST} = 5.8I_{FL}$ when started on full voltage.

Solution

a. I_{ST} (line side) $= [(0.65)^2 \times 5.5 \times 39] + 11 = 91 + 11 = 102$
b. I_{ST} (line side) $= [(0.65)^2 \times 5.8 \times 13.6] + 11 = 33 + 11 = 44$

Note that for the purposes of illustration the two components of current were added arithmetically. Although this procedure is not strictly correct because these components are somewhat out of phase and the polyphase system of currents is unsymmetrical, the direct additions are not significantly different from the correct values.

Figure 2·26 indicates how the solution can be represented diagrammatically.

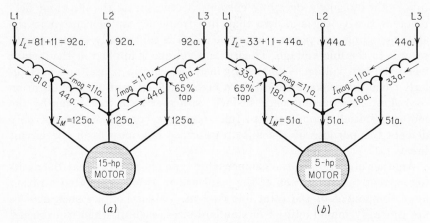

(a)   (b)

Fig. 2·26   Diagrams illustrating solutions to Example 5. Note that the L1 and L3 starting currents are shown as arithmetical sums of two transformer components; the L2 currents are the vector differences of the motor and transformer components.

QUESTIONS AND PROBLEMS

1. What is the purpose of an *accelerating resistor* in the control circuit of a d-c motor? Upon what does its ohmic value and physical size depend?

2. Explain why an accelerating resistor in the control circuit of a motor that is started and stopped frequently must be physically larger than one used for a continuous-duty motor.

3. The control circuit of a 10-hp 230-volt 36-amp shunt motor is to be provided with an accelerating resistor that is to be cut out in two steps. If the resistance of the armature (including brushes) is assumed to be $6\frac{1}{4}$ per cent of V/I, calculate the ohmic value of each of the two sections of the accelerating resistor if the maximum current is not to exceed $1.75I_{FL}$ and the first section is to be cut out when the current drops to I_{FL}.

4. What would be the approximate inrush current to the motor of Prob. 3 if no accelerating resistance is used during the starting period?

5. Why is it often permissible to start small d-c motors, up to perhaps 2 hp and particularly when started without load, by connecting them directly to the line?

6. Discuss the difference between universal and induction motors with regard to use of power supply, speed range, torque characteristics, ratio of weight to horsepower, and inrush current when started.

7. Discuss the differences between single-phase and polyphase induction motors with regard to horsepower output and operating characteristics such as power factor, efficiency, starting torque, and overload capacity.

8. Distinguish between squirrel-cage and wound-motor induction motors with regard to construction, starting torque, speed control, and starting methods.

9. The inrush current to a squirrel-cage motor is 100 amp when started directly from the line. What will be the inrush line current if the machine is started

(a) by the *star-delta* method? (b) by the line-resistance method that reduces the motor voltage to $0.58V$? (c) by the autotransformer method that reduces the motor voltage to $0.65V$?

10. In wound-rotor motors, how are the starting torques, maximum torques, and full-load speeds affected by rotor-resistance change?

11. Describe the *part-winding* method of starting induction motors, and explain why the inrush current is limited by this scheme.

12. Point out important differences between *three-point* and *four-point* manual faceplate starters for d-c motors.

13. What important advantages are possessed by the compound starter for d-c motors?

14. Explain the purpose and action of the *auxiliary contact* in the control circuit for the wound-rotor motor of Fig. 2·13.

15. Explain the purpose of the *interlock contact* in the control circuit for the d-c motor of Fig. 2·16. What operating changes result when the auxiliary contacts are placed in the alternate position shown?

16. Referring to Figs. 2·17 and 2·18, why is the released position of the operating handle marked OFF-RUN?

17. The inrush current to a 75-hp 220-volt 185-amp 60-cycle squirrel-cage induction motor is $6.5I_{FL}$ when connected to a rated-voltage source. Assuming a starting power factor of 0.4, calculate the ohmic value of each of three line resistors that will reduce the inrush current to $3I_{FL}$. Determine also the starting power factor under this condition.

18. For the motor of Prob. 17, calculate the reactance and inductance of each of three line reactors that will reduce the inrush current to $4I_{FL}$. Determine also the starting power factor under this condition.

19. The voltage drop across each of three reactors in a line reactor starter is 242 volts at the instant of starting when used in connection with a 500-hp 550-volt 480-amp 3-phase squirrel-cage induction motor; the inrush current under this condition is $3.5I_{FL}$ and the motor voltage is $0.682E_L$. (a) Calculate the value of each of the line reactors. (b) What will be the inrush current if the motor is connected directly across the 550-volt source?

20. A 50-hp 440-volt 3-phase 63-amp 1,160-rpm squirrel-cage motor takes $4.1I_{FL}$ amp at the instant of starting when connected to the 65 per cent taps of a two-winding (open-delta) autostarter (compensator). Calculate the *motor side* and *line side* currents (a) if the 50 per cent taps are used, (b) if the 80 per cent taps are used. Assume a magnetizing current of 27 amp.

21. If the motor of Prob. 20 develops a starting torque of $1.2T_{FL}$ on the 65 per cent taps, calculate the starting torques (a) on the 50 per cent taps, (b) on the 80 per cent taps.

22. What important operating differences exist between line resistor and compensator starting?

23. Briefly describe the actions in compensator starting when the motor is "switched" from reduced voltage to full voltage, i.e., by opening the circuit (open-circuit transition) when the voltage change is made.

CHAPTER 3

Automatic Starters and Control
Circuits for D-C Motors

Types of starter for automatic acceleration. The principles of d-c motor acceleration were discussed in Chap. 2 (see Figs. 2·1 and 2·2), where it was shown that, except for rather small motors, accelerating resistors must be connected in series in the armature circuit during the starting period and be cut out (or short-circuited) progressively as the rotor comes up to speed. The following reasons are given for the practice of applying reduced voltage across the armature terminals: (1) to limit the inrush currents that are commutated and pass across commutator-brush sliding contacts, because excessive currents would tend to produce arcing and commutator burning; (2) to minimize line disturbances caused by high inrush currents that produce excessive line-resistance drops; (3) to provide smooth acceleration so that driven machinery and equipment will not be subjected to undue mechanical stresses caused by the sudden application of high torques.

There are many types of automatic starter for d-c motors, but all can be classified under two *general* methods of acceleration, namely, (1) *current-limit* acceleration and (2) *definite-time* acceleration. In the first of these a set of series or current relays is made to function, i.e., pick up and drop out, for changing values of armature current as the motor accelerates; the relays, in turn, operate to energize contactors as they successively cut out, i.e., short-circuit, an equal number of resistors. The definite-time acceleration method, on the other hand, employs a set of timed relays that function in sequence at a definite rate, regardless of load current; the operation of the relays then permits properly interconnected contactors to short circuit progressively a group of resistors. Note particularly that the two systems of acceleration are completely different in their behavior, since the load (or load current) will determine how rapidly a motor is brought up to full speed in one of them, while in the other the preset timing of the relays will repeatedly cause the motor to operate on a given acceleration-time cycle independently of the load conditions.

Since the rate of current change varies with the load, it should be clear that a current-limit acceleration starter will bring a lightly loaded motor up to speed much more rapidly than it will when a heavy load must be started. Although this may be desirable in some installations in contrast to the definite-time starter, the latter has the advantage that its timed acceleration is often useful when a driven machine must always repeat the same cycle of operation in a manufacturing process or when several motors in a mechanical system must perform in a given timed sequence.

Current-limit acceleration starters. Three types of automatic starter that function to accelerate a motor on the basis of current change are (1) the series-relay type, (2) the cemf type, and (3) the lockout type.

Series-relay Starter. A wiring diagram showing a shunt motor connected to a four-step series-relay type of starter is given in Fig. 3·1. In addition to the START-STOP push-button station the control circuit consists of three accelerating relays $1AR$, $2AR$, and $3AR$; four contactors M, $1A$,

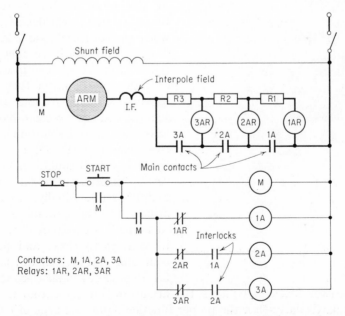

Fig. 3·1 Shunt motor connected to a series-relay current-limit type of starter.

$2A$, and $3A$; and three resistors $R1$, $R2$, and $R3$. The series relay is constructed with a lightweight, well-balanced armature and a rather short magnetic air gap, so that its action in opening and closing contacts is very fast. Moreover, the N.O. electrical interlocks on the contactors must be set and maintained to close *after* the main power-circuit contacts close.

When the START button in Fig. 3·1 is pressed, the M contactor will be energized to close the main contact M and start the motor with the three acceleration resistors $R3$, $R2$, and $R1$ and acceleration relay $1AR$ in series in the armature circuit. Also, two N.O. interlocks M will close, one to seal the M contactor coil and the other to provide an energizing circuit for the accelerating contactors. Since the armature inrush current is high when the motor starts, the first acceleration relay $1AR$ picks up to open its N.C. contact in the $1A$ contactor circuit; then, after the motor speeds up sufficiently and the armature current drops to normal, $1AR$ drops out and closes its contact. This action permits contactor $1A$ to operate, close its contact across $R1$ and $1AR$, and short-circuit the first

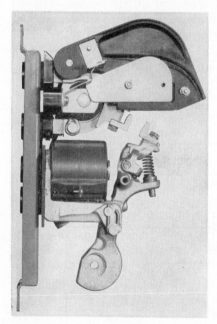

block of resistance. A second current inrush now passes through $R3$, $R2$, and $2AR$ to open the N.C. $2AR$ contact *before* the $1A$ interlock closes in the $2A$ contactor circuit. As the motor accelerates further, the armature current decreases and when normal again causes $2AR$ to drop out and close its contact where the $2A$ contactor is energized. The operation of the latter then closes its contact across $R2$ and $2AR$ to short-circuit the second block of resistance. A third current inrush, this time through $R3$ and $3AR$, opens the $3AR$ contact before the $2A$ interlock closes in the $3A$ contactor circuit. Finally, when the armature current diminishes again to normal, $3AR$ drops out, causes its contact to close, and permits contactor $3A$ to pick up; contact $3A$ then closes to short-circuit the

Fig. 3·2 100-amp d-c contactor. (*Square D Co.*)

third resistor block and places the armature directly across the line.

Note particularly that the proper functioning of this type of current-limit acceleration starter depends upon (1) *current variations*, which, in turn, cause *series acceleration relays* to pick up and drop out and (2) operation of accelerating relays that are very much faster than contactors.

Figure 3·2 depicts a 100-amp d-c contactor that clearly shows the operating coil, the N.O. contact, and the raised blowout shield.

Counter EMF Starter. For rather small motors, up to perhaps 5 hp, the cemf type of starter is economical and performs reasonably well. As Fig. 3·3 shows, it consists essentially of a push-button START-STOP station,

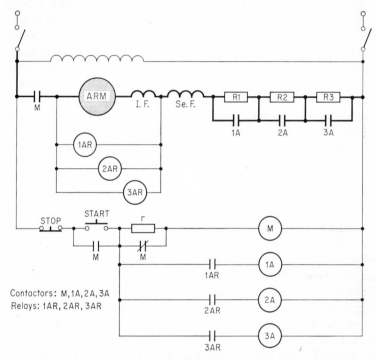

Fig. 3·3 Compound motor connected to a cemf type of starter.

a main contactor, and voltage-type accelerating relays, contactors and resistors. Note particularly that the relays are connected across the armature, where the voltage rises progressively as the motor comes up to speed. Since the relays are adjusted to pick up at a succession of increasing voltages (in the four-step starter shown, this might be at 50, 75, and 90 per cent of rated values), a set of contacts closes to energize the normal-voltage contactors, which, in turn, short-circuit the resistors. Relays are used across the armature to time the operation of the contactors because they are easily adjusted to operate at definite preset voltages. An alternate scheme that eliminates the need for relays is to connect the contactors across the armature and have them function directly to accelerate the motor. Except for small single-contactor starters, this arrangement is, however, generally unsatisfactory because it is rather difficult to adjust several contactors to close accurately at different voltages.

When the START button in Fig. 3·3 is pressed, contactor M is energized at full voltage through N.C. interlock M. This action first causes line contact M to close and then inserts resistor r in the coil circuit as the N.C. interlock M is opened. The latter procedure makes the contactor pick up at maximum mmf (when the air gap between armature and core is wide) and, after resistor r is added, allows the coil to function at a permissible lower current and mmf (when the air gap is shortened) with reduced heating. As the motor starts with the three steps of resistance in the armature circuit, the voltage drop across the accelerating relays is low. However, when the motor accelerates and the cemf rises sufficiently to operate relay $1AR$ (at perhaps 50 per cent volts), contact $1AR$ closes in the $1A$ contactor circuit; this causes contact $1A$ to close and short-circuit the first step of resistance $R1$. With further acceleration of the motor, relay $2AR$ picks up at its adjusted potential, say 75 per cent volts, and closes its contact $2AR$ in the $2A$ contactor circuit, with the result that the second step of resistance $R2$ is short-circuited. Finally, when the motor has accelerated to increase the armature emf to about 90 per cent of line voltage, relay $3AR$ is actuated to close its contact in series with the $3A$ contactor coil, under which condition $R3$ is short-circuited and the armature operates normally at full line potential.

Lockout Acceleration Starter. Another type of current-limit d-c starter employs a special contactor that is constructed with two operating coils, one, called the *holding coil*, that tends to keep the contact open and another, the *closing coil*, that attempts to close the contact. During operation, both coils are connected in series in the armature circuit and create fluxes in two independent magnetic circuits so that two opposing magnetic forces act on a spring-loaded pivoted iron bar that carries one of the contact elements. The two mmfs and the magnetic circuits are designed so that, when the armature current is high, the contact is *locked out* by the superior electromagnetic force exerted by the holding coil. Then, after the motor armature has accelerated sufficiently to bring the current down, the closing-coil mmf overpowers the holding-coil mmf, under which condition the iron bar moves to close the N.O. contact. The unique design of the contactor that produces the lockout-closing actions described is a *closing magnetic circuit* that is made up largely of iron and is easily saturated, and a *holding magnetic circuit*, containing a large adjustable air gap, which does not saturate easily. Thus, when the armature current is diminishing, the holding-circuit flux drops rapidly because the magnetic current is unsaturated while the closing-circuit flux is only slightly reduced, since its magnetic circuit is saturated.

A circuit diagram for a *lockout acceleration* starter connected to a compound motor is illustrated in Fig. 3·4. Note that there are three lockout

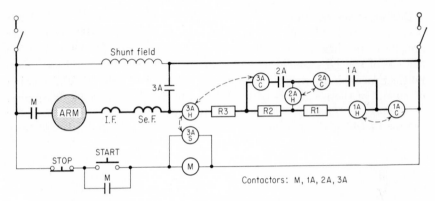

Fig. 3·4 Compound motor connected to a lockout current-limit type of starter.

contactors $1AH$-$1AC$, $2AH$-$2AC$, and $3AH$-$3AC$-$3AS$. In each case the H and C letters following the $1A$, $2A$, $3A$ symbols stand, respectively, for holding and closing coils. The third contactor $3A$ has a shunt coil S wound directly over the C coil in addition to the H and C coils; its function is described below.

When the START button is pressed, the M contactor is energized to close the main contact M and the interlock M that seals the control circuit. The motor therefore starts with resistors $R1$, $R2$, and $R3$ in the armature circuit and with contactor coils $1AH$-$1AC$ and $3AH$ excited by the armature current. Contactor $1A$ is thus locked out by the initially high inrush current, and contactor $3A$ is also held open even though its shunt coil $3AS$ is excited, since it is connected across the M coil. The latter condition prevails because $3AS$ creates a weak field that is capable of keeping the pivoted bar pulled up only after the contactor is actuated and the air gap is greatly reduced. After the motor accelerates sufficiently and the armature current drops, the holding-coil mmf $1AH$ yields to its closing coil $1AC$, which then causes that contactor to pick up and close contact $1A$. With $R1$ now short-circuited, increased armature current passes through $2AH$-$2AC$ and keeps that contactor locked out until the motor speed increases further. When the armature current falls again to permit contactor $2A$ to pick up, its contact $2A$ short-circuits resistor $R2$. With the circuit thus established through $3AH$-$3AC$, $2AC$, and $1AC$ a third large current inrush keeps contactor $3A$ dropped out (as contacts $1A$ and $2A$ remain closed). Finally, when motor acceleration reduces the armature current again, contactor 3A closes its contact to place a short-circuit around all resistors and all closing and holding coils. With the shortened air gap in contactor 3A, the latter remains picked up by its shunt coil 3AS, still connected across the line.

Definite-time acceleration starters. Many types of d-c starters have been developed in which contacts connected across accelerating resistors are made to close at successively longer time intervals. These involve the use of special solenoid or motor-driven timers, dashpot relays, pneumatic timers, inductive time-limit contactors, and others. A discussion of several of these schemes with appropriate circuit arrangements follows.

Mechanical Timers. Figure 3·5 illustrates a circuit for a d-c motor starter in which the mechanical timing relay TR may be operated by a

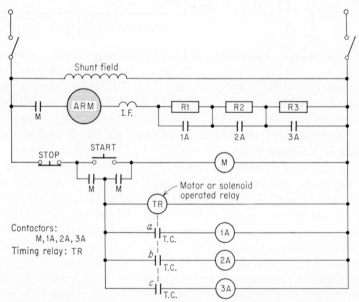

Fig. 3·5   Shunt motor connected to a mechanical-timer type of starter.

synchronous motor (similar to that used in electric clocks) or a solenoid which, in turn, actuates a shaft that closes a set of contacts in a definite and adjustable time sequence. Contactors are thus made to pick up with definite and increasingly longer time delays to short-circuit a group of accelerating resistors.

When the START button is pressed, contactor M is energized and is sealed in by its interlocks M. Timing relay TR, operated by either a solenoid or a motor, also picks up to close contacts a, b, and c in sequence and with adjusted time delays which increase progressively. The relay contacts, in turn, then cause contactors $1A$, $2A$, and $3A$ to operate progressively to short-circuit resistors $R1$, $R2$, and $R3$. It should be understood in this connection that relay TR is designed to have several contacts (three in this starter) that are arranged on a common shaft which, when turned, performs a contact closing function.

Another type of mechanical timing involves the use of a dashpot relay, a photograph of which is shown in Fig. 3·6. As illustrated the relay consists of a plunger which, when the coil of the former is energized, moves

Fig. 3·6   Oil dashpot timing relay.

slowly through a bath of oil and closes a contact at the end of its stroke. The dashpot is usually provided with a bypass near its upper limit of travel so that the contact is permitted to close with a snap action. In addition, a valve is included in its construction to allow the oil to flow freely as the plunger falls when the relay is deenergized.

In Fig. 3·7, dashpot timing relays are represented by $TR1$, $TR2$, and $TR3$. Contactor M and timing relay $TR1$ are energized by pressing the START button. This closes the main M contact to start the compound motor with resistors $R1$, $R2$, and $R3$ in the armature circuit and closes the two series-connected M interlocks to bypass the START button and *then* to provide a path for the remaining contactors and relays. After relay $TR1$ times out and N.O. contact $(TR1\text{-}T.C.)$ closes to actuate contactor $1A$, resistor $R1$ is short-circuited by contact $1A$ to accelerate the motor further; relay $TR2$ is energized simultaneously to begin timing for its part in cutting out the second resistor $R2$. Similar actions continue when relay $TR3$ times out and resistor $R3$ is short-circuited. An interesting feature in this circuit is the addition of two interlocks on

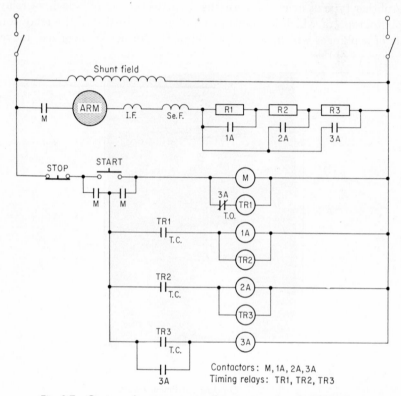

Fig. 3·7   Compound motor connected to starter using dashpot relays.

contactor $3A$; one of them N.O., is connected across $TR3\text{-}T.C.$ and the other, N.C., is in series with relay $TR1$. Thus, when contactor $3A$ operates one contact closes to bridge $TR3\text{-}T.C.$ in preparation for its opening and then opens another in the $TR1$ circuit to drop out that relay. These operations disconnect relays $TR2$ and $TR3$ and deenergize contactors $1A$ and $2A$. Since interlock $3A$ continues to maintain contactor $3A$ and power contact $3A$ effectively short-circuits all three accelerating resistors, the motor operates normally with all but two contactors (M and $3A$) dropped out.

An extremely popular type of mechanical timing for motor starters and many kinds of sequencing in automated systems is provided by the pneumatic timer. Illustrated by Fig. 3·8, the relay is constructed in two parts,

Fig. 3·8   Pneumatic timing relays.

the lower section consisting of the magnetic core, its exciting winding, and several sets of *instantaneously operated contacts* and the pneumatic *timing head* with its *timed contacts* in the upper part. Internally, the timer consists of an upper and lower chamber separated by a silicon rubber diaphragm, and timing is accomplished by permitting air to move through a small orifice from one chamber to another. The size of the orifice, adjusted from the outside by turning a screw (shown at the top of the photograph), determines how long it takes for the air to pass from one side of the diaphragm to the other. When the coil is deenergized, the core of the electromagnet is pushed up by a spring to exhaust the air from the lower to the upper chamber through a check valve. To begin the timing function the coil is energized; this pulls the core down electromagnetically against the force of a spring, relieves the thrust against the bottom of the rubber diaphragm, and permits air to travel slowly by atmospheric pressure through the orifice from upper to lower chamber. A bar fastened to the slowly moving diaphragm trips the contact mechanism after it reaches its final downward position to close and/or open several sets of contacts.

A circuit similar to that of Fig. 3·7, comprising pneumatic timers instead of dashpots, can be used to start a motor by definite-time acceleration.

Time-delay Contactors. A unique scheme for obtaining time delay without the use of relays is to construct the contactors with special copper sleeves, slipped directly over the cores and uninsulated therefrom, before the exciting coils are installed. Under steady-state conditions, i.e., when a constant direct current energizes the coil of such an electromagnet, the contactor behaves normally to set up a magnetic force to attract its armature, and the copper sleeve is electrically idle. However, when the coil is *disconnected* from the power source, the copper sleeve, acting like a short-circuited secondary of a transformer, carries an induced current whose mmf establishes flux that opposes decay of magnetism. The armature, therefore, is *not* released instantly upon deenergization of the coil but stays picked up until the flux is reduced sufficiently to permit dropout. Obviously, time delay for such a contactor to close N.C. contacts or open N.O. contacts in a control circuit is fixed by the size of the copper sleeve and is not subject to adjustment.

A circuit diagram illustrating how *time-delay contactors* are employed in connection with d-c motor acceleration is given in Fig. 3·9. In this

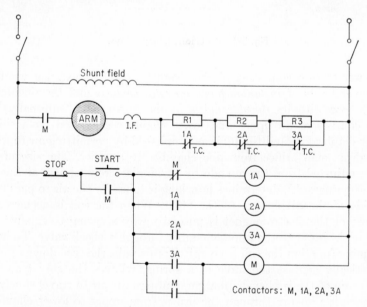

Fig. 3·9 Shunt motor connected to a starter with time-delay contactors.

arrangement the contactors are equipped with N.C. contacts that are connected across the appropriate accelerating resistors and are held open when the coils are energized. Then, when the contactors are deenergized in sequence, each one drops out in turn to permit the accelerating resistors $R1$, $R2$, and $R3$ to be short-circuited and bring the motor up to speed.

In this circuit it is necessary to keep the START button depressed long enough to permit contactors $1A$, $2A$, $3A$, and M to pick up, open the N.C. contacts across $R1$, $R2$, and $R3$, and close main contact M. The motor will then start properly, after which the START button may be released because contactor $1A$ will be sealed by interlock M; also, interlock $3A$ will be bypassed by another interlock M to keep contactor M picked up. With the energization of contactor M, the N.C. contact in the $1A$ circuit will open; this causes contactor $1A$ to drop out with a time delay (as described above) to close contact $1A$-T.C. across $R1$. Deenergization of contactor $1A$ also opens an interlock in the $2A$ contactor circuit which proceeds to drop out with a time delay; this is followed by the short-circuiting of resistor $R2$, a further increase in motor speed, and the opening of the $2A$ interlock in the $3A$ contactor circuit. Finally, when contactor $3A$ is disconnected from the line, it, too, releases its armature with a time delay so that the last resistor $R3$ is short-circuited.

An extremely important point to be noted in this type of starter, in addition to the fact that no relays are needed, is that the time-closing contacts across the accelerating resistors *do not interrupt current;* these contacts are opened before they carry current. The air gaps in such contactors, are, therefore, short, and the mmfs to operate them are necessarily comparatively low.

Inductive Time-limit Contactors. Another type of contactor that exerts a time delay on the closing of its N.O. contact (or the opening of an N.C. contact) is constructed with two coils. Designed like the lockout acceleration contactor (discussed on page 59 and illustrated in Fig. 3·4), it has two coils whose mmfs act simultaneously on a pivoted armature; one of them, the holding coil, tends to oppose pickup, while the other, the closing coil, attempts to pull the armature up. As constructed, the holding coil magnetizes a circuit consisting largely of iron—its air gap is extremely short, i.e., several thousandths of an inch, and it can prevent pickup by the closing coil even when it is excited at about 1 per cent of line voltage. Also, its circuit is highly inductive and, when the holding coil is short-circuited in a properly connected circuit, will permit the current to decay slowly. The closing coil, on the other hand, acts on a magnetic circuit that has a large air gap, about $\frac{3}{8}$ in., and with line voltage applied across its coil is unable to pick up the contactor when the holding coil carries current.

Figure 3·10 represents a circuit in which three *inductive time-limit* contactors are used to provide the necessary time delays for short-circuiting

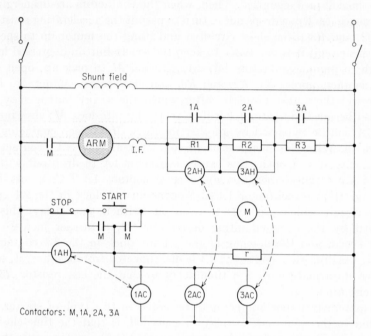

Shunt field

1A 2A 3A

ARM I.F. R1 R2 R3

2AH 3AH

STOP START

M

M M

1AH r

1AC 2AC 3AC

Contactors: M, 1A, 2A, 3A

Fig. 3·10 Shunt motor connected to a starter with inductive time-limit contactors.

the acceleration resistors. With the closing of the line switch, holding coil $1AH$ is energized through protecting resistor r; this keeps contactor $1A$ open until the current in $1AH$ drops to zero. When the START button is pressed, contactor M will pick up to close the main contact M and the two interlocks M. The inrush of armature current starts the motor and, at the same time, produces voltage drops across $R1$, $R2$, and $R3$; the latter then permits current to pass through $2AH$ and $3AH$ to lock out contactors $2A$ and $3A$ until the currents in the respective holding coils drop to zero. Closing coils $1AC$, $2AC$, and $3AC$ are also excited, but this has no immediate effect upon the three contactors because, as indicated, they are held open by the holding coils.

The short-circuiting of $1AH$ by one of the M interlocks causes the current in that coil to decay with a time delay, and when its value is nearly zero, closing coil $1AC$ actuates contactor $1A$, which, in turn, closes contact $1A$ across resistor $R1$. Since $2AH$ is now short-circuited, its current falls with a time delay until, at a low value, closing coil $2AC$ permits contactor $2A$ to pick up. This action also short-circuits $3AH$, and its current, like the others, drops with a time delay. At a value close to zero closing coil $3AC$ picks up contactor $3A$, closes contact $3A$ across $R3$, and puts the armature directly across the line.

It should be pointed out that, when each of the contactors closes, the magnetic air gap of the closing coil becomes less than that of the already small holding-coil gap. The result of this change in magnetic-circuit reluctance is to strengthen the action of the closing coil with respect to the holding coil so that a definite and strong pull is developed by each contactor in the closed position.

Time-current acceleration starters. To combine the advantages of both current-limit and definite-time acceleration, the *time-current* relay was developed by the Electric Controller and Manufacturing Division of the Square D Company. In this unique relay design, motor acceleration proceeds at a definite time rate, although it does so more slowly as the mechanical load on the motor increases. Thus, more time is allowed between steps when loads are heavier, with the result that the motor has an opportunity to develop higher values of cemf before the accelerating resistors are short-circuited; current peaks are thereby reduced.

A sketch illustrating the general construction of the time-current acceleration is given in Fig. 3·11. With the series coil of the relay connected

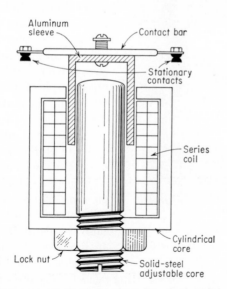

Fig. 3·11 Cross-sectional sketch of a time-current acceleration relay.

in series with the armature circuit, an inrush of current creates a sudden increase in magnetic flux that passes through the adjustable solid-steel core and the outer cylindrical steel shell. As magnetism rapidly builds up, a current is induced in the aluminum sleeve surrounding the upper portion of the central steel plug. Since, by Lenz's law, the induced current creates a flux that opposes the very action that produces it, a momentary force of repulsion exists between the core and the aluminum sleeve. The latter, carrying the upper movable contacts, therefore, jumps up to open two interlock contacts that are connected in the coil circuit of a contactor; under this condition the contactor is dropped out. However, as the armature current subsides with motor acceleration, the sleeve falls slowly, being retarded both by a cushion of air and by eddy currents that are induced in the opposite direction. Moreover, with

increasing values of armature current the rate of fall of the sleeve diminishes. When the aluminum piece finally reaches its lowest point of travel, both interlock contacts close, an accelerating contactor picks up because its coil is now energized, and a contact short-circuits a resistor. Time adjustment is effected by screwing the steel plug up or down in a loosened lock nut. Moving the core down to expose less iron to the aluminum sleeve reduces the time interval for a given coil current, and vice versa. A photograph of a completely assembled unit is shown in Fig. 3·12; clearly seen are the upper part of the aluminum cup and two contacts.

A circuit diagram illustrating the use of the time-current acceleration relay is given in Fig. 3·13. With the pressing of the START button, the coil of the M contactor is excited to close the main M contact and two M interlocks. The inrush current

Fig. 3·12 Completely assembled time-current acceleration relay. (*Square D Co.*)

through the armature circuit, which includes the series coil of accelerating relay $1AR$, causes the N.C. contact $1AR$ in the $1A$ contactor circuit to open. After a time delay, the falling sleeve permits the $1AR$ contact to close; contactor $1A$, therefore, picks up, closes its main $1A$ contact, and short-circuits coil $1AR$ and resistor $R1$. This is followed by a second inrush of armature-circuit current through relay coil $2AR$, the sudden rise of the sleeve of the latter relay, and the instant opening of the N.C. $2AR$ contact *before* the $1A$ interlock closes in the $2A$ contactor-coil circuit. When contact $2AR$ closes after the proper time delay, contactor $2A$ is energized and its contact bridges the $2AR$ coil and resistor $R2$. The armature current rises again, now through relay coil $3AR$, and the N.C. contact $3AR$ opens when the cup jumps up; contact $2A$ in the $3A$ contactor-coil circuit then closes. Upon reclosure of contact $3AR$, contactor $3A$ is actuated to short-circuit the

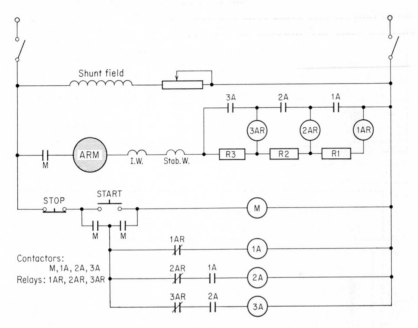

Fig. 3·13 Shunt motor with stabilizing field connected to a starter with time-current accelerating relays.

final accelerating resistor $R3$. The shunt motor with its series stabilizing field and a shunt-field rheostat for speed adjustment (to be discussed subsequently) is thus connected for normal operation.

Neo Timer acceleration starter. An interesting timing device that can be applied to starters for a-c as well as d-c motors and is capable of being adapted readily to give a wide range of time intervals between the operation of any number of accelerating contactors is the Neo Timer. Developed by the Electric Controller and Manufacturing Division of the Square D Company, it consists of two timing relays, a control relay, a neon tube, a capacitor, resistors, and a tap-switch rheostat. A circuit diagram showing how it is applied to a compound motor is given in

Fig. 3·14. Before its behavior is described, the following points should be noted: (1) Each of the two timing relays has two N.O. and two N.C. contacts, with the N.O. contacts labeled $1TR_a$, $1TR_b$, $2TR_a$, and $2TR_b$ and the N.C. contacts identified by $1TR_c$, $1TR_d$, $2TR_c$ and $2TR_d$; (2) timing is effected through a series RC circuit consisting of a capacitor

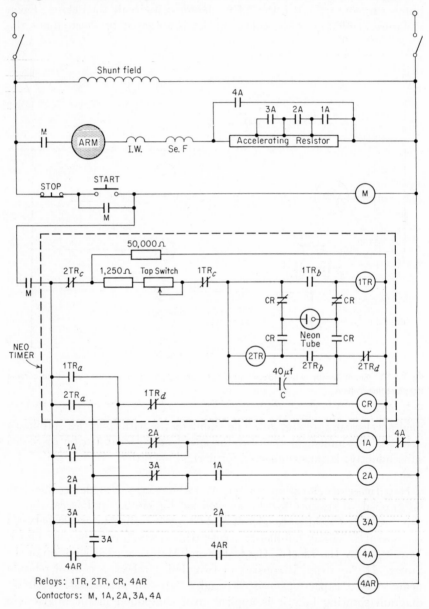

Fig. 3·14   Compound motor connected to a Neo Timer type of starter.

C, a fixed resistor (1,250 ohms), and a tap switch resistor, with adjustable time intervals of about 1 to 15 sec; (3) the neon glow-discharge tube is connected so that, when the capacitor C reaches a potential of about 90 volts, it breaks down and conducts current with zero voltage drop.

Operation begins when the START button is pressed. This energizes the M contactor which closes the main contact M, the sealing interlock M, and an auxiliary contact M that connects the power source to the Neo Timer and the control circuit. As the motor starts with all the accelerating resistor in the armature circuit, the capacitor C begins to charge through $2TR_c$, two resistances, $1TR_c$ and $2TR_d$, and contact $4A$. After a definite time (determined by the RC constant of the timer) the neon tube breaks down, i.e., conducts, under which condition the $1TR$ relay is actuated; the latter is then connected to the line through two N.C. CR contacts. This action closes the $1RT_a$ contact leading to the $1A$ contactor, which in turn cuts out part of the acceleration resistance; contactor $1A$ also makes its own holding circuit through interlock $1A$. In addition, relay $1TR$ opens $1TR_c$, opens $1TR_d$ in the CR relay circuit, and closes $1TR_b$ across the neon tube.

The capacitor now begins to discharge through the $1TR$ relay. After the charge is sufficiently dissipated, relay $1TR$ drops out to open $1TR_a$ and $1TR_b$ and close $1TR_c$ and $1TR_d$. Control relay CR now picks up through contacts $1A$, $2A$, and $1TR_d$, and this action opens two N.C. CR contacts and closes two N.O. CR contacts in the Neo Timer. With the contact changes indicated, relay $2TR$ is connected to the line through the neon tube as was relay $1TR$ when operations were first started.

The capacitor starts its second period of charge and, when its voltage is high enough, causes the neon tube to break down again. The timing relay $2TR$ is energized with the result that $2TR_c$ and $2TR_d$ open and $2TR_a$ and $2TR_b$ close; also, contactor $2A$ is excited through $2TR_a$ and interlocks $3A$ and $1A$, is sealed by its own interlock $2A$, and opens the N.C. contact in the $1A$ contactor circuit. The last action causes the CR relay to drop out with the result that two N.C. contacts close and two N.O. contacts open in the Neo Timer. Obviously, when contactor $2A$ operated, a second step of accelerating resistance was short-circuited.

After the capacitor goes through its third period of charge and discharge, accelerating contactor $3A$ picks up, closes its interlock sealing contact $3A$, and cuts out a third step of accelerating resistance.

Finally, the Neo Timer passes through a fourth series of actions, timed by the charging and discharging of the capacitor, after which relay $4AR$ is connected to the line through $2TR_a$ and another interlock $3A$. This causes two interlocks $4AR$ to close, connects contactor $4A$ to the source, and cuts out the entire acceleration resistor. The operation of contactor $4A$ also opens an N.C. interlock $4A$ (to the right of contactor $1A$) with the result that contactors $1A$, $2A$, and $3A$ are deenergized in turn and the

Neo Timer is open-circuited; main contacts 1*A*, 2*A*, and 3*A*, therefore, open.

Should there be a power failure during operation, the capacitor will discharge through the 50,000-ohm shunting resistor; the Neo Timer will then revert to normal and prevent any change in the timing period when starting operations are resumed.

Reversing of d-c motors. The direction of rotation of d-c motors can be reversed in either of two ways; these are (1) reversing the current through the armature, and this *includes the interpole field* if there is one, and (2) reversing the current through all fields (shunt and/or series) but *not the interpole field*. Since field reversing generally involves breaking and making highly inductive circuits and when, as in compound motors, shunt and series windings must be changed, armature reversing is the preferred method. Alternate schemes used occasionally are (1) to have two series fields in a small series motor, one of them connected in series with the armature to give one direction of rotation and the other replacing the first one to provide the opposite rotation, and (2) having the armature of a separately excited motor connected to its own d-c power supply whose polarity may be reversed when the rotation of the motor is to be changed. The discussion that follows concerns armature reversal only, although the other methods will be considered in appropriate places later.

The fundamental motor connections showing the forward *F* and reverse *R* contacts in the armature circuit is given in Fig. 3·15. It indicates that

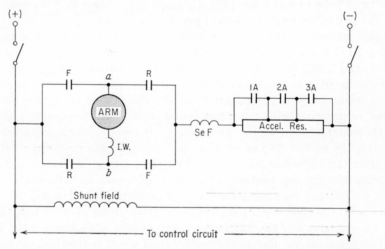

Fig. 3·15   Fundamental circuit for armature reversing of a compound motor.

the current will be from *a* to *b* when the FORWARD contactor is energized and the *F* contacts close and from *b* to *a* when the REVERSE contactor is picked up and the R contacts close. Obviously, some scheme must be provided that prevents the R contactor from being actuated when the motor is running in a forward direction with the F contacts closed, or vice versa. This is because the accelerating resistors are cut out when the motor is operating normally, so that having both sets of contacts closed simultaneously would be equivalent to short-circuiting the power lines and the cemf of the armature, an extremely serious possibility. One of the most common methods that avoids the condition indicated is to interlock the contactors mechanically. This locks out one pair of contacts when the other pair is closed and permits armature-current reversal only after the armature is disconnected completely from source and the accelerating resistors are again inserted. Frequently, electrical interlocks are provided in the control circuit that make it impossible for one contactor to pick up while the other is energized. Sometimes special push buttons are used that have two sets of contacts, one pair, the back contacts, being normally closed and the other pair, the front contacts, being normally open. Wiring connections in the control circuit can then be made to interlock the contactors electrically so that only one contactor will function at a time. Moreover, many circuit arrangements include both mechanical and electrical interlocking to incorporate an added measure of safety, timing relays that prevent reversal until the motor has come to a complete stop, special antiplug relays and zero-speed plugging switches (to be discussed later), and others.

A control-circuit diagram showing two mechanically interlocked contactors (represented by dashed lines that join them) connected to a push-button station and interlocks F and R that lead to accelerating relays and contactors (omitted) is illustrated by Fig. 3·16a. Also included are

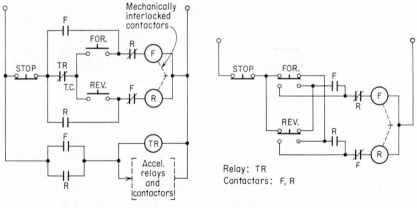

(a) Circuit using single-pole push-buttons   (b) Circuit using two-pole push-buttons

Fig. 3·16 Control circuits for armature reversing of d-c motors.

Fig. 3·17 A FOR.-REV.-STOP push-button station shown with the front cover removed.

electrical interlocks to provide extra protection, with N.C. *F* contact in series with the REVERSE contactor coil and the N.C. *R* contact in series with the FORWARD contactor coil. A further addition is a timing relay *TR* to ensure that the motor does not reverse until it has come to a complete stop; with this relay the N.C. *TR* contact opens instantly when the motor starts in either direction and makes both buttons inoperative until the *TR* relay times out after the STOP button is pressed.

In Fig. 3·16*b* the FORWARD AND REVERSE push buttons are shown to have back and front contacts. Note particularly that, when the FORWARD button is pressed, connection to the FORWARD contactor coil is made through the back contact of the REVERSE button and vice versa. Thus, if the *F* contactor is picked up and the REVERSE button is pressed, the back contact of the latter opens first to release the FORWARD contactor and then closes the front contact to energize the REVERSE contactor coil. A photograph of a FOR.-REV.-STOP push-button station of the type described here is illustrated in Fig. 3·17 with the front cover removed.

Figure 3·18 shows a push button with clearly visible back and front contacts in dissassembled and assembled views.

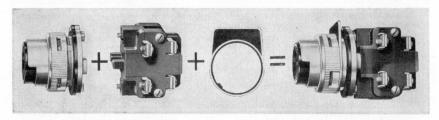

Fig. 3·18   A push button with back and front contacts shown disassembled and assembled.

Jogging. As defined by NEMA *"jogging (inching)* is the quickly repeated closing of a circuit in order to start a motor from rest for the purpose of accomplishing small movements of the driven machine."* Starting and stopping of the motor are then under the direct and immediate control of the operator, whose primary objective is generally to move tools, equipment, or processed material into position very slowly. Motors are jogged frequently to align and space machine parts, position a tool accurately, thread cloth, webbing, sheet steel, or other such material into a long processing line, move a small portable crane in incremental steps, and many others. Since d-c motors operate for short periods and are not expected to reach normal speeds *during jogging operations,* control circuits are designed to prevent the accelerating resistors from being cut out. This means, therefore, that the accelerating contactors are not permitted to pick up when the JOG button is pressed and that the maximum speed is limited to a low value by the comparatively large armature-circuit resistance. Furthermore, in some installations jogging is restricted to one direction of rotation while in others it involves both forward and reverse motions.

A simple control circuit that provides for jogging in one direction only is given in Fig. 3·19. Note particularly that the M contactor coil is

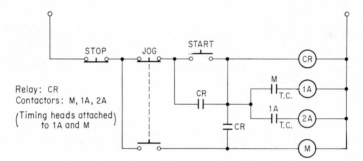

Fig. 3·19   Control circuit that provides for jogging a d-c motor in one direction of rotation

energized but *not* sealed when the JOG button is pressed and that the accelerating contactors cannot pick up under this condition because the control relay CR is inoperative. However, when the START button is pressed, control relay CR is actuated to close the two CR interlocks; the latter then permit the M contactor to pick up and seal the two coils (M and CR). Since a path is now provided to the accelerating contactors $1A$ and $2A$ through the CR contacts, they will be energized after the closing of the properly timed contacts that are associated with M and $1A$.

The control-circuit diagram of Fig. 3·20 illustrates how a d-c motor can be jogged or operated normally in both directions of rotation. Note particularly that the forward and reverse contactors F and R are mechanically interlocked and that provision is also made for electrical interlocking by employing N.C. contacts F and R. When the JOG FOR. or JOG REV. button is pressed, the corresponding contactor F or R is energized but is not sealed in; neither do the accelerating contactors $1A$ and $2A$ pick up because contacts CRF and CRR are open. For normal running on the other hand, the FOR. or REV. button is pressed. This will first actuate either the forward control relay *CRF* or the reverse control relay *CRR* and will in turn cause the proper interlocks *CRF* or *CRR* to close. The forward or reverse contactor then picks up, is sealed in, and relay *TR* starts to time. When contacts *TR*-T.C. and *1A*-T.C. close after preset time delays, accelerating resistors are short-circuited.

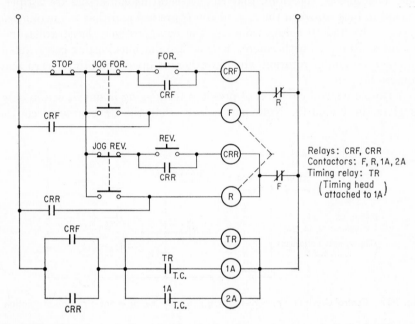

Fig. 3·20 Control circuit that provides for jogging a d-c motor in both directions of rotation.

Dynamic braking. Another function that is frequently provided by a control circuit is that of bringing a motor to a quick stop. Three general methods used to accomplish this are (1) by coupling a mechanical brake directly to the motor, (2) by plugging the motor to a stop, and (3) by dynamic braking, which makes use of the stored energy in the rotating system to stop the motor. A fourth method, called regenerative braking, can be applied to a motor driving an overhauling load, like a hoist, when it is desired to check the lowering speed, not to perform a complete stop. The third of these will now be considered, with the others deferred for later discussion.

The *dynamic-braking* principle involves generator action in a motor to bring it to rest. If the armature terminals are disconnected from the source and immediately transferred to a resistor, *the field being kept energized*, the motor will come to a quick stop. The reason for this braking action can be explained in two ways. While the motor is operating normally, the polarity of the source, although opposed by a slightly lower cemf, determines the direction of the current in the armature winding. However, when the armature terminals are switched from the source to a resistor, the existing *oppositely directed cemf reverses the current direction in the armature* and attempts to cause motor reversal. Remember that the cemf is a *generated voltage* that results from the rotation of the armature, which, in turn, depends upon the stored mechanical energy in the revolving system. The motor, therefore, proceeds to slow down as it tries to run in the opposite direction and in doing so uses up and causes the energy of rotation to diminish. Obviously, when the motor comes to rest, there is no longer any stored mechanical energy, so reversal cannot take place. Another explanation is that the cemf sends a current into the resistor when the changeover is made, and this involves a conversion

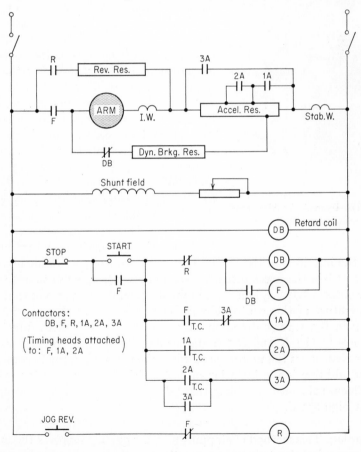

Fig. 3·21   Shunt motor connected to a starter that provides dynamic braking and jog reversing.

of mechanical energy that is stored in the rotating system into electrical energy for the resistor, i.e., I^2R. During this period of energy conversion the motor continues to slow down until, at zero speed, the interchange is complete. The rapidity of the stop depends on the rate of energy conversion, and this is determined by the value of the resistor; for small values of the latter, the energy is dissipated rapidly, and this results in a quick stop.

A control-circuit diagram that provides for dynamic braking is shown in Fig. 3·21. It will be noted that the armature and a portion of the accelerating resistor are shunted by a dynamic-braking resistor and an N.C. *DB* contact. Moreover, the dynamic-braking contactor is equipped with two coils that act to oppose each other to *open* the N.C. contact. The arrangement of the coils on the contactor as shown in Fig. 3·22 indicates that the upper *retard coil* tends to keep the contact closed while the lower one attempts to open it. When the START button is, therefore, pressed, the *existing flux produced by the retard coil* prevents the lower main coil from actuating the contactor to open the *DB* contact *until the main flux is well established on the saturation curve*, at which instant the contactor quickly opens with a snap action. This disconnects the dynamic-braking resistor from the armature circuit,

Fig. 3·22 Dynamic braking contactor with retard coil. (*Square D Co.*)

closes the interlock in the F contactor-coil circuit to permit the motor to start, and lets the accelerating contactors operate in succession to close main contacts $1A$, $2A$, and $3A$ and finally open the first two of these. An added feature of this control circuit is an arrangement that provides for jogging in the reverse direction only. Observe the two N.C. directional interlocks in the R and DB contactor circuits and the special *reversing resistor* that limits the motor speed when the motor is jogged in reverse. A further addition is the inclusion of a stabilizing field, connected outside the influence of control contacts, that helps to stabilize shunt-motor operation at high speed.

Plugging. This method of stopping a motor quickly involves the principle of reversing the armature connections while it is running in a given direction, under which condition a countertorque is developed that retards rotation. The motor, therefore, slows down, and when it has stopped and is ready to reverse, an automatically actuated switch functions in the control circuit to disconnect the power source. During the plugging period the armature cemf adds to the impressed voltage, in contrast to their opposition for normal operation. Therefore, to prevent the armature current from becoming excessive, particularly at the instant the motor is plugged when the armature-circuit voltage is nearly twice rated value, a so-called plugging resistor must be inserted.

Plugging is frequently used, not only to bring a motor to a complete stop rapidly, but in drives that are reversed repeatedly, as in certain machine tools, cranes, and rolling mills. In such cases, master controllers, to be discussed later, are generally employed to perform a plug-reverse operation. This is accomplished by moving the handle from *full-speed forward to full-speed reverse*, or vice versa, in which event the motor plugs to a stop and then, with accelerating resistors properly inserted and cut out automatically by timing relays, comes up to speed in the opposite direction.

Fig. 3·23 illustrates how a *nonreversing* shunt motor, running normally,

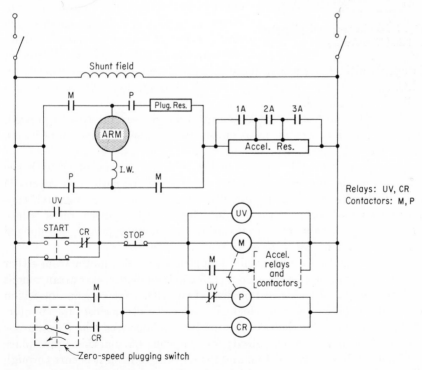

Fig. 3·23 Shunt motor connected to a starter and a zero-speed plugging switch that provides for plugging to a stop.

is brought to a quick stop by pressing the STOP button. Coupled directly to the shaft of the motor is a so-called *zero-speed plugging switch*, depicted in Fig. 3·24, whose N.O. contact closes at

the instant rotation of the armature begins. This is accomplished by having a revolving pulley transmit a force to a friction belt whose ends are fastened to a lever that carries one of the contact points. The motor is started in the usual way by pressing the START button. This energizes the M contactor and the undervoltage relay UV, and they in turn, close the two main contacts M for the armature, two interlocks M for the accelerating relays and contactors, as well as a plugging contactor P and a control relay CR, and a UV interlock for sealing purposes; also the N.C. UV contact in the P contactor circuit is opened. At the instant the motor starts to revolve, the contact in the plugging switch closes, but this action has no immediate effect on the P contactor (the UV contact

Fig. 3·24 Zero-speed plugging switch. (*Allen-Bradley Co.*)

is open), although the CR relay picks up, seals itself in, and opens an N.C. contact in the START button circuit. Acceleration now proceeds until full speed is reached.

The motor does *not* coast to a stop when the STOP button is pressed. Instead, the M contactor drops out to reinsert the acceleration resistors, and then with the deenergization of the UV relay the P contactor picks up when the N.C. UV contact closes. The motor now plugs with the addition of the plugging resistance (the armature current is reversed) and quickly comes to a stop at which instant the plugging-switch contact opens to disconnect the P contactor from the power source. Since the CR relay remains energized during the plugging period, the opened N.C. CR contact makes the START button inoperative.

A circuit diagram that permits plugging a *reversing* motor from either direction of rotation is given in Fig. 3·25. Included in this arrangement is a zero-speed plugging switch with two N.O. contacts, one of which closes when the motor runs clockwise while the other closes for counterclockwise rotation. If the FOR. or REV. button is pressed, the P contactor and the CR relay are energized; for the FOR. button this takes place through the N.C. REV. button, and for the REV. button it occurs through the N.C. FOR. button. (This is desirable because the F contactor drops out when the REV. button is pressed and vice versa.) Assume that the

FOR. button is pressed. The F contactor picks up through the closed CR contact, and the motor accelerates normally through the proper timing of the accelerating relays and contactors. Also, with the closing of the FOR. contact in the plugging switch, the reverse relay RR is energized to

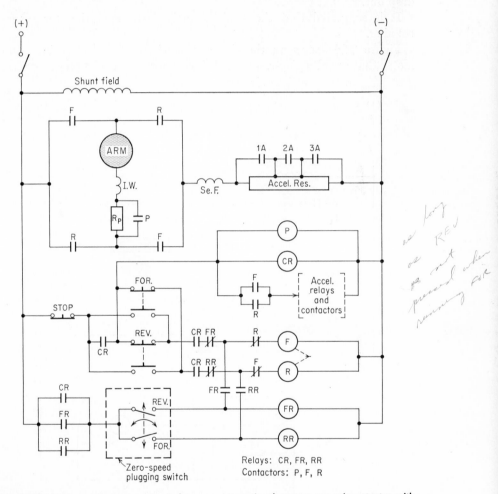

Fig. 3·25 Compound motor connected to an automatic plug-stop reversing starter with lockout protection.

be sealed in by one RR interlock and to close another interlock that leads to the N.C. F interlock that was opened by the F contactor when it operated. If, next, the STOP button is pressed, the motor plugs to a stop because the F contactor drops out to close the N.C. F interlock in the R contactor circuit. This permits the R contactor to pick up, and since the P contact across the plugging resistor R_P is open (P has been disconnected from the line), a countertorque is developed with reduced armature current. When the motor finally comes to rest and the FOR. contact in the plugging switch opens, the RR relay and the R contactor drop out.

It is suggested that the student analyze the operation for reverse running.

Figure 3·26 shows another interesting circuit for a reversing shunt motor that permits *plug-reversing*. It incorporates a special plugging relay,

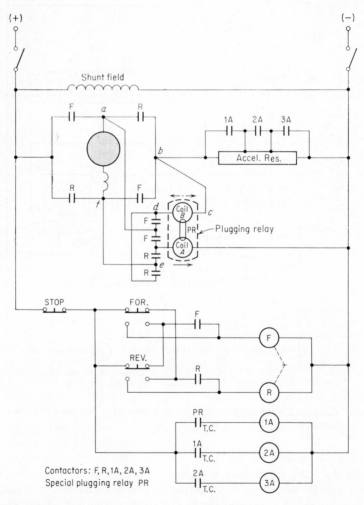

Fig. 3·26 Shunt motor connected to an automatic plug-stop reversing starter using a special plugging relay.

illustrated by Fig. 3·27, which has two coils whose mmfs act on a common magnetic circuit that includes an armature and four N.O. contacts.

As connected, coil A always carries current in the *same* direction but the current in coil B can be in either direction. Also, when the current directions in A and B are the same or if no current passes through B, the relay will be picked up; however, with current in B opposite to that of A the relay will not be actuated.

For forward running, the FOR. button is pressed. This causes the forward contactor F to pick up through the back contact of the REV. button and permits the motor to start with the accelerating resistors in the armature circuit; all F contacts will, of course, close under this condition. Since armature terminal a is positive, current passes through the A coil of the plugging relay from *left* to *right;* however, there will be very little current in coil B at the instant the motor starts (the voltage drop across the armature is extremely low), although it will increase rapidly from *left* to *right* as acceleration continues. The PR relay will, therefore, pick up and close the PR-T.C. contact in the $1A$ contactor circuit after a time delay. Then, as contactors $2A$ and $3A$ are also energized when contacts $1A$-T.C. and $2A$-T.C. close,

Fig. 3·27 Plugging relay. (*Clark Controller Co.*)

the motor comes up to speed in the usual way. Pressing the STOP button or touching the FOR. or REV. button lightly will drop out the F contactor to permit the motor to *coast* to a stop.

To plug-reverse the motor, the REV. button is pressed down tightly. As the shorting bar passes through the OFF position in moving from the back contact to the front contact, the F contactor is deenergized and the R contactor is energized. The latter, therefore, picks up to close the main R contacts and three interlocks. With coil A still connected across the line as before, its current continues to pass from *left* to *right*, the high cemf at the armature with terminal a positive causes current to pass through coil B from *right* to *left* as a path is made from a to b to c to d to e to f. Thus, the PR relay does *not* pick up and the motor plugs to a stop, at which instant the cemf drops to zero. At this point the plugging relay begins to function as for reverse running, causes contact

PR-T.C. to close after the proper time delay, and permits the motor to accelerate in the reverse direction.

Acceleration-resistor calculations. The design of the acceleration resistors for a d-c starter generally involves (1) computations for the resistance of the individual steps and (2) proper selection of the physical sizes of the units based on current ratings and duty cycles. The latter topic, requiring the use of standard NEMA resistor classifications, will be deferred to a later discussion, and calculations for the former will now be given as it applies to an illustrative example.

Figure 3·28 shows a resistor bank of the type under consideration, with nonbreakable grids assembled in a steel frame and taps brought out to a terminal board.

Fig. 3·28   Nonbreakable multiple resistor frame assembly with wiring to a terminal board.

Example 1. It is desired to determine the values of the four individual resistances in a five-step starter, given the following information for a shunt motor: 15 hp, 230 volts, 56-amp full-load current, resistance of motor (armature) is 8 per cent of V/I, inrush currents to be limited to

150 per cent of full load value, resistance steps to be cut out at rated current. (Calculations are to be made neglecting the shunt-field current.)

Solution

Max. inrush current $= 1.5 \times 56 = 84$ amp

$R_M = 0.08 \times {}^{230}\!/_{56} = 0.33$ ohm

Referring to Fig. 3·29,

$$(R1 + R2 + R3 + R4) = {}^{230}\!/_{84} - 0.33 = 2.74 - 0.33 = 2.41 \text{ ohms}$$

When the motor accelerates and the current drops to 56 amp, the cemf is

$$E_{c_1} = 230 - (56 \times 2.74) = 77 \text{ volts}$$

At this instant the first resistor $R1$ is short-circuited, a second inrush of

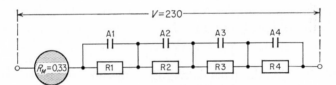

Fig. 3·29   Illustration for Example 1.

84 amp follows, when

$$(R2 + R3 + R4) = \frac{230 - 77}{84} - 0.33 = 1.82 - 0.33 = 1.49 \text{ ohms}$$

and

$$R1 = 2.41 - 1.49 = 0.92 \text{ ohm}$$

The motor continues to accelerate, and when the current again reaches 56 amp, the cemf is

$$E_{c_2} = 230 - (56 \times 1.82) = 128 \text{ volts}$$

At this instant the second resistor $R2$ is short-circuited, a third inrush of 84 amp follows, when

$$(R3 + R4) = \frac{230 - 128}{84} - 0.33 = 1.22 - 0.33 = 0.89 \text{ ohm}$$

and

$$R2 = 1.49 - 0.89 = 0.60 \text{ ohm}$$

When the motor speeds up until the current is again 56 amp, the cemf is

$$E_{c_3} = 230 - (56 \times 1.22) = 162 \text{ volts}$$

At this instant the third resistor $R3$ is short-circuited, a fourth inrush of

84 amp follows, when

$$R4 = \frac{230 - 162}{84} - 0.33 = 0.48 \text{ ohm}$$

and

$$R3 = 0.89 - 0.48 = 0.41 \text{ ohm}$$

When the current again drops to 56 amp, the fourth resistor $R4$ is short-circuited, and a fifth inrush of 84 amp follows.

Dynamic-braking resistor calculations. To bring a motor to rest by dynamic braking, it is often desirable to have the armature take a peak current that is close to the commutation limit at the instant the braking action is initiated. This means that the rotating energy of the system will be converted to electrical energy very rapidly with a resulting quick stop. To accomplish this it is obviously necessary to make the dynamic-braking resistance rather low. (In Fig. 3·21 it includes a portion of the acceleration resistance because the latter is part of the dynamic-braking loop.)

Example 2. Using the given data of Example 1, calculate the resistance of a dynamic-braking resistor if the peak current is to be limited to 250 per cent of rated value. (Assume that braking is to be initiated when the motor is operating at rated load.)

Solution

$$I_{\text{peak}} = 2.5 \times 56 = 140 \text{ amp}$$
$$E_{c_{\text{initial}}} = 230 - (56 \times 0.33) = 211.5 \text{ volts}$$
$$R_{db} = \frac{211.5}{140} - 0.33 = 1.18 \text{ ohms}$$

Remembering that dynamic-braking action is at a maximum at the instant it is initiated and that it diminishes as the motor slows down because the cemf drops progressively, it is possible to hasten motor stopping if a part of the dynamic-braking resistor is short-circuited at some lowered value of current. Indeed, this is sometimes done in several steps by progressively cutting out sections of the dynamic-braking resistor and in a manner similar to that used in accelerating a motor. In such procedures, called *graduated* braking, the currents follow a stepped pattern, rising to a peak at each reduction in resistance; moreover, quick steps are possible with low current peaks.

Example 3. For the given data of Example 1, determine the ohmic values of three sections of a dynamic-braking resistor if graduated braking is to be accomplished between 150 and 100 per cent of rated currents,

Solution

$$I_{peak} = 1.5 \times 56 = 84 \text{ amp}$$

Referring to Fig. 3·30,

$$(R1 + R2 + R3) = \frac{230 - 0.33 \times 56}{84} - 0.33$$
$$= 2.52 - 0.33 = 2.19 \text{ ohms}$$

$$E_{c_1} \text{ (at } I = 56) = 56 \times 2.52 = 141 \text{ volts}$$

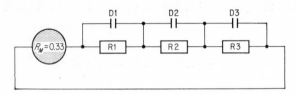

Fig. 3·30   Illustration for Example 2.

To raise the current to 84 amp by cutting out one step of resistance,

$$(R2 + R3) = \frac{141}{84} - 0.33 = 1.68 - 0.33 = 1.35 \text{ ohms}$$

Therefore,

$$R1 = 2.19 - 1.35 = 0.84 \text{ ohm}$$
$$E_{c_2} \text{ (at } I = 56) = 56 \times 1.68 = 94.3 \text{ volts}$$

To raise the current to 84 amp by cutting out a second step of resistance,

$$R3 = \frac{94.3}{84} - 0.33 = 1.12 - 0.33 = 0.79 \text{ ohm}$$

Therefore,

$$R2 = 1.35 - 0.79 = 0.56 \text{ ohm}$$

Plugging-resistor calculations. As was previously mentioned, the voltage across the armature circuit at the instant a motor is plugged is almost twice rated value because the impressed and cemfs are aiding then. If plugging is initiated when the armature current is I_{FL}, this plugging voltage is

$$V_P = V + E_c = V + (V - I_{FL}R_M)$$

But

$$R_M = p \frac{V}{I_{FL}}$$

where p is the per unit resistance of the motor

Therefore,

$$V_P = V(2 - p)$$

If the total resistance during plugging is

$$R_t = R_P + R_M + R_{\text{accel}}$$

$$R_P = \frac{V(2 - p)}{I_{\text{inrush}}} - R_M - R_{\text{accel}} \tag{3.1}$$

Example 4. What should be the magnitude of a plugging resistance for the motor of Example 1 if the inrush current is to be limited to $1.5I_{FL}$?

Solution

$$R_P = \frac{230(2 - 0.08)}{84} - 0.33 - 2.41 = 2.51 \text{ ohms}$$

QUESTIONS AND PROBLEMS

1. Give several reasons for the use of reduced armature-voltage starters for d-c motors. Which of the reasons given do not apply to the starting of squirrel-cage induction motors?
2. Name two general classifications of automatic starter for d-c motors. Under each classification list several particular types of starter.
3. What important mechanical constructions and adjustments are necessary in the *series-relay* automatic starter of Fig. 3·1 if it is to function properly? Why are two interlocks M used in the circuit?
4. In the cemf starter circuit of Fig. 3·3, it would be possible to eliminate the three relays $1AR$, $2AR$, and $3AR$, replacing them by contactors $1A$, $2A$, and $3A$; this simplification would reduce the number of components. Why would such a change be undesirable?
5. In the lockout current-limit starter of Fig. 3·4, describe how the holding and closing coils function to provide proper timing. What purpose is served by the shunt coil on contactor $3A$?
6. What is meant by *time-closing* (T.C.) and *time-opening* (T.O.) contacts? What useful function does the $3A$-T.O. contact have in Fig. 3·7?
7. Explain the action of the *copper sleeves* on contactors $1A$, $2A$, and $3A$ in Fig. 3·9. Is the timing of such contactors adjustable?
8. Referring to Fig. 3·10, explain the actions of the holding and closing coils in contactors $1A$, $2A$, and $3A$ to provide timing. What design features are essential for proper operation?
9. What is meant by a time-current acceleration relay? What advantages does such a relay have over those of the simple current-limit or time-limit classifications? Explain the action of the relay illustrated by Fig. 3·11.
10. Modify the circuit of Fig. 3·13 so that contactors $1A$ and $2A$ may be permitted to drop out after the motor comes up to speed.
11. Explain the operation of the Neo Timer circuit of Fig. 3·14. Before doing so, indicate how timing is adjusted.
12. Modify the Neo Timer circuit of Fig. 3·14 to include a fifth contactor $5A$ for a six-step starter.
13. List several methods for reversing the direction of rotation of a d-c motor. Which is the preferred method? Why?

14. When a control circuit is designed to provide *jogging*, it is essential that the acceleration relays and/or the contactors do not function while the operator is jogging. Why?

15. Using Fig. 3·19 as a guide, draw a complete diagram for the control of a non-reversible shunt motor so that the latter can be jogged or made to accelerate normally to full speed. Provide three acceleration relays and contactors.

16. Using Fig. 3·20 as a guide, draw a complete diagram for the control of a reversible compound motor so that the latter can be jogged or made to accelerate normally to full speed. Provide three acceleration relays and contactors.

17. The reversing control circuit of Fig. 3·20 shows N.C. *F* and *R electrical interlocks* in addition to mechanically interlocked *F* and *R* contactors. The former are not essential for proper protection in d-c control circuits but must be used to protect the operating coils of a-c contactors in a-c control circuits. Why?

18. List four general types of braking for electric motors. Under each type indicate suitable applications and give examples which require two kinds of braking.

19. Carefully explain the operation of the dynamic-braking contactor that employs a *retard* coil (Fig. 3·22) in addition to the operating coil.

20. A 25-hp 230-volt 92-amp shunt motor has an armature-circuit resistance that is 8 per cent of V/I. Calculate the ohmic value of a dynamic-braking resistor which will limit the braking current to 200 per cent of rated value, assuming that braking is initiated when the motor is delivering a 25-hp load.

21. Design a control circuit for a compound motor in which the automatic starter will have two accelerating contactors and include jogging and dynamic-braking features.

22. Using the given data of Prob. 20, calculate the ohmic values of two sections of dynamic-braking resistor if the maximum current peaks are not to exceed $1.75I_{FL}$ and the first resistor section is to be cut out at I_{FL}.

23. What is a *zero-speed plugging switch* and how does it function in a control circuit that is designed to plug a motor?

24. Discuss the relative merits of dynamic braking and plugging.

25. Design a control circuit for a compound motor that can be plugged to a stop using a zero-speed plugging switch. Include the motor connections, acceleration resistors, and main and plugging contactors, but omit acceleration contactors and relays.

26. For the motor of Prob. 20 determine first the acceleration resistance that will limit the inrush current to $1.25I_{FL}$ and then calculate the magnitude of a plugging resistor that will not permit the current to exceed $1.75I_{FL}$.

27. Calculate the values of the plugging currents in Probs. 20 and 26 if the plugging resistance in each case is reduced to zero.

28. The following information is given in connection with a 230-volt d-c contactor: the core is circular with a diameter $D = 1.75$ in.; the exciting coil has 12,000 turns of wire with a hot resistance of 640 ohms; the air gap between end of pole core and armature is $\delta = 0.875$ in. Calculate the pull exerted by the contactor at the instant the coil is energized. (Neglect fringing of flux in air gap.)

NOTE: The pull of an electromagnet is given by the equation $P = B_g{}^2A/72 \times 10^6$, where B_g and A are, respectively, the flux density in maxwells per square inch and area in square inches of the air gap.

Also, ampere-turns required to overcome air-gap reluctance is given by the equation $NI = 0.313 B_g\delta$.

CHAPTER 4

Master Controllers and Control
Circuits for D-C Motors

Master switches—drum and cam types. Two extremely common types of switches that perform essentially the same functions as push buttons are *drum-* and *cam-type master switches*. Designed to be operated manually by rotating a lever, they control the coil currents in relays and contactors and, in general, the behavior of most circuit components.

As Fig. 4·1 clearly illustrates, a drum switch is constructed to open

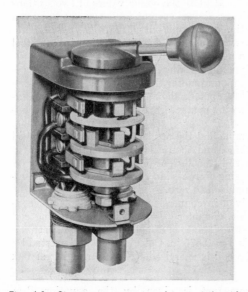

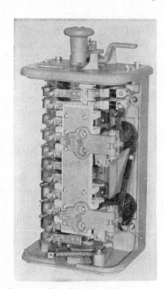

Fig. 4·1 Six-contact reversing drum switch with cover removed.

Fig. 4·2 Multiple-contact drum switch with cover removed.

and close contacts on segments or surfaces on the periphery of a rotating cylinder or sector. Depending upon the simplicity or complexity of the circuit arrangements, such switches are made in a variety of ways, with few or many contacts, for nonreversing or reversing service, and for use in d-c or a-c circuits. Figure 4·2 shows a multiple-contact drum switch that is designed for a reversing d-c motor (also see Fig. 2·15). As can be observed, conducting segments of various lengths are spaced in accordance with desired sequencing requirements and arranged to slide under stationary spring-loaded contact fingers. As the drum is rotated, segments and contact fingers touch at various designated positions to establish conducting paths to electrical devices. In practice, the unit is completely enclosed by a gasketed cover to protect it from dust and dirt.

An exposed view of a cam switch (Fig. 4·3) shows how electrical contacts are opened and/or closed by a mechanical action of a group of cams. Unlike the drum switch whose contact sequencing arrangement is somewhat difficult to alter once it is assembled, the cam switch lends itself readily to modification; if a closing-opening change is necessary after the switch is constructed, it can be accomplished by substituting another cam having a new contour or, under certain conditions, by renotching an existing cam. Note in Fig. 4·3 that small rollers fastened to stationary arms ride on the rims of cams and permit contacts to close or open in the proper sequence as the shaft is turned. An excellent view

Fig. 4·3　Multiple cam switch with cover removed.

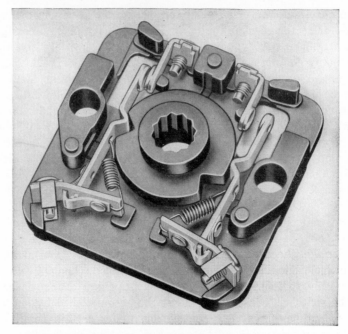

Fig. 4·4   Bakelite cam and contact section of a cam switch.

Fig. 4·5   Completely assembled cam switch for a reversing-motor drive. (*Square D* Co.)

of one bakelite cam section and two contacts is given in Fig. 4·4; it shows how the contacts that are fastened to the spring-loaded levers are made to move as the cam is rotated.

A completely assembled unit for a motor used in a reversing drive is illustrated by Fig. 4·5.

Advantages of master switches. In addition to the points already made, drum and cam switches have many advantages that make them useful in control systems, and especially where numerous functions such as acceleration, deceleration, reversing, braking, speed adjustment, and others must be provided. They not only are built for heavy-duty service but are capable of withstanding considerable abuse. This is because arc-blowout protection and excellent heat-resisting insulation are incorporated; moreover, heavy contact pressure can be maintained to prevent contact burning and poor electrical continuity. An extremely important aspect of these devices is that they can be arranged with a multiplicity of contacts which can be opened and closed to perform almost any desired sequencing and timing operations. This means, of course, that they can be readily adapted to the most complicated circuits and often with a minimum of wiring connections.

A number of circuit diagrams will now be given to illustrate the use of master switches in control circuits for d-c motors.

Nonreversing master. The circuit diagram of Fig. 4·6 represents a four-point nonreversing master connected to a compound motor. Note particularly that the master has one N.C. contact and four N.O. contacts in the OFF position. Also, when the handle is moved successively to positions 1, 2, 3, and 4, the N.C. contact opens and remains open in all four positions, while the N.O. contacts close in successive positions; contact *a* closes in position 1, contact *b* closes in position 2, contact *c* closes in position 3, and contact *d* closes in position 4, and having done so they continue closed in all higher numbered positions. These operating conditions are interpreted from the ✕ designations under the position numbers. Moreover when a master contact is closed or open, the relay or contactor to the right is either picked up or dropped out depending on whether and how other series contacts affect circuit continuity. Thus, for example, in the OFF position the UV relay is energized and sealed in by its interlock and remains picked up as the master is moved through the four positions. However, should there be a power failure while the master is in *any* ON position, the UV relay will drop out and will not be reenergized with a return of power until the handle is moved to the OFF position.

Analyzing Fig. 4·6 further, relay $1AR$ will be actuated in the OFF position of the master, and its contact in the coil circuit of contactor $1A$ will be opened. When the handle is moved to position 1, the M contactor operates (note the ✕ under point 1), and the armature circuit that includes the three accelerating resistors is connected to the line. The motor therefore starts, voltage drops exist across relays $2AR$ and $3AR$ causing them

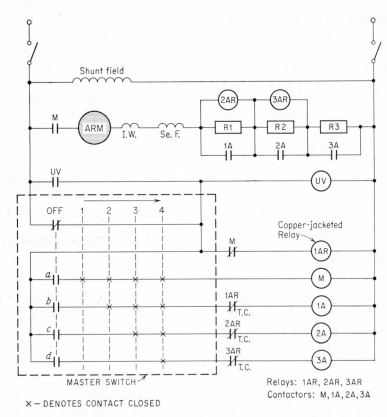

Fig. 4·6 Nonreversing controller connected to a compound motor.

to pick up, and contacts $2AR$-T.C. and $3AR$-T.C. open. Simultaneously, the N.C. M interlock in series with the $1AR$ relay opens, and since the latter is copper jacketed (see Fig. 3·9), it will drop out with a time delay; when it does, contact $1AR$-T.C. will close and contactor $1A$ will be prepared for the next step. In position 2 (note the two ✕'s under point 2) the $1A$ contactor is energized, resistor $R1$ is short-circuited, and the motor accelerates to a higher speed. At the same time the *highly inductive* $2AR$ *relay* is bridged, the current through its coil will decay *slowly* and, after the flux has dropped sufficiently, will close its contact and prepare contactor $2A$ for the next step. When the handle is moved to position 3 (note the three ✗'s under point 3), contactor $2A$ picks up, resistor $R2$ and relay $3AR$ are short-circuited, and the motor increases its speed. Again, relay $3AR$ drops out after a time delay, closes its $3AR$-T.C. contact after a time delay, and sets up the circuit for the final step. In position 4 (note the four ✗'s under point 4) contactor $3A$ is energized, contact $3A$ short-circuits resistor $R3$, and the motor accelerates to its normal operating speed.

It should be pointed out that exactly the same time-delaying procedure as described above would result if the master handle were moved immediately to position 4. This is because the successive dropouts of the three time-limit relays $1AR$, $2AR$, and $3AR$ depend only upon *when* contact M opens and contacts $1A$ and $2A$ close after the handle is moved to the last point. Moreover, the operator can select any intermediate motor speed by quickly moving the master to the desired point; in each case the relays will time out properly to accelerate the motor smoothly. Obviously, accelerating resistors that are used for continuous-duty service in adjustable-speed motor installations must be chosen on that basis rather than the short-duty limited-heating periods.

Reverse-plugging master. The master that permits motor reversing has a *center* OFF point and two sets of positions, one set for FORWARD or HOIST and the other set for REVERSE or LOWER. The same contacts are associated with both directions of motor rotation and may or may not be closed and/or opened in the same sequence.

A complete circuit diagram of a five-point reversing master controller connected to a series motor that operates a hoist is given in Fig. 4·7. This is a standard arrangement used by one manufacturer of control equipment and is designed for plug-reversing of motors rated at 50 to 125 hp. Note particularly that it incorporates time-current types of accelerating relay (see Figs. 3·12 and 3·13), a series brake that is spring set and electrically released (to be discussed subsequently), and a special plugging relay PR with its rectifier. The acceleration resistors are, of course, designed for continuous-duty service, so that hoisting and lowering speeds can be adjusted to desired operating conditions.

With the master in the OFF position all contactors are dropped out and the motor is at rest with the series brake clamping the armature tightly. For hoisting, the handle is moved to the first hoist point; this causes two master contacts to close so that the *directional* contactors $1H$ and $2H$ (double-pole type) and the main contactor M are energized. Current then passes through a series circuit consisting of the armature, the series field, the series brake, and the entire accelerating resistor. The brake releases instantly, and the motor starts with reduced torque. However, since points *a* and *b* are, respectively, at plus and minus potentials, the rectifier blocks current through the *plugging relay PR*, Fig. 4·8, so the latter does not pick up and the N.C. *PR* contact remains closed. Also, the *M* interlock in the *P* contactor circuit closes to prepare the master for the second position. At point 2, the *P* contactor picks up and cuts out one step of resistance

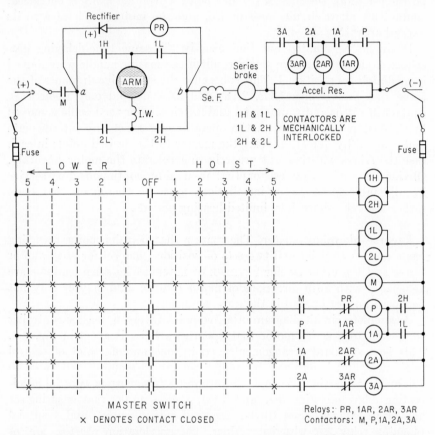

Fig. 4·7 Reversing controller connected to a series motor. Designed for reverse-plugging using a plugging relay and rectifier.

as a high inrush current passes through timing relay $1AR$ successively to open the N.C. $1AR$ contact and close the *P* contact in the 1A contactor circuit. When the armature-circuit current drops sufficiently, the $1AR$ relay drops out; then, in master position 3 contactor 1A operates to short-circuit a second section of the accelerating resistor through relay $2AR$. As the master is then moved progressively to points 4 and 5, contacts 2A and 3A close as their respective contactors are energized after proper time delays. In fact, if the master is moved at once to point 5, all contactors except P (which operates instantly on position 2) pick up with proper time delays in much the same way as in any automatically operated push-button starter.

If, next, the handle is returned to the OFF position, all contactors drop out, the series brake resets mechanically, and the motor comes to rest.

To plug-reverse the motor so that it comes to a quick stop and then accelerates normally in the reverse direction, the master is rapidly moved

Fig. 4·8   Special plugging relay used with a rectifier (see Fig. 4·7). (*Square D Co.*)

to point 5 LOWER. Under this condition the directionals $1L$ and $2L$ pick up, their main contacts close and, because the armature is still rotating in the same direction, cause the cemf in the armature to make points b and a, respectively, plus and minus. Now relay PR is actuated, since the rectifier is connected to properly polarized terminals. This means that N.C. contact PR in the P contactor circuit opens and remains open until, at a low motor speed when the cemf drops sufficiently, the PR relay drops out. The contact of the latter then closes, and the motor proceeds to speed up in the lowering direction.

The operating conditions described will, of course, repeat themselves to plug-reverse from LOWER to HOIST if the master is quickly moved to position 5 on the opposite side.

Another circuit arrangement that performs in essentially the same way as the foregoing but uses two plugging relays (no rectifier) is given in Fig. 4·9. As previously described, a five-point reversing master controller

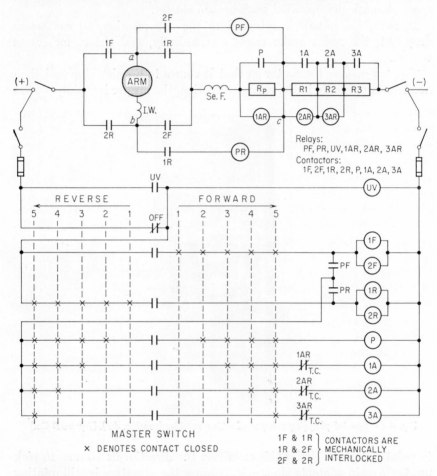

Fig. 4·9 Reversing controller connected to a series motor. Designed for reverse-plugging using two plugging relays.

is used with the usual complement of contactors and relays; in addition, a special plugging resistor R_P is provided that serves primarily to limit the armature current and torque during plugging operations. Note particularly that, when the directional contactors pick up *during normal acceleration*, interlock 2F (in the plugging-forward relay circuit PF) or interlock 1R (in the plugging-reverse relay circuit PR) closes. Either of these actions then causes the corresponding relay (PF or PR) to pick up and close its associated contact and permits the P contactor to operate *instantly* when the master is moved to point 2; this means, of course, that there is no time delay in the short-circuiting of R_P. In practice, therefore, position 1 is designed to make the motor develop only sufficient torque to pull up slack in cables or tighten gears, but not to start a reasonably large load. Then, as the master is advanced toward position 5, contacts P, $1A$, and $2A$ short-circuit the resistors and the accelerating relays, the latter drop out with the proper time delays, and the motor accelerates to full speed.

Assume that, with the motor running forward and the master in position five, the handle is moved quickly to the fifth point REVERSE. As the master passes through the OFF point, all contactors ($1F$, $2F$, P, $1A$, $2A$, and $3A$) and accelerating relays ($1AR$, $2AR$, and $3AR$) drop out and so does plugging-forward relay PF. Then, at REVERSE points 1 to 5 contacts $1R$ and $2R$ close as the corresponding contactors pick up. Since rotation is still forward, the cemf in the armature raises the potential of point a to approximately twice the value at point b. Now then, if the resistance of R_P is selected to incur a voltage drop that is approximately equal to the cemf, the potential of terminal c would be the *same* as that of point b. Under this condition relay PR will *not* be energized, its contact that leads to contactor P (and $1A$, $2A$, and $3A$) will remain open, and the motor will plug with accelerating resistors $R1$, $R2$, and $R3$ in the armature circuit. However, as the motor decelerates and the cemf drops, the potential difference between points b and c increases until, at approximately zero speed, the voltage across relay PR is sufficient to cause the latter to pick up. When this happens, contact PR closes, contactor P is energized, contact P short-circuits R_P, and accelerating relays and contactors proceed to operate with the usual time delays to bring the motor up to speed in the reverse direction.

Obviously, returning the master to point 5 FORWARD will produce a similar sequence of actions to plug the motor to a stop and then speed it up in a forward direction.

Reversing master for armature shunting drive. When a motor drives a high-inertia load at normal or above-normal speed, it is generally desirable to slow down the mechanical system before a stop is attempted. This is because the stored kinetic energy is reduced at the lower speed, under which condition the stop is smoother and softer and less jarring on the machinery. Shunt and compound motors can be slowed down quite readily by first applying full field, i.e., by turning the field rheostat to the all-out position, and then introducing resistance into the armature circuit. Series motors, on the other hand, do not lend themselves so well to speed reduction by the insertion of series resistance unless the load current is sufficiently high to cause an appreciable voltage drop across the added resistor.

The speed of a lightly loaded series motor can be reduced considerably by connecting a resistance across—in parallel with—the armature. This is because the series field is strengthened (more flux is produced) by the additional shunt-resistance current that must pass through the series-field winding. Moreover, if the available acceleration resistance is also connected in series with the parallel combination, i.e., armature and shunted resistor, a reasonably large voltage drop is incurred in the series resistor to lower the armature voltage and further reduce the speed.

A sketch representing a series motor, with one resistor connected across its armature and another in series with the parallel combination, is given in Fig. 4·10. Note particularly that the current I_S in the series field is

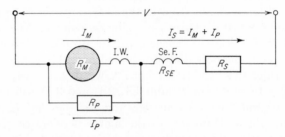

Fig. 4·10   Series motor with armature-shunt and series resistances.

the sum of the motor current I_M and the parallel-resistance current I_P and that this current, apart from the comparatively low value of I_M for a lightly loaded motor, can be made rather large by using a low value of R_P. Under this condition, the series-field flux and the series-resistance voltage drop $I_S R_S$ will both be high and together will contribute to a measurable speed reduction. To emphasize further the foregoing relationships, it can be stated that the motor speed S is directly proportional to the cemf E_c developed in the armature and inversely proportional to the flux ϕ; in equation form, this can be written as

$$S = k\,\frac{E_c}{\phi}$$

But for the circuit shown, the cemf is

$$E_c = V - (I_M + I_P)R_S - I_M R_M$$

from which

$$S = k\,\frac{V - I_M R_S - I_P R_S - I_M R_M}{\phi} \tag{4·1}$$

Equation (4·1) indicates that, even for small load currents I_M, the series-motor speed can be greatly reduced by making R_P small; this results in a high voltage drop for $I_P R_S$ and a greatly strengthened field flux ϕ.

A circuit diagram of a five-point master connected to a series motor with provision for armature shunting is given in Fig. 4·11. Observe that

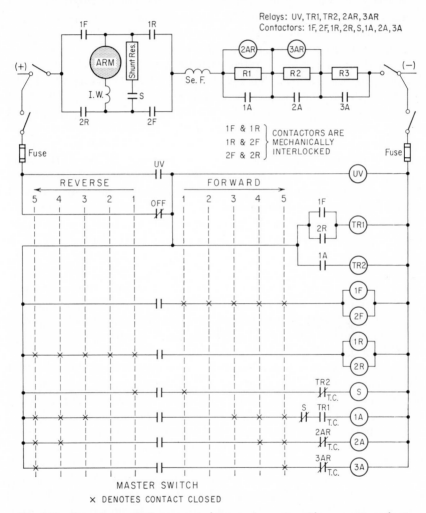

Fig. 4·11   Reversing controller connected to a series motor with an armature shunt.

a *shunt* contactor S is included, so that, in position 1, it is energized to close its N.O. contact in the shunt-resistance circuit. This, as was described above, will give the controller a slow-speed point, both on acceleration and when the motor is brought to a stop as the handle is moved to the OFF position through point 1. These actions will, of course, occur for *forward* and *reverse* directions of rotation.

Note that contactor S has an N.C. interlock in the $1A$ contactor circuit and that in position 1 this contact is open. Therefore, should the operator move to point 3 very quickly, contactor $1A$ cannot pick up until the S contactor drops out and closes its opened interlock. This means that the main S contact (in the parallel branch) is opened *before* the $1A$ contactor is energized, with the result that no part of the high inrush current passes through the shunt resistor.

When the motor is decelerated with the return of the master to the *first position*, it is essential that the S contactor be prevented from closing until *after* the accelerating contactors have dropped out; if such a condition did not prevail, the motor would develop an excessively high torque of retardation. A unique feature of the controller that causes the S contactor to pick up after $1A$ is deenergized is a circuit containing an N.O. interlock $1A$ in series with a timing relay $TR2$, with the time-closing contact of the latter in S. Thus, when the handle is quickly moved back to position 1, the $1A$ contactor drops out first, interlock $1A$ in the $TR2$ circuit opens next, and after a short delay, $TR2$-T.C. closes to permit S to pick up. In this last position, therefore, the motor slows down with its shunted armature resistor and all accelerating resistors in the armature circuit.

Reverse-plugging master for two-motor (duplex) drive. In some heavy-duty installations such as bridge and gantry cranes and large man trolleys of ore bridges and unloaders, it is frequently necessary to employ one or more pairs of motors to power the load. The motors are generally located at convenient driving points on the moving structure and are mechanically coupled and electrically interconnected so that all exert proper shares of the total torque. Since similarly rated series motors operate well together and their speed-torque characteristics are ideally suited for these applications, they are nearly always used.

An important aspect of the control system for a *duplex* (two-motor) drive is that one master regulates both motors simultaneously. This means, therefore, that there must be only one set of contactors and accelerating relays and these, being multiple-contact components, always act together on the motor armatures and similar sections of the accelerating resistors. In addition, it is generally necessary to provide for the contingency that the failure of one of the motors will still permit limited operation of the equipment with the other motor. This is usually accomplished by mounting a group of double-throw knife switches on a separate panel so that the closing of the switches in different combinations will provide for normal and emergency operation.

A complete circuit diagram of a duplex reverse-plugging drive connected to a five-point reversing master is given in Fig. 4·12. Note especially that the arrangement of the various components follows the general pattern of the one-motor drive of Fig. 4·7, with the exceptions that (1) two sets of acceleration resistors are used, (2) each of the directional contactors (*F* and *R*) has four contacts, (3) each of the other contactors (*M*, *P*, 1*A*, 2*A*, and 3*A*) has two contacts, (4) there are two sets of

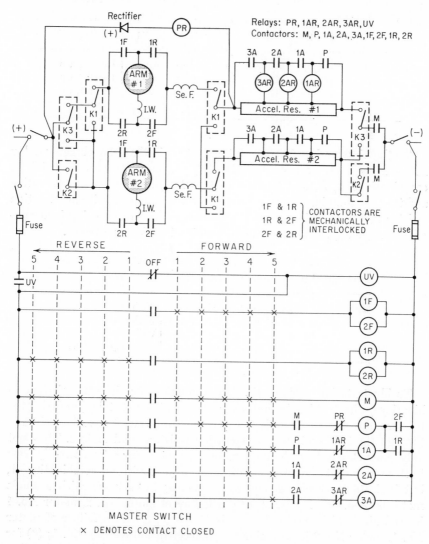

Fig. 4·12 Duplex reverse-plugging control system connected to series motors.

accelerating resistors, and (5) a group of three knife switches, $K1$, $K2$, and $K3$, are included for two-motor or single-motor operation.

A table listing four possible modes of motor operation and acceleration-resistor use and their corresponding knife-switch combinations is given herewith.

Combination	Motor		Accel. resistance		Knife switch		
	1	2	1	2	$K1$	$K2$	$K3$
A	Run	Run	In	In	Up	Up	Up
B	Run	Off	In	Out	Up	Down	Up
C	Off	Run	In	Out	Down	Down	Up
D	Off	Run	Out	In	Up	Down	Down

Another interesting feature of the system is that the plugging relay PR and the rectifier (see also Fig. 4·7) are used for plug-reversing in combinations A, B, and C when motors 1, 2, or 1 and 2 are operative and that for combination D, where motor 2 is employed with acceleration resistor 2, the plugging relay PR is inoperative but plugging is accomplished manually. A table of these combinations is given below.

Combination	Motors	Resistances	Plug
A	1 and 2	1 and 2	With PR
B	1	1	With PR
C	2	1	With PR
D	2	2	Manually (no PR)

To illustrate further how the system functions for the four switch combinations, Fig. 4·13 has been drawn. The simplified sketches, omitting the master controller and its relays and contactors, trace the various circuits through the knife switches, acceleration resistors, and motors. They not only show clearly how the duplex control system operates normally with both motors and the two sets of acceleration resistors, combination A, but indicate three emergency arrangements, in each of which a single motor is used. Specifically: (1) for a failure of motor No. 2, motor No. 1 is used alone with acceleration resistor No. 1; (2) for a failure of motor No. 1, motor No. 2 is used alone with acceleration resistor No. 1; (3) for a failure of motor No. 1 and acceleration relays and/or acceleration

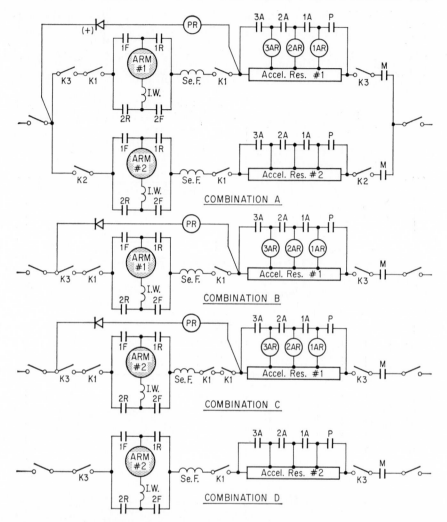

Fig. 4·13 Sketches showing switch, motor, and acceleration resistor combinations for duplex reverse-plugging control system (see Fig. 4·12).

resistor No. 1, motor No. 2 and acceleration resistor No. 2 are used; in this combination, it is necessary for the operator to perform all timing and plugging functions manually because acceleration and plugging relays are omitted.

Reverse-plugging master with dynamic braking. A compound motor connected to a four-point master designed for reverse-plugging and dynamic braking is given in Fig. 4·14. Before the operation of the controller and its circuit are discussed, brief comments will be made concerning several protective and economy features.

Since the shunt field of a motor is wound around a good magnetic circuit with a comparatively large number of turns of wire, its inductance

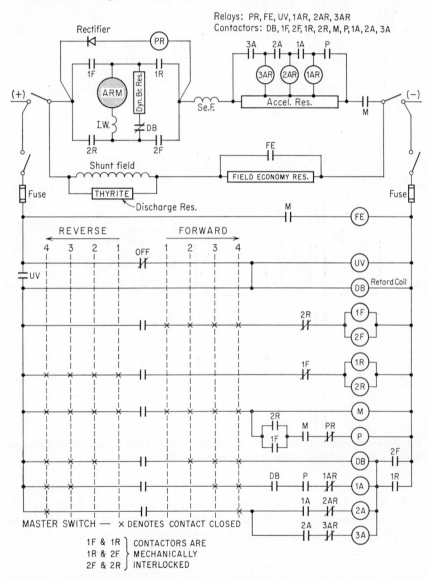

Fig. 4·14 Reversing controller connected to a compound motor. Designed for reverse-plugging and dynamic braking.

is high. This means that an extremely large voltage will be *induced* in the winding of an energized field if, *unprotected*, it is suddenly disconnected from the source. This is because energy that is stored in the magnetic field cannot be dissipated instantly when the switch is opened. However, as the flux decays with time, the voltage developed in the winding permits the current to change gradually to zero. Since the rate of current change is a maximum at the instant the switch is opened, the initial transient emf, represented by $L(di/dt)$, may reach a value of as much as 10,000 to 25,000 volts, sufficiently high to puncture the insulation. This must obviously be prevented, and especially so in the larger machines, where the effect of inductance is more serious than in the smaller sizes.

To limit the induced voltage to a reasonably low value, it is customary to connect a so-called *discharge resistor* in parallel with the field, as indicated in Fig. 4·15a. When this is done, the field current decays slowly

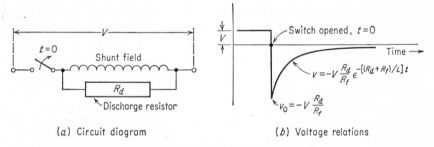

(a) Circuit diagram   (b) Voltage relations

Fig. 4·15   Shunt-field protection and transient voltage.

(di/dt is reduced), and the voltage appearing across the field after the switch is opened will be

$$v = -V \frac{R_d}{R_f} \epsilon^{-[(R_d + R_f)/L]t}$$

As shown in Fig. 4·15b, the voltage will drop exponentially to zero but will have its maximum value at $t = 0$ when the exponential term becomes zero; it will then be

$$v_0 = -V \frac{R_d}{R_f} \tag{4·2}$$

Equation (4·2) indicates, therefore, that a discharge resistance R_d whose ohmic value is four times the field resistance R_f will prevent the maximum field voltage from exceeding 1,000 volts, a safe value, in a 250-volt circuit. Such a discharge resistor will, of course, consume power if connected continuously to the field; in some cases, therefore, a special discharge switch is used which connects the discharge resistor to the field just before the source is disconnected.

A material often employed for the discharge resistor is *Thyrite*. Made by combining silicon carbide with a binder under high pressure and firing the mixture at an elevated temperature, it has nonlinear characteristics. At low voltages its resistance is rather high, under which condition the current is negligible. However, when the field circuit is disconnected and the current in the Thyrite is the initial value for an induced emf of, say, 1,000 volts, the transient is effectively limited. Figure 4·14 includes a Thyrite resistor for the reasons given.

When a motor is at rest and the field is kept energized at rated voltage, the temperature of the field winding tends to become excessive; this is because the stationary armature produces no fanning action for cooling purposes. To minimize the heating effect under this condition, a so-called *field economy resistor* is connected permanently in series with the field, and the former is paralleled by an N.O. contact that is closed by a field-economy relay *FE* at the instant the motor starts. This arrangement is shown in Fig. 4·14, where an *M* interlock in the *FE* relay circuit closes when the master is moved to the first point and the *M* contactor picks up. Thus, if the field-economy resistance is made equal to the shunt-field resistance, the field heating will be reduced to one-fourth of the full-field condition; this is because the current is diminished by one-half and the power loss varies as the square of the current.

Figure 4·14 includes another innovation in that a four-point master is used with four contactors, i.e., *P*, *1A*, *2A*, and *3A*; in previous circuit diagrams (Figs. 4·7, 4·9, and 4·11), five-point masters were employed. The arrangement shown here is acceptable when a definite last speed point is not particularly important, in which case accelerating relay *3AR* is made to drop out with a time delay on the fourth master point so that contactor *3A* can operate to short-circuit the last resistance step.

In the OFF position of the master, the undervoltage relay *UV* will be picked up and will be sealed in for all higher speed points by its interlock. Also, the retard coil of the dynamic-braking contactor will be energized so that, as was previously explained (see Figs. 3·21 and 3·22), the latter will not open its contact until the flux of the operating coil is well-established; this will occur in position 2. The motor is started, forward or reverse, by moving to the first master point. Several actions occur when this is done. As the *M* contactor picks up, it closes an interlock in the *FE* relay circuit to energize the latter, which, in turn, short-circuits the field-economy resistor; it also closes another interlock in the *P* contactor circuit so that one step of accelerating resistance is cut out. However, since the N.C. contact remains closed, the dynamic-braking resistance continues to parallel the armature and shunts a part of the inrush current to lessen the initial torque somewhat; this is, therefore, a low speed point.

With the operation of the DB contactor on point 2, the armature shunt resistance is removed to increase the motor speed further. Contactor $1A$ then picks up in position 3 (after accelerating relay $1AR$ times out) to short-circuit a second step of resistance. Moving the handle to point 4 causes contactor $2A$ to pick up after relay $2AR$ drops out, and this is quickly followed by the energization of contactor $3A$ as the current in relay $3AR$ drops sufficiently to permit it to close its contact. Motor operation now continues normally with all accelerating resistance short-circuited.

When the handle is returned to position 1, the DB contactor drops out to close its main contact in the motor circuit. The motor thus proceeds to slow down because, as before, the dynamic-braking resistance shunts the armature. A quick dynamic-braking stop can then be made by moving to the OFF point, since, in this position, power is removed from the armature and the latter dissipates its energy into the dynamic-braking resistor.

As in the circuits of Figs. 4·7, 4·9, and 4·12, this motor can be made to plug and reverse by rapidly moving the master to the fourth point on the opposite side.

QUESTIONS AND PROBLEMS

1. Discuss the constructional and operating differences between drum- and cam-type master switches.
2. What important advantages do master switches possess in comparison with faceplate starters when used for motor-starting purposes? Which type can be used for reversing service?
3. Referring to Fig. 4·6, what change should be made if it is desired that resistor $R1$ be short-circuited at the instant (without time delay) the master handle is moved to position 2?
4. With the motor of Fig. 4·6 running at full speed in position 4, explain the actions of the controller when the master handle is moved backward toward position 1.
5. Referring to Fig. 4·7, what change should be made if it is desired that the first block of accelerating resistance be cut with a time delay when the master handle is moved to position 2?
6. In the reversing controller circuit of Fig. 4·9, carefully explain how the plugging relays PF and PR function to insert plugging resistor R_P and accelerating resistors $R1$, $R2$, and $R3$ in the armature circuit during a plugging operation.
7. A 50-hp 230-volt 180-amp 540-rpm series motor has an armature resistance that is $6\frac{1}{4}$ per cent of V/I and a series-field resistance of 0.06 ohm. (**a**) At what speed will the motor operate when the load torque is reduced so that the motor takes 25 per cent of rated current under which condition the flux is reduced to 35 per cent of its full-load value? (**b**) At what speed will the

motor operate if it develops the same torque as in (a) with a 0.3-ohm resistor connected in series? (c) Referring to Fig. 4·10, what will be the motor speed if a shunt resistor R_P is connected across the armature so that, with the same series resistance R_S and the same armature current as in (b), the series-field current is increased to rated value?

8. Referring to Fig. 4·11, draw a set of simple schematic sketches showing how the series motor is connected to the shunt resistor and the accelerating resistors in master positions 1 to 5.

9. Explain why a high inrush current cannot pass through the shunt resistor in Fig. 4·11 when the master handle is quickly moved to position 3.

10. Explain the actions of the $TR1$ and $TR2$ timing relays in the master control circuit of Fig. 4·11.

11. Referring to Fig. 4·12, trace the armature and accelerating-resistance circuits for each of the four knife-switch combinations listed in the table on page 104, noting particularly that such circuit tracings agree with the sketches of Fig. 4·13.

12. In the duplex reverse-plugging control system of Fig. 4·12, why is it possible to use only one set of accelerating relays and one set of contactors? Why would two independently operated sets of contactors, one set for each motor, be unsatisfactory in system such as this?

13. The resistance of the shunt field of a 230-volt motor is 45 ohms. (a) If a discharge resistance of 400 ohms is connected permanently across the field, what maximum voltage will be induced in the field when the switch is opened and what will be the power consumption in the resistor? (b) If the maximum induced emf is to be limited to 1,500 volts, calculate the ohmic value of the resistor and its power consumption.

14. The field resistance and inductance of a 115-volt shunt motor are, respectively, 28.8 ohms and 48 henrys. If the field current rises exponentially in accordance with the equation $i = V/R_f[1 - \epsilon^{-(R_f/L)t}]$ after the field switch is *closed*, calculate (a) the current at time equal to the time constant L/R, (b) the time required for the current to build up to 95 per cent of its normal value. If a discharge resistance of 115.2 ohms is connected across the field at the instant the main switch is *opened*, calculate (c) the current at time equal to the time constant $L/(R_f + R_d)$, (d) the time required for the current to decay to 5 per cent of its normal value, (e) the peak voltage induced in the field.

15. The shunt-field resistance of 230-volt motor is 36 ohms. (a) What will be the normal power loss in the field? (b) If a *field-economy* resistor of 50 ohms is connected in series with the field as in Fig. 4·14, what will be the power loss in the field winding and field-economy resistor when the motor is at rest?

16. Referring to Fig. 4·14, explain (a) the purpose of the UV relay, (b) the function of the N.O. interlock contact M in the P contactor circuit, (c) the action of the *retard coil* on the dynamic-braking contactor, (d) why dynamic braking and plugging are provided in this control circuit, (e) why a four-position master can be used in a circuit that has five accelerating points.

CHAPTER 5

Protective Devices

Kinds of motor protection. In addition to providing a motor with such primary functions as acceleration, deceleration, reversing, braking, and speed control, a properly designed control system must include one or more kinds of protection. As used in practical installations, electrical equipment is frequently subjected to rather severe service, and this makes it imperative that safeguards be established to prevent failure or complete breakdown. The usual sources of trouble against which it is generally necessary to take protective measures are overloads of a sustained nature, suddenly applied overloads, short circuits, acceleration and/or deceleration that are too rapid, field loss in a shunt motor, contact arcing when loads are disconnected or highly inductive circuits are opened, loss of one phase in a polyphase system, and others.

Two common sources of trouble that are likely to cause motor failure are (1) excessive temperature rise and (2) extreme mechanical forces that result from high speed or vibration; either of these may reduce the dielectric strength of the insulation and eventually destroy it, or parts of the machine may be thrown out of alignment so badly that shutdown is necessary. Protective devices that are used to avoid the difficulties indicated take many forms, but most of them are designed to act in the control circuit, where a motor is disconnected from the line if its temperature exceeds a safe value or proper speed adjustments are made to forestall trouble. In most cases such devices are relays or circuit breakers of unique design that often employ special kinds of materials; moreover, these components are made to operate either instantaneously or with adjusted time delays, depending upon the protection desired.

Temperature and overload protection. Most motors are constructed to operate within established NEMA standards and, depending upon the classes of insulation used, have definite temperature-rise limitations. This implies, therefore, that the heating of a machine will depend not only on the magnitude of the load but upon whether or not the load is sustained, the duty cycle, the number and frequency of its starts, and time required for acceleration, the ambient temperature and cooling ability of the rotating structure, and other factors. It should be recognized, however, that, although the maximum load that may be imposed upon a motor before it stalls is often several times the nameplate value, its practical upper limit must be restricted to prevent excessive heating and, in d-c machines, to avoid poor commutation.

The most serious effect of an overload is a temperature rise that tends to weaken the insulation and thus shorten its life. Indeed, studies have shown that, as a general rule, the expected insulation life is halved for every 10°C rise in temperature above the recommended upper rating. Since temperature rise depends upon heating and the latter is a function of such conditions as the type of load (continuous or intermittent), the ambient temperature, and the ability of the machine to cool itself, it should be clear that protective devices must be designed to trip a control circuit on the basis of temperature rather than some particular overload current. An exception to this practice is the circuit breaker, which generally operates to trip instantly to avoid the damaging effects of a short circuit or a sudden current increase of extremely high value.

In practice, motors are frequently subjected to the following types of overload: (1) those that are applied continuously by the driven machinery; (2) those that simulate mechanical overloads because of reduced line voltage; (3) mechanical loads that retard acceleration, often causing motors to stall or reach full speed slowly; (4) intermittent-duty drives that start and stop frequently and cause repeated high inrush currents; (5) high ambient-temperature conditions that have the effect of increasing actual motor temperatures even though load currents may not be excessive.

Overload relays. Although many designs and constructions of overload relay are used in control circuits for the protection of electric machines, they can be classified on the basis of two general principles of operation. One of these, the *thermal principle*, is responsible for the opening of an N.C. contact when the line current (or a definite fraction thereof) passes through a temperature-sensitive element such as a bimetallic strip or when a low-melting-point alloy is softened by the passage of current through a short conductor. Such *thermal overload relays* have *inverse-time* characteristics, i.e., they respond more rapidly to increased currents, and vice versa. The second or *electromagnetic principle* causes an N.C. contact to open when current passes through a coil of wire to produce a magnetic force that acts on a contact-making mechanism. This type of magnetic overload relay operates instantaneously when the overcurrent reaches some definite value and may have a time-delay attachment such as a dashpot or variable magnetic assembly that gives it inverse-time characteristics. In all these, however, motors are shut down by deenergizing the contactor coils which, in turn, open the contacts to disconnect the motor from the line.

Overload relays can be constructed to be reset manually or automatically. In the first of these it is usually necessary for the operator to reclose the control-circuit contacts by pressing a button or by moving a lever, after which the motor can be restarted in the usual way. This is generally the preferred arrangement because a shutdown often implies a source of trouble which should be corrected. In the automatically reset type, the contacts close immediately after the circuit is opened and the motor stops; a new start can then be made without the need for reclosing the overload relay. It should be pointed out, however, that automatic-reset overload relays must not be used in control circuits which have maintained-contact pilot devices such as thermostats, pressure switches, level indicators, and the like, because motors would "pump"; that is, they would repeatedly attempt to restart against the overload.

Thermal overload relays. Several types of thermal overload relay are available for the protection of motors, but all operate to open a contact in the control circuit when a temperature-sensitive element receives sufficient heat by conduction, radiation, or convection from a load-carrying conductor. When properly designed with regard to heat-storing capacity and heat-dissipating properties, such relays can be made to follow quite closely the inverse-time heating characteristics of a motor. It is in this way that a machine can be protected against high-temperature failure and still permit operation under momentary or short-duty overload conditions.

A cutaway view of an extremely popular type of thermal unit is illustrated by Fig. 5·1. As a section of a block assembly with spring latch and contact, this detachable part contains a heater (plainly visible), melting pot, and ratchet wheel. Continued overcurrent through the heater raises its temperature and finally melts a *eutectic* alloy (between the ratcheted stem and the cup), allowing the ratchet wheel to rotate; a spring latch engaging the ratchet is then released, causing the relay contact to open. A few moments after the motor power circuit has been broken, the melted alloy resolidifies and the relay can be reset.

A completely assembled melting-pot overload relay of different construction is shown in Fig. 5·2. Here, with the dust cover lifted up, can be seen the ratchet wheel (with the melting-pot unit to one side) and the spring latch that actuates the contact. In this arrangement, a heater unit of the style illustrated in Fig. 5·3 is fastened over the

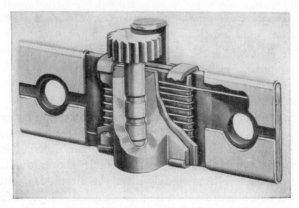

Fig. 5·1   Melting-pot unit for a thermal relay.

Fig. 5·2   Completely assembled melting-pot overload relay (see **Fig.** 5·3 for heater elements). (*Allen-Bradley Co.*)

stem and, carrying line current, transfers the developed heat to the melting pot.

A wiring diagram illustrating how a thermal relay is represented and connected to protect a compound motor is given in Fig. 5·4. Note particularly that the heater carries armature current and acts to open an

N.C. contact in the control circuit where all contactors and relays drop out when an overload exists sufficiently long to overheat the motor. The contact is, in this respect, similar to the STOP push button which is normally pressed to stop the motor. An added protective feature of this circuit is the use of a so-called *antiplug relay* whose purpose it is to prevent plugging. Being a potential relay that is connected across the armature terminals and with N.C. contacts in the N.O. FOR. and REV. push-button circuits, the contacts will be open while the armature is rotating and a voltage exists across the antiplug relay coil. Thus, with the motor running forward (or reverse) the reverse button (or forward

Fig. 5·3   Heater elements for melting-pot overload relay (see Fig. 5·2 for assembled relay without heater).

button) is inoperative; that is, the motor must come to a complete stop, or almost so, before the direction of rotation can be changed.

Another type of thermal overload relay, employing a bimetallic strip that responds to heat, is illustrated in the cutaway view of Fig. 5·5. Clearly visible are the inner U-shaped heater through which line current passes and the similarly shaped outer bimetal piece. The deflection of the latter does not open the contact directly but acts to release a mechanical latch which, in turn, permits a spring-loaded bar to move with a snap action. This arrangement eliminates sluggish, arc-forming operation and avoids irregular, nonrepetitive tripping.

The solder-pot principle of tripping an overload relay (Figs. 5·1 and 5·2) is applied to a special construction that is used primarily for the protection of *a-c motors*. Called an *inducto-therm* relay, Fig. 5·6, its operation does not depend upon a resistor type of heater element but utilizes the heating effect created by a current *induced* in a copper tube. As previously explained, the contact is held in a closed position by a reset

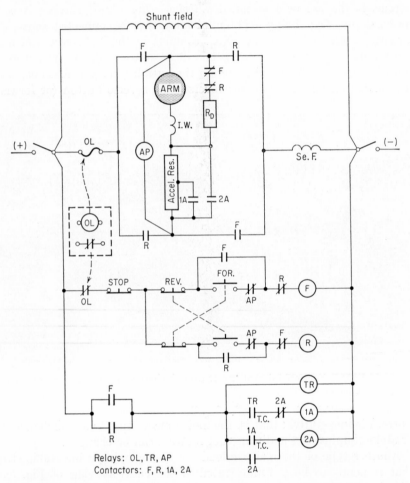

Fig. 5·4 Compound motor connected to a controller provided with thermal overload relay, antiplugging protection, and dynamic braking.

lever which engages a soldered ratchet wheel. An *alternating-current* overload through the magnetizing coil of the relay, clearly visible in Fig. 5·6 with the cover removed, increases the *induced current* in the copper heater tube in direct proportion to the overload current, which, continuing sufficiently long, will melt the eutectic alloy, release the ratchet wheel, and trip the relay. The contact must be reclosed manually by pushing a reset lever as soon as the copper tube cools and permits the solder to solidify. The accuracy of the relay depends only upon the melting point of the eutectic alloy, which is a definite value and remains constant. Moreover, since the heat is generated exactly where it is utilized, heat-transfer inaccuracies are avoided.

The tripping value of the inducto-therm relay is adjustable over a wide range of current ratings by simply changing the position of the threaded

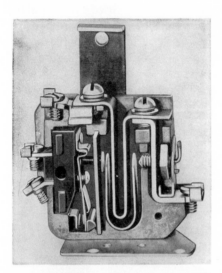

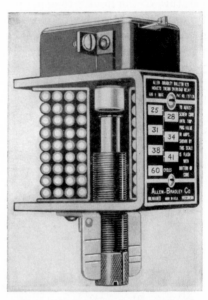

Fig. 5·5 Cutaway view of bimetallic type of thermal overload relay.

Fig. 5·6 Inducto-therm overload relay for polyphase a-c motors.

iron core within the magnetized coil; this alters the inductive coupling between the coil and the copper tube. Thus, when the core is lowered, the magnetic flux is weakened and a larger overload current is required to melt the solder. This adjustment, therefore, eliminates the necessity for an assortment of heater elements to provide correct overload protection. (See Fig. 5·3 for d-c motor protection.) The relay is designed for the protection of squirrel-cage induction motors whose locked-rotor current is not less than five times rated value and for slip-ring motors.

It should be understood that there is always some time lag between the heating effect of an overload current and the tripping action of a thermal relay. Although this delay is precisely what is desired for normal overloads, it is objectionable when the current suddenly reaches an extremely high value. Under such conditions, approaching those of a short circuit, it is necessary that fuses or fast-tripping circuit breakers be used, and these, when installed in a control circuit, are always in the line ahead of the relays. In accordance with the National Electric Code, the rating of such fuses or circuit breakers shall not be more than four times the rated current of the motor.

Magnetic overload relays. In contrast to the thermally actuated overload relay which operates only after a sustained current raises the temperature of the device to some predetermined upper limit, the magnetic type functions to open or close a contact whenever the current rises to a certain value. In some special cases the relay can be used for a similar purpose, operating when there is a definite current drop. As previously pointed out, the *magnetic overload relay* exerts a magnetic force on a plunger when an overcurrent passes through a coil of wire. Figure 5·7

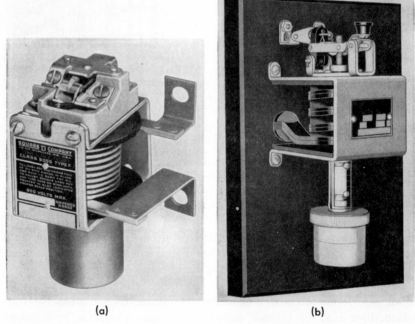

(a)   (b)

Fig. 5·7 Magnetic types of overload relay with dashpot time-delay attachment. (a) Unmounted; (b) mounted on slate base.

illustrates one construction which clearly shows the heavy, spirally wound current coil at the top and the cup that holds the dashpot at the bottom. Attached to the plunger is a piston, immersed in the oil of the cup and arranged with several bypass holes. As the current increases in the relay coil, the force of gravity is overcome and the plunger and piston move upward. When this happens, oil is passed through the bypass holes, with the result that the motion of the plunger to open (or close) the contact is delayed. To provide time-delay adjustment, the rate of oil flow can be altered by turning a valve disc that changes the size of the bypass holes. Inverse time-limit overload is afforded because the rate of upward travel of the core and piston depends directly upon the degree of overload.

The foregoing relay will not trip on the usually high motor starting currents or harmless *momentary* overloads. In these cases, line current drops to normal value before the operating coil is able to lift the core and piston far enough to actuate the contact. On the other hand, if the overcurrent continues for a prolonged period, the core is pulled far enough to perform its protective function. Moreover, since the tripping time decreases as the line current increases, the relay displays the desired inverse-time characteristic.

Another type of magnetic overload relay is depicted in Fig. 5.8. In this construction the inverse-time operation depends on the change of flux in the magnetic circuit due to a steel cylinder in a specially designed *vari-time* core. Note particularly that no oil or valve is used, so that change in fluid viscosity or aperture clogging cannot affect response. Normally, the cylinder remains at the bottom of the core and introduces a long air gap in the magnetic circuit. This results in a comparatively high reluctance which prevents the flux from building up to a value that is high

Fig. 5·8 Magnetic type of overload relay with special steel cylinder-core (*vari-time*) attachment for time-delay adjustment. (*Clark Controller Co.*)

enough to trip the relay. However, if an overload condition occurs, the flux in the magnetic circuit and the corresponding force increase sufficiently to raise the core slightly. With the upward movement of the core the reluctance continues to decrease to raise the flux and force further until the latter becomes strong enough to pull in the armature and open the contact. Under short-circuit conditions the flux rises immediately to a relay-tripping value, independent of the vari-time core action, and disconnects the motor from the line almost instantaneously.

A typical inverse-time characteristic for the kinds of overload relay described is given in Fig. 5·9. Drawn in *per cent full-load current versus*

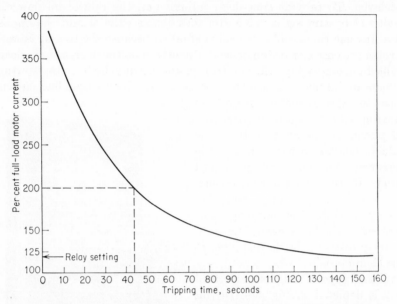

Fig. 5·9 Typical inverse-time characteristic of overload relay.

tripping time in seconds, it follows the motor-heating curve only approximately. Although differences between tripping and heating characteristics exist because of the large variety of motor constructions, service conditions are, for the most part, quite satisfactory. In practice it is, of course, essential that a motor be kept in operation as long as its temperature is below a permissible maximum, but it should be disconnected from the power source as soon as that temperature is reached. These functions are performed reasonably well by properly designed overload relays. For example, if a relay having the characteristic shown in Fig. 5·9 is set to trip at 120 per cent of full-load current, a 50-hp 230-volt 185-amp d-c motor will remain in operation a comparatively long time—more than 2 min—if the current reaches 1.2 × 185 = 222 amp; for a 200 per cent load, i.e., 370 amp, the tripping time will be about 44 sec, while a current of 400 per cent or 740 amp, will trip the relay instantaneously.

Instantaneous-trip overload relays. A special type of magnetic overload relay that trips instantaneously under conditions of extremely high overcurrents is sometimes used where it is desirable to take a motor off the line as soon as a predetermined load condition is reached. This safety precaution is often essential, in addition to normal overload protection that is provided by inverse-time relays, to guard against damage to motor and driven machinery when overloads come on rather quickly. As indicated, *instantaneous-trip overload relays* offer no protection to motors that carry sustained currents that may cause excessive temperature rise but do quickly disconnect the power source when gears bind, a driven mechanical system jams, or an extremely heavy load is suddenly applied. A typical application would be a woodworking machine, where a "jam-up" of material would cause a sudden high current. Such a relay would quickly disconnect the feed motor to prevent damage. After the cause of the jam has been removed, the motor can be restarted immediately. Another use of this relay is in conveyors, where a motor is stopped before mechanical breakage results from the effects of a jam.

Figure 5·10 shows an instantaneous overload relay in which the heavy series current coil and the tripping mechanism are openly displayed. Its

Fig. 5·10 Instantaneous-trip magnetic-type overload relay.

operating mechanism consists of a solenoid coil through which the motor current flows and a movable iron core within the coil. Mounted on top of the solenoid frame is a snap-action switch.

Under normal operating conditions the magnetic pull developed by the mmf of the coil is insufficient to lift the core. However, on overcurrent, the core is lifted to the upper position, where a precision snap switch is actuated to open the contact.

An interesting control circuit illustrating the use of an instantaneous-trip overload relay in connection with a two-motor milling-machine drive is given in Fig. 5·11. In this machine tool the *spindle* motor operates in

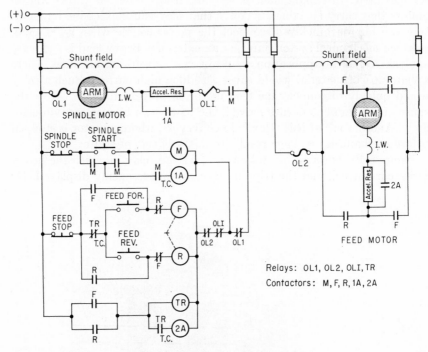

Fig. 5·11 Wiring diagram of a typical two-motor milling-machine control circuit with overload protection interlocking.

one direction while the *feed* is driven by a reversing motor. Note particularly that the tripping of the *OL*1 inverse-time overload relay will stop both machines while the feed motor only will be disconnected from the line if its inverse-time overload relay *OL*2 trips. However, should the cut be too heavy on the spindle, the current will rise to actuate the instantaneous-trip relay *OLI* to stop the feed motor only, thus permitting the milling cutter to set itself clear. This overload-relay interlocking is frequently used on multimotor drives when one or more motors must be protected against failure by the overloading of another. The reversing control for the feed motor was given in Fig. 3·16a and explained on page 73.

Combination thermal-magnetic overload relay. An overload relay that combines the advantages of the thermal principle for inverse-time tripping and the magnetic principle for instantaneous tripping is illustrated in Fig. 5·12. Used only for the protection of *a-c motors*, it has (1) a U-shaped bimetallic element which, when sufficiently deflected,

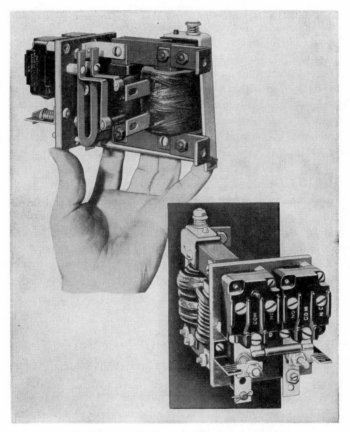

Fig. 5·12 Combination thermal-magnetic overload relay with inverse-time and instantaneous-trip elements. (*Square D* Co.)

acts on one quick-snap switch and (2) a pivoted strip of steel—an armature—which, when pulled up against a highly magnetized core, opens a contact in another precision snap switch. The basic design embodies a current transformer, the primary of which is connected in one of the line leads, with the secondary short-circuited by the bimetallic element. In this respect the device is similar to a current transformer with the heat unit replacing the indicating instrument. Under normal operating conditions the bimetallic element of the overload relay is heated by the secondary current, which, in turn, is a definite measure of the primary motor current. If an adjustment is made for, say, 125 per cent of full-load current, the *curled* bimetallic piece strikes a button that opens the contact.

However, should the current rise *suddenly* to an extremely high value, say twelve times rated current, the thermal element will be too slow to provide the necessary protection; on the other hand, the flux density in the iron core would become so great that sufficient magnetism will "spill over," that is "leak," to the externally mounted strip of steel. Since the latter is pivoted and free to move against the force of a spring, it does so, transmitting its motion through a connecting rod that opens the *instantaneous-trip* contact. Normal adjustment of the inverse-trip current is obtained by changing the spring loading on the trip lever, with calibration provided from 110 to 125 per cent of rated full-load current. The instantaneous-trip adjustment is made in a similar manner by turning another screw that sets the tension of a restraining spring, with a range of calibration points from twelve to twenty-five times the full-load value of the primary coil. In practice, both N.C. contacts are permanently wired *in series* and connected in the operating coil of the control circuit. Moreover, the inverse-time trip contact is automatically reset when the current drops sufficiently to permit the thermal unit to flatten out, while the instantaneous trip must be reset manually by pressing a button on the snap switch.

A simple circuit diagram is given in Fig. 5·13 to illustrate the application of the relay in a 3-phase line-start squirrel-cage motor installation

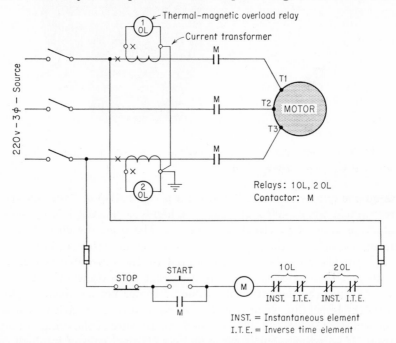

Fig. 5·13   Three-phase line-start squirrel-cage motor connected through a starter provided with thermal-magnetic overload relays.

Since one design and construction are used for all motor ratings up to 100 hp at 220 volts and 200 hp at 440 volts, it is necessary to connect each of them to the secondary of a properly rated current transformer. This means, therefore, that the current in the primary coil of the overload relay must not exceed 5 amp because the latter is the normal rating or burden of a standard current transformer. Thus, for a 25-hp 220-volt 64-amp motor, each current transformer would have a rating of 80 to 5, which means that, with a motor current of $1.25 \times 64 = 80$ amp, the coil of the overload relay would carry 5 amp.

Arcing protection. The dropping out of a relay or contactor while there is a circuit current is generally accompanied by arcing at the contact when the surfaces separate. Since the arc exists at an extremely high temperature and tends to burn away some of the contact material and even, under certain conditions, attempts to weld together the two separating pieces, it is essential that arcing be extinguished very rapidly. This can be done in one of several ways, but the method usually employed is to elongate the arc artificially until it can no longer be sustained by the available voltage.

To understand why an arc is formed and how it is extinguished, it should be recognized that for every contact material, such as copper, silver, and special alloys, there exist certain current-voltage values above which arcing will occur and below which it will not. For example, a minimum drop of 50 volts across an arc will sustain a current of 1 amp or more between silver contacts, or a minimum drop of 150 volts will prolong an arc across similar contacts if the current is 0.5 amp or more. For copper contacts, the 50- and 150-volt drops must have minimum current values, respectively, of 1.2 and 0.7 amp. Thus, when the surfaces of the contact first separate, the current-to-voltage ratio is high enough to keep the arc going. As the two pieces move farther apart, the resistance of the arc increases; this results in a reduction in circuit current and a voltage rise across the arc. However, since the current drops comparatively more than the voltage rises, a point is soon reached where the arc can no longer be sustained; under this condition the current drops to zero. It should, therefore, be noted that any arc can be extinguished if it is sufficiently elongated. Specifically, a separation of about 4 in. would be necessary if a current of 150 amp is to be interrupted in a 230-volt motor circuit.

Since it is obviously impractical to move the contact surfaces apart more than a fraction of an inch in relays and contactors, it is necessary to stretch the arc *artificially*. This is generally done electromagnetically, that is, by superimposing a strong magnetic field, electrically produced, on the arc and in such a manner that the latter, acting like a flexible, stretchable current-carrying conductor, is lengthened and forced upward. The arc, therefore, rises by motor action, aided by the natural tendency of a rising "band of heat," and extinguishes itself when the available voltage is too low for the reduced current at the simulated wide spacing.

Figure 5·14 shows an exploded view of a contactor with provision for

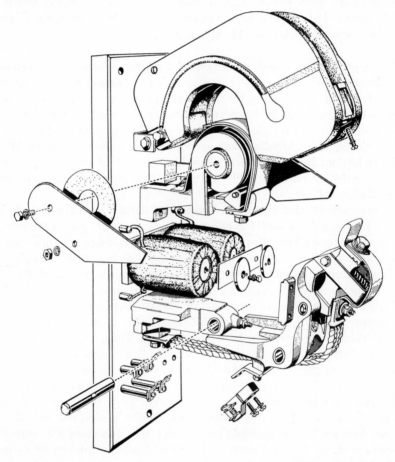

Fig. 5·14   Exploded view of a contactor, showing the blowout coil, pole plates, and arc chute. (*Square D Co.*)

"blowing out" the arc. The so-called *blowout coil*, clearly seen in the upper part of the drawing, is made up of a few turns of heavy copper strap wound on edge and over an iron core of circular cross section. Fastened to the core ends are two steel plates (shown separated in the sketch) whose ends straddle the two contact pieces. Thus, motor current passing through the winding of the blowout coil would produce a magnetic field at *right angles* to the direction of the arc. Then, with the arc-current direction and the magnetic polarities of the steel plates correct, the arc will be blown up and into the asbestos or ceramic enclosure, the arc chute. (In Fig. 5·14, if the current or arc direction is toward the plate mounting, i.e., front to rear, the magnetic polarity of the right plate must be *north* and the left plate *south*.) The two heat-resisting barrier shields that form the arc chute prevent the flaming arc from reaching other parts of the structure and, in addition, have a cooling effect. Moreover, since the reduced temperature increases the voltage necessary to sustain the arc, the latter is extinguished more rapidly.

The contactor of Fig. 5·14 incorporates a rather unique feature that tends to squeeze the arc flame away from the arc-chute barrier plates and toward the center as it moves upward. Designated by its manufacturer, the Square D Company, as a *Line-Arc* contactor, it includes two copper conductors mounted on the outside of barrier plates that line up parallel to the path of the formed arc. With both wires carrying the arc current in the *same* direction but opposite to that of the arc itself, magnetic forces exist to keep the flame away from the inside surfaces. Referring to Fig. 5·15, note that the arc is directed *in*, away from the

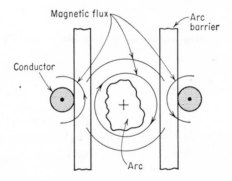

Fig. 5·15 Sketch illustrating the *Line-Arc* effect that tends to squeeze the arc together.

observer, and that the conductor currents are both directed *out*, toward the observer. With flux lines as indicated, the arc would be forced away from the two barrier plates, i.e., toward the center. This effect, therefore, tends to reduce the burning action of the arc and provides a good measure of added protection.

As previously stated, a blowout coil is usually connected in series in the circuit in which the arc is formed when the contact surfaces separate. Since the blowout force is proportional to both the arc current and the flux created by it, i.e., I and ϕ, and ϕ is also directly proportional to I, the action to extinguish the arc is a function of the *square* of the current. Thus, for high current interruption the blowout force is extremely powerful.

In some cases, especially on contactors or relays (for example, field-acceleration relays to be discussed later) designed for comparatively small currents, a *shunt* blowout coil is connected to the control circuit or a small alnico permanent magnet is used. When this is done, a constant blowout flux is provided which is independent of the load current. Generally, however, the blowout force is adequate in these low-current devices. Figure 5·16 illustrates such an arrangement in connection with

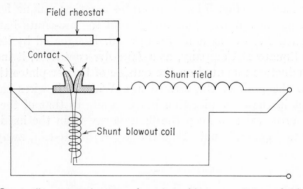

Fig. 5·16   Circuit illustrating the use of a shunt blowout coil in a field-acceleration application.

a field-acceleration application. As will be explained subsequently, the contact opens and closes rapidly as a shunt motor accelerates during the starting period with a definite field-rheostat resistance cut in. In doing so, the constant flux produced by the shunt blowout coil repeatedly blows out the arc as it forms when the contact opens.

Field-failure protection. Although extreme weakening or complete loss of the main field of a shunt or compound motor is not particularly common, its occurrence may result in serious damage under certain conditions of operation. This is because the speed of such a machine tends to rise rapidly when the flux is diminished. When the motor is coupled to a load which can neither be lost nor reach a very low value, an open main field will merely reduce the flux to a residual or subnormal state, in which case the developed torque is incapable of sustaining rotation; the motor will, therefore stall, and the current will rise to the tripping point of the overload relay. If, on the other hand, the application does permit the motor to be unloaded or, perhaps, overhauled as in the case of a hoist, loss of the field will not reduce the torque to the stalling condition but will allow the armature to accelerate quickly to a mechanically dangerous high speed or will produce destructive commutation.

To prevent the actions described when the shunt field is lost in an installation that is subject to overspeeding, a so-called field-loss relay must be used. As Fig. 5·17 illustrates, it is a simple N.O. single-contact device with a series operating coil that is connected in the shunt-winding circuit and the contact wired in the control circuit where its opening would stop the motor. In Figs. 4·6 and 4·14, for example, the contact would be in series with the undervoltage relay coil *UV*. A wiring diagram of a shunt motor connected to a push-button starter and including a field-loss relay is given in Fig. 5·18. Note also that it incorporates a Thyrite discharge resistor and an overload relay.

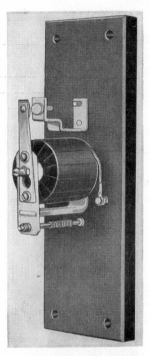

Fig. 5·17   Field-loss relay.

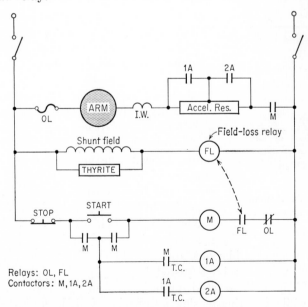

Fig. 5·18   Shunt motor connected to a starter including a field-loss relay.

Field-acceleration protection. To adjust the speed of a shunt or compound motor so that it operates *above* base value, it is customary to insert a rheostat in series with the shunt-field winding. The resistance must, however, be added *after* the motor has attained base speed because it is essential that the motor operate at full field strength while the accelerating resistors are being cut out and the speed of the motor is increasing. This is because the inrush currents are minimized when the motor is started with full field, since the developed torque depends upon both the armature current and the flux.

Where it is desirable to have a motor reach some above-rated speed after the START button is pressed, it is, of course, necessary to preset the field-rheostat. However, since the latter must be kept in its all-out position during the initial stages of accelera-

tion, i.e., until the acceleration resistors are short-circuited, it is required that some device be incorporated in the control circuit to accomplish this automatically. Furthermore, if a motor is running at base speed with the field rheostat cut out, a sudden resistance increase in the field circuit will result in an extremely high armature current because the motor must develop considerable additional torque to speed up to the new upper value. This rapid rise in current must also be avoided by having some device weaken the field *slowly* or in a multiplicity of steps. Both of the requirements described, that is, (1) short-circuiting the field rheostat until the motor reaches base speed and (2) weakening the field slowly above base speed, are fulfilled by a *field-acceleration relay*, one design of which is illustrated in Fig. 5·19. Its action, therefore, limits armature-current peaks, prevents the tripping of overloads and commutator flashover, and causes the motor to accelerate smoothly to the weakened field speed.

Fig. 5·19 Field-acceleration relay with tapped series winding.

The relay of Fig. 5·19 is constructed with a series exciting coil having an intermediate tap and a magnetic blowout and arc shield as previously described. The pickup value can be adjusted by varying the air gap with the slotted nut on a brass stud projecting from the core through the armature, while a spring-tension adjustment will determine the dropout current. The double-coil design permits the relay to close on comparatively low currents while the motor is accelerating up to base speed, but to close only at the higher currents when acceleration proceeds above base speed. An explanation of its operation in a control circuit will now be given.

Referring to Fig. 5·20, note that a field-acceleration relay *with a tapped exciting coil* is connected in series in the armature of a circuit similar to

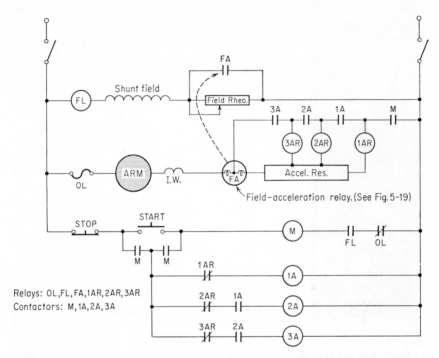

Fig. 5·20 Shunt motor connected to a controller including a field-acceleration relay with tapped series winding.

that given in Fig. 3·13. When the motor is started, the armature current passes through *all* the turns of the accelerating relay which is designed to close its N.O. contact across the field rheostat at comparatively low values of current. The field rheostat is, therefore, short-circuited during the accelerating period, i.e., while the motor is coming up to the rated speed. Note particularly that, when contactor 3*A* picks up, its contact short-circuits a portion of the field-acceleration relay winding in addition to the last acceleration resistor. This means, of course, that the accelerating relay will now develop fewer ampere-turns for given values of armature current. Thus, when the armature current drops sufficiently, the field-acceleration relay will drop out, open its contact across the field rheostat, and, assuming that the latter is set to increase the motor speed, will permit the machine to accelerate further. In doing so, however, the armature current rises enough to actuate the accelerating relay (even with the fewer turns), the contact closes again, and the rheostat is short-circuited a second time. The armature current then subsides with a repetition of the actions indicated. In fact, the relay may open and close the contact several times before the motor reaches its ultimate speed. Obviously the arcing at the contact will be blown out by the blowout attachment every time it opens.

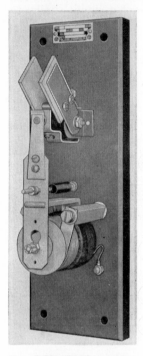

The operation described can be accomplished with another type of acceleration relay depicted in Fig. 5·21. This unit, as observed, has a many-turn shunt coil (rear) and a series coil (front) wound on edge with few turns of heavy copper. Following Fig. 3·9, the performance of which was previously described, note that in Fig. 5·22 the series coil is placed in the armature circuit and the shunt coil with a $3A$ interlock is connected across the source. The polarities of the coils are additive; that is, their mmfs produce fluxes in the same direction during acceleration.

At the instant the motor starts, contactors $1A$, $2A$, $3A$, and M are energized, N.C. contacts across the accelerating resistors are open, and N.O. interlock $3A$ in the shunt-coil circuit is closed. This means that the accelerating relay will be picked up by the mmf of its *shunt wind-*

Fig. 5·21   Field-accelera-tion relay with series and shunt windings.

ing and the contact FA across the field rheostat will be closed. Acceleration now proceeds in a normal manner as contactors $1A$, $2A$, and $3A$ drop out with time delays to permit contacts $1A$-T.C., $2A$-T.C., and $3A$-T.C. to close. However, with the deenergization of the last contactor $3A$, its interlock in the shunt-coil winding opens and the acceleration relay is now under the control of its *series* winding. Since the relay with fewer turns will now be actuated only when the armature current exceeds about $1.25I_{FL}$, the FA contact closes and opens several times as the motor accelerates, during which period the current rises and falls with the torque demands. When the motor reaches the upper speed as determined by the setting of the rheostat, the FA contact will be open.

Field-deceleration protection. When the speed range of a shunt or compound motor is greater than 2 to 1 and there is a possibility of rapid deceleration from a high to a low speed (by quickly cutting out field-rheostat resistance), the cemf may momentarily exceed the impressed voltage. Under this condition the *reversed armature current*, "pumping" back into the line, will often increase so quickly that destructive com-

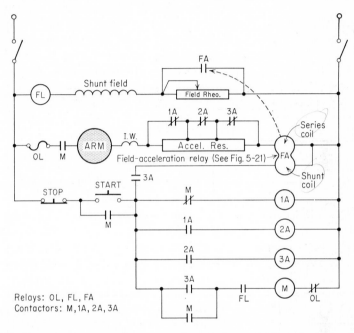

Fig. 5·22 Shunt motor connected to a controller including a field-acceleration relay with series and shunt coils.

mutation or flashover may occur. The actions described can be avoided by using a so-called *field-deceleration relay* which automatically weakens the field during the slowdown period. In construction, this relay is similar to that illustrated by Fig. 5·21, having series and shunt exciting coils and an N.C. contact. (In the field-acceleration relay the contact is normally open.)

A circuit diagram showing the application of the field-deceleration relay (in addition to other protective features such as field economy, field discharge, field loss, and field acceleration) is given in Fig. 5·23. It should be noted particularly that the shunt and series coils are connected so that the polarities of their mmfs are normally *bucking*, i.e., they tend to produce fluxes that are oppositely directed; also, the associated N.C. contact is across a separate field-deceleration resistor. This means that under stable operating conditions the *FD* relay will be dropped out, its *FD* contact will be closed, and the shunt-field winding is excited through the inserted field-rheostat resistance. However, when the speed of the motor is suddenly reduced by strengthening the field (cutting out field-rheostat resistance), the armature current reverses momentarily with

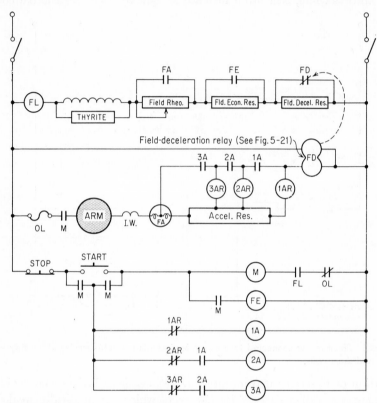

Relays: OL, FL, FA, FE, FD, 1AR, 2AR, 3AR
Contactors: M, 1A, 2A, 3A

Fig. 5·23 Shunt motor connected to a controller including a field-deceleration relay.

the result that the series- and shunt-coil mmfs become additive. The relay is, therefore, actuated, its contact across the field-deceleration resistor is opened, and the cemf is brought down quickly to reestablish normal current flow through the armature. With the reclosing of the *FD* contact the motor settles down to its reduced-speed setting.

Thermistor-type overload relay. One of the more recent overload relay developments makes use of a special semiconductor resistor, chemically similar to ceramic oxides. The unique property of the material, called a *thermistor*, is that the resistance is comparatively low at normal temperatures, remains nearly constant up to some critical point, and has an extremely large positive temperature coefficient of resistance within a narrow range above the critical temperature. Moreover, by varying the composition of the "doping" of the ceramic disc, about the size of an aspirin tablet, the critical range can be controlled to permit selection of the temperature at which protection will be maintained. Another good feature is that its power-carrying capacity is self-limiting because the device has the tendency to increase its resistance as the temperature rises.

Since the device is small enough to be placed in direct contact with the motor winding and has good thermal response, it eliminates the delay factor in transferring heat to actual sensing elements; it does, in this respect, overcome the shortcomings of remote current-sensitive devices for motors that are hermetically sealed. The element is encapsulated in an epoxy resin having compatible thermal, electrical, and mechanical properties and is generally capable of carrying the coil current of industrial-type relays.

In the circuit diagram of Fig. 5·24, three thermistors are shown at the surfaces of the three stator-winding phases of an a-c motor. Assuming

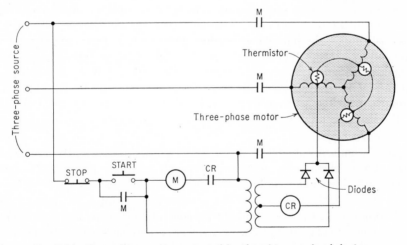

Fig. 5·24   Three-phase motor protected by *thermistor* overload devices.

that the temperature and resistances of the thermistors are low, sufficient current will flow through the *CR* relay to actuate it when the START button is depressed; this is rectified direct current after passing through the transformer secondary and the diodes. With the closing of the *CR* contact, the M contactor coil is energized and sealed in, the three line contacts close, and the motor starts. Should excessive motor temperature cause a large increase in thermistor resistance, the CR relay will drop out because of reduced current, the M contactor will be deenergized, and the motor will stop. Temperature and resistance of the thermistors must then drop to rather low values before the circuit can be reenergized and the motor restarted. This is accomplished by adjusting the relay so that its pickup current is more than twice the dropout current.

Overload protection for 3-phase motors when single-phasing. A 3-phase motor operating under *ideal* conditions will draw three equal line currents for all load conditions; this implies that the three line voltages are exactly equal and that the motor windings and magnetic circuits are identical. Ideal operation is, however, rarely possible in practice, since balanced voltages in particular are difficult to maintain in industrial electrical systems. Significantly, 2 and 3 per cent unbalances in voltage will result in unbalanced line currents that are, respectively, as much as 15 and 25 per cent, where the unsymmetrical system of voltages is generally caused by connecting one or more single-phase loads to a distribution circuit that normally delivers power to 3-phase motors.

It is also important to understand that overload relays in a control circuit are not intended to protect a motor against internal faults such as short circuits and grounds, nor are they capable of doing so. The function of the overload relay, whether thermally or magnetically operated, is to open the power lines within specified time limits under conditions of locked-rotor current or less. Controllers, on the other hand, are generally designed to interrupt currents that are as much as ten times the full-load rating of the motor. Since fault currents may be considerably more than this value, fuses or circuit breakers are usually installed ahead of the controller to clear such faults as may occur in the motor or in the line wiring.

Most standard controllers for 3-phase motors are presently equipped with two overload relay elements, although adequate protection is not always provided under unbalanced current conditions unless there is a unit in each of the three lines. The arbitrary adoption of the two-relay-unit standard was made originally by control-equipment manufacturers because early relay designs were expensive and occupied considerable space within the cabinet. This arrangement is still retained, although it is felt that two relays are not sufficient when there is extreme unbalance. The latter kind of aggravated circuit condition may occur when a 3-phase motor is powered through its own Y-delta or delta-Y bank of transformers, one of whose primary line wires is opened. When this occurs while the motor is running, the machine "single-phases" and draws about

73 per cent more current through each of the two intact primary line wires than on 3-phase operation. Also, one of the three *motor leads* then carries twice as much current as either of the other two, and, if an overload relay is omitted in this line, a motor burnout is possible.

Referring to Fig. 5·25*a* for Y-delta transformer connections, assume that primary line *L*1 is open and that transformer primaries *B* and *C*

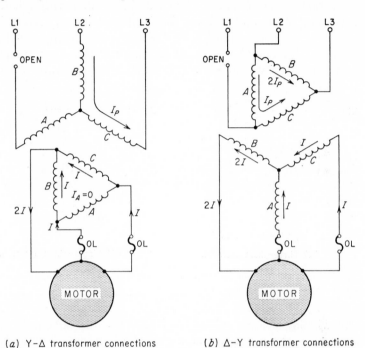

(*a*) Y–Δ transformer connections   (*b*) Δ–Y transformer connections

Fig. 5·25 Current relations during single-phase operation of 3-phase motor with one primary line open.

are energized. Under this condition transformer secondary *A* carries *no* current while the currents in *B* and *C* are each *I* amp and *in the directions shown*. The current in the unprotected line will, therefore, be 2*I*. It should be clear that with this combination of currents, one motor-winding phase is subject to extreme heating and possible damage (its current is 2*I*) should the overload relays fail to trip.

Similarly, a 3-phase motor is not fully protected if it is "single-phasing" through a delta-Y bank of transformers as in Fig. 5·25*b*. Assuming again that primary line *L*1 is open, the current in the primary of transformer *B* will be twice as much as in the series combination of *A* and *C*. By transformer action, therefore, secondary current *B* will also be twice the values *A* or *C*. Here again the unprotected line will permit one of the motor-winding phases to overheat, with possible damage, in the event the overloads do not trip.

Operating conditions simulating those described above have indeed been responsible for the failure of large numbers of oil-pump motors in isolated oil fields where individual installations were served through Y-delta banks of transformers and controllers with two overload relays.

A less serious situation arises when a 3-phase motor with a delta-connected winding is running and one line is opened. If the current in the two intact line wires is assumed to be I amp, one winding phase will carry $\frac{2}{3}I$ amp, and $\frac{1}{3}I$ amp will pass through each of the other two phases. These relations are illustrated in Fig. 5·26. By comparison, the current in

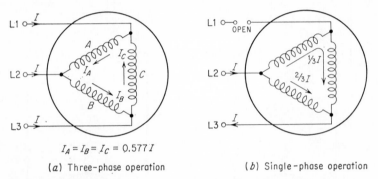

$$I_A = I_B = I_C = 0.577\,I$$

(a) Three-phase operation   (b) Single-phase operation

Fig. 5·26 Current relations in a delta-connected motor for balanced 3-phase and unbalanced single-phase operation.

all three winding phases is $I/\sqrt{3} = 0.577I$ (for the same assumed line currents) when the motor is running normally from a 3-phase source. The relations indicated should, therefore, make it clear that the B phase will be overloaded to the extent of about 16 per cent $[(0.667/0.577 - 1)100]$. However, since the currents in phases A and C are materially reduced (they are $0.33I$), the high-current phase will ordinarily dissipate its excess heat through the other two sections and the overload will usually trip before any part of the motor exceeds the allowable temperature rise.

Undervoltage protection with time delay. Two general kinds of low-voltage device are usually provided in control circuits that operate motors. In one of these, the *low-voltage release*, a power failure or a substantial voltage dip interrupts the power flow but does *not* prevent the reestablishment of the main circuit when voltage is restored. In another, *undervoltage protection*, power interruption is *maintained* after a failure or reduction of voltage.

Circuits in which momentary push-button pilot devices are normally

used will, of course, provide undervoltage protection because all magnetic components are deenergized when the voltage dips sufficiently, and
these can be energized again only by manually actuating the proper
push button after the circuit voltage returns to normal. Motors that
drive such equipment as pumps and blowers are often permitted to
restart automatically when power is reestablished after the circuit is disconnected from the line by the action of a protective device or because
of reduced voltage. A low-voltage release is employed in such cases, since
this avoids the inconvenience of manually restarting the motor and,
furthermore, often prevents possible danger in the event the machine is
not brought up to speed immediately after voltage is restored. In the
interest of safety, on the other hand, it is usually necessary to employ
undervoltage protection to *prevent* automatic starting in master-switch
controllers (Chap. 4); the operator is then required to return the handle
to the OFF position before a so-called *undervoltage relay* resets the circuit
for a proper start.

A modification of the foregoing practices is to keep the motor connected to the line for several seconds *after* a voltage dip or failure so
that, with a restoration of power within a timed period, operation can
continue automatically without the need for restarting. Special circuits
that will function in this way are especially convenient and desirable
in systems that suffer frequent interruptions of short duration. Figure
5·27 illustrates an arrangement with two *off-delay* timing relays that will

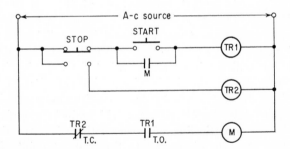

Fig. 5·27 Undervoltage protection circuit with time delay provided by timing relays.

provide *undervoltage protection with time delay*. With the pressing of the
START button, the motor is energized when the M contactor picks up
through the $TR1$-T.O. contact which closes instantaneously. Then, should
there be a power failure or a voltage dip, the M contactor will, of course,
drop out but will remain connected to the line for several seconds, i.e.,
until the $TR1$ relay times out and opens its $TR1$-T.O. contact. If voltage
is restored *after* $TR1$-T.O. opens, a new manual start will be necessary,
but if this occurs *before* the opening of that contact, the M contactor

will pick up automatically and permit the motor to continue operation, after slowing down slightly. When the STOP button is pressed to stop the motor, the $TR2$ relay is energized and instantly opens its $TR2$-T.C. contact to drop out the M contactor. This is followed by the deenergization of the $TR1$ relay which is timed to open its $TR1$-T.O. contact *before* $TR2$-T.C. closes.

Another scheme that eliminates the need for timing relays and a special STOP button but uses instead a rectifier, a capacitor, and a control relay is given in Fig. 5·28. In this arrangement, the motor makes a normal

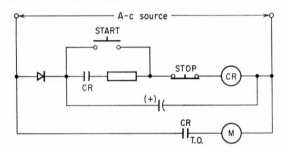

Fig. 5·28 Undervoltage protection circuit with time delay provided by a rectifier, a capacitor, and a control relay.

start when the START button is pressed, since the M contactor is energized through the CR-T.O. contact which closes at the instant the control relay CR picks up. The latter is sealed in through a resistor and a CR contact and, at the same time, is placed in parallel with a charged capacitor. A power failure or a voltage dip will, as before, drop out the M contactor but will not immediately affect the CR relay inasmuch as this unit is energized by the slowly discharging capacitor. Then, should power or voltage be restored *before* the capacitor loses sufficient charge to drop out the relay, the M contactor will again be energized and the motor will start automatically. However, if there is a sustained outage period and the capacitor discharges to the dropout point of the relay, a new motor start will be necessary. When the STOP button is pressed, the relay is instantly deenergized because it is disconnected from the charged capacitor as well as form the power source; the motor, therefore, comes to a stop and can be restarted only by pressing the START button.

QUESTIONS

1. What is meant by a thermal overload device and how is it connected into a control circuit to protect a motor against excessive temperature rise?
2. Why is the "life" of a motor reduced if its normal operating temperature is

higher than the manufacturer's recommendation? What operating conditions are likely to cause excessive temperature rise?

3. Distinguish between thermal and electromagnetic overload relays with regard to principle of operation, sustained overload protection, and suddenly applied, extremely high overload protection.

4. Why are manually reset types of overload relay preferable to those that reset automatically?

5. What is meant by "pumping" when referring to adverse motor operation?

6. Describe the construction and operation of a *melting-pot* overload relay. What is meant by eutectic alloy?

7. Referring to Fig. 5·4, describe the operation of an *antiplug* relay. Why is it sometimes desirable to include antiplugging protection in a control circuit?

8. Describe the construction and operation of the inducto-therm overload relay.

9. Explain the operation of the electromagnetic overload relay shown in Fig. 5·7. Will such a relay trip on momentary overloads that normally cause it to trip if the overload persists? Give reasons for your answer.

10. Explain the operation of the *vari-time* type of magnetic overload relay illustrated by Fig. 5·8. What unique constructional feature makes this relay function without the need for fluids and bypass holes as in Fig. 5·7?

11. Why is it desirable that an overload relay have an inverse-time characteristic?

12. What is an instantaneous-trip overload relay? Under what operating conditions should such a relay be used? Does it offer motor protection against normal overcurrents? If used, should additional overload protection be provided? Explain carefully.

13. Explain the operation of the two-motor machine-tool interlocking control circuit of Fig. 5·11 in which normal overcurrent and instantaneous-trip relays are used.

14. Describe the construction and operation of the thermal-magnetic overload relay illustrated by Fig. 5·12. Can it be used for the protection of motors of various current ratings?

15. Give three methods which may be employed to extinguish an arc after it forms when contact surfaces separate. Does arc formation apply to both direct and alternating current?

16. Describe the action of a *blowout coil* to extinguish an arc.

17. What important advantage does the so-called *Line-Arc* contactor (Figs. 5·14 and 5·15) have? Explain its operation.

18. Indicate the merits of series- and shunt-type blowout coils.

19. Under what operating conditions is it desirable to provide *field-failure* protection? How is this accomplished in a motor control circuit?

20. Why is it often desirable to incorporate a *field-acceleration* relay in a control circuit? Explain the operation of such a relay in Fig. 5·20.

21. Explain the operations of the field-acceleration relays in Figs. 5·20 and 5·22, pointing out particularly the differences in their construction and the manner in which they prevent rapid motor acceleration.

22. Referring to Fig. 5·23, explain why the *field-deceleration* relay has two coils whose mmfs normally act in opposition. How does this relay function to prevent regeneration when field-rheostat resistance is quickly cut out while the motor is operating at high speed?

23. Describe the action of the *thermistor* as an overload relay. Can such a device be used for d-c motor protection? If so, illustrate by a sketch how this can be done.

24. Explain why a 3-phase motor, supplied with power through a Y-delta or delta-Y bank of transformers, is not adequately protected by *two* overload relays. What are the practical objections to the use of three overloads for 3-phase motors?

25. Describe the operation of the circuit of Fig. 5·27 to provide under-voltage protection and yet permit automatic restarting during a momentary power failure or voltage dip.

26. What advantage does the circuit of Fig. 5·28 have over that of Fig. 5·27 for undervoltage protection with time delay?

CHAPTER 6

Automatic Starters and Control
Circuits for Polyphase Motors

Types and characteristics of polyphase induction motors. All electric motors consist essentially of two major sections, the *stator* and the *rotor*. The construction of each part is basically a slotted *laminated* core of good electrical steel with an embedded winding; when current passes through the winding, magnetic flux is produced. The steel serves primarily to provide effective *magnetic circuits* for the various fluxes, while the windings of copper, aluminum, or other conducting material act as *electric circuits* for the currents. This implies, therefore, that an electric motor is a combination of electric and magnetic circuits properly linked to develop a mechanical turning force or torque.

The stationary part of the polyphase induction motor—the *stator*—is a cylindrical structure, usually a stack of laminations (14-mil sheets for frequencies up to 60 cps), that contains a carefully insulated copper winding. The frame of Fig. 6·1 illustrates how the winding coils, pre-formed and insulated, are inserted in the stator of a moderate-size motor. In the construction shown, the stator slots are open to permit the free entry of the formed coil sides and, in addition, make the winding reasonably accessible when individual coils must be replaced or serviced in the field. With the open slot, however (see Fig. 6·2a), iron is removed from that portion of the magnetic circuit near the air gap, where it contributes greatly to good performance; this means, therefore, that easier manufacturing and servicing conditions are attained with some sacrifice in performance. The partially closed slot (Fig. 6·2b), on the other hand, makes winding somewhat more difficult because the coils must be taped and insulated *after* they are placed in the core. Also, since the coils usually have several turns of wire, the winding procedure requires that the individual conductors be fed into the slots between the narrow openings formed by overhanging teeth. The latter design, generally emphasized in the smaller ratings, does, nevertheless, have the advantage that

Fig. 6·1   Partially wound stator for a polyphase induction motor.

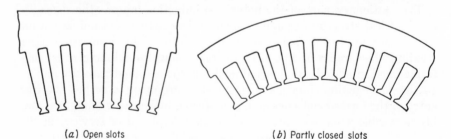

(*a*) Open slots   (*b*) Partly closed slots

Fig. 6·2   Two general types of slots for induction-motor stators.

iron is provided in the magnetic circuit where it is most useful. For motors of equal output performance, this makes it possible to have a core that is shorter axially than one with open slots.

Two general types of rotor construction are employed for induction motors, namely, the *squirrel cage* and the *phase wound*. Both designs have slotted, laminated cores tightly pressed on the shaft, but the squirrel cage consists of an *uninsulated winding* of bars of aluminum or copper that are joined together at both ends by rings of similar conducting material, whereas the phase-wound rotor contains a *completely insulated copper winding*, very much like that found on the stator. Although either can be used with a given stator, the torque and speed characteristics can be altered with the wound rotor and remain fixed by the design constants in the squirrel-cage rotor. Figure 6·3 shows a cutaway view

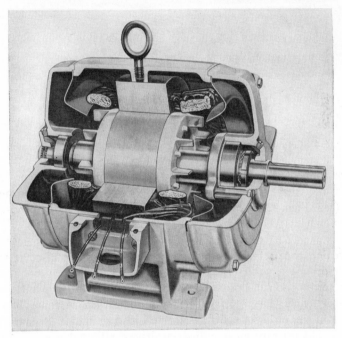

Fig. 6·3   Cutaway view of a typical 3-phase squirrel-cage induction motor.

of a typical 3-phase squirrel-cage induction motor of modern design in which the smooth, structurally solid appearance of the rotor, with projecting cooling fins, is clearly visible. The rotor bars pass through the laminated core below the surface, and the ends become parts of the two heavy end rings. Moreover, the bars are not parallel to the shaft axis; i.e., they are *skewed*, because this feature permits the motor to operate more quietly and prevents cogging, a condition that tends to make the machine run at a subsynchronous speed. The cutaway view of Fig. 6·4 illustrates the general constructional details of a 3-phase wound-rotor induction motor. Although the stator is, in most respects, similar to Fig. 6·3, it will be observed that the rotor has an insulated winding whose

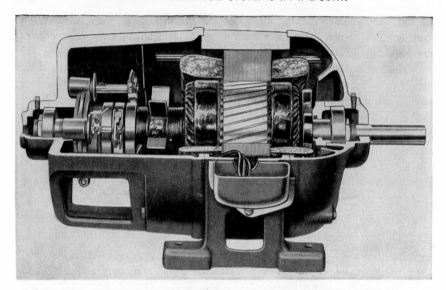

Fig. 6·4   Cutaway view of a typical 3-phase wound-rotor induction motor.

ends are connected to brass slip rings seen at the left; carbon brushes, riding on the rings, are, during operation, connected to a resistance controller (see Figs. 2·5, 2·13, and 2·14) for starting and speed-control purposes.

Operating characteristics of NEMA classified squirrel-cage motors. Whether or not a squirrel-cage motor can be started by connecting it directly across the full-line voltage (see Fig. 5·13) will depend upon the rotor design and the *permissible* maximum inrush current. The latter is usually of the order of six times the full-load value for general-purpose machines that operate most industrial drives, although some high-speed 2-pole motors for special applications may take inrush currents that are as much as ten times their normal ratings.

Class A motors, for example, have *deep-slot rotors with comparatively large slot areas*, and this makes it possible for the conductors to have relatively high thermal capacities. Moreover, during the accelerating period the current is not distributed uniformly in the rotor bars but is crowded into the upper part of the conductor (near the rotor surface) where the flux density is high. This phenomenon tends to produce a high *effective* rotor resistance and low reactance, a condition that results in good starting torque as well as an extremely fast start. When such a motor reaches full speed and the input current and flux have dropped to normal values, the current in the rotor bars distributes itself more or less uniformly to reduce the rotor resistance effectively and improve the operating efficiency. Thus, the deep-bar rotor design, with its excellent cooling and quick-start properties, makes it suitable for line-voltage starting and many severe-service applications.

Class B motors, with *deep-slot rotors* resembling those used in the class A designs, develop somewhat less starting torque than the latter because they are characterized by smaller values of field (stator) excitation. It is for this reason that they do not lend themselves so readily to full-voltage starting, although they have in other respects those performances characteristics that are desirable for general-purpose drives.

Class C motors, with *double-cage rotor constructions*, Fig. 6·5, behave somewhat like the class A designs from the standpoint of inrush current

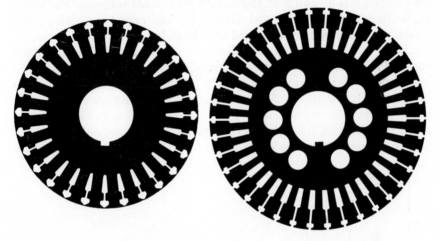

Fig. 6·5   Typical double-cage rotor laminations for squirrel-cage induction motors.

and starting torque and can be started on full voltage. Here again the slots extend deep into the punchings, but there are, as noted, two enlarged sections. Often called rotors with "double-deck" slots, they start with the rotor current crowded to the upper cages, where, with small cross-sectional areas, the resistances are high. Such motors, therefore, develop good starting torques and accelerate rapidly. Then, at full speed, the fluxes penetrate into the rotors to make the inner cages useful, under which condition both cages of a given rotor are in parallel; the rotor resistances are thereby greatly reduced, and the motors develop fine running characteristics. During the starting period, however, the currents are confined to the outer cages, and with the lower cages virtually isolated, the thermal capacities of the rotors are somewhat limited. This means that these motors are not adapted to high-inertia loads that prevent rapid accelerations and should be used only where the starting-torque requirements are high but of a static nature.

Class D motors develop extremely high values of starting torque doing so with considerable slip (low speed). This is because the rotor slots extend only a moderate distance into the punchings and the conductors have small cross-sectional areas. The deep-slot effect (classes A and B) is, therefore, slight, while the rotor resistances are comparatively high during the running as well as the starting period. Moreover, because of the elevated rotor resistances under load, efficiencies are low. Another objection to the small cross sections of the squirrel cages is that, having poor thermal capacities, the rotors tend to become quite hot under high-inertia loads. Such motors are, however, especially suitable for installations in which energy is stored as the rotors speed up and where the stored energy is given up at the instant peak loads come on. Particular applications are shearing and stamping-press operations, where the mechanical machines are equipped with large heavy flywheels.

Class F motors are designed to draw less starting and full-load running currents than the class B type and sometimes replace the latter. Used in installations where the available system power is limited, they draw such comparatively low values of current that power circuits are not adversely affected. The motors are generally large, in sizes above 25 hp, and when they are started on full voltage, the starting torques are only about 125 per cent of normal ratings. These characteristics result because extremely

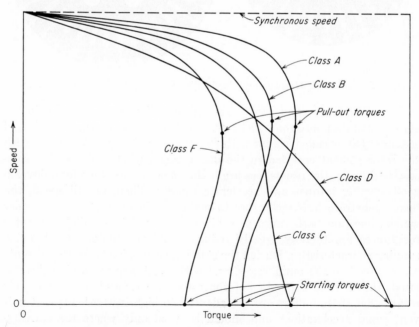

Fig. 6·6   Typical speed-torque characteristics for five classes of squirrel-cage motor.

high-resistance double-cage rotors are used which, as was previously pointed out, also limit the thermal capacities during the accelerating periods. Additionally, class F motors have very low overload capacities and, because of the high rotor resistances, rather poor efficiencies. They are generally employed in systems having restricted power supplies and to applications whose starting loads are light.

Figure 6·6 shows typical speed-torque curves for the foregoing five classes of motor. In analyzing and comparing them it will be well to note the speed variations with load, the starting torques, and the breakdown (pullout) torques.

Analysis of wound-rotor motors. The chief advantage of the polyphase induction motor with a thoroughly insulated winding in the rotor—the wound-rotor motor—is that a set of variable resistors can be connected externally and in series with the rotor winding (1) for accelerating torque adjustments and (2) to control the speed. The curves of Fig. 6·7 indicate

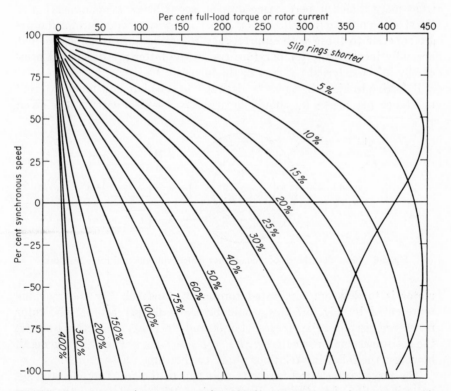

Fig. 6·7   Speed-torque (or rotor current) curves for a typical wound-rotor motor with different values of secondary (rotor) resistance.

how different *relative* values of externally connected resistances affect the relation between speed and torque and also show that starting torques are readily adjusted to meet different conditions of loading.

In practice it is customary to have the controller insert values of resistance in the rotor circuit that limit inrush currents to approximately 200 per cent of normal at the various accelerating points. Such resistances may, of course, be reinserted to make adjustments below synchronism, although the effect upon the speed at light loads is generally small. Moreover, since the slopes of the speed-torque curves increase with the greater values of rotor resistance, small variations in torque produce undesirably large speed variations. Figure 6·7 also indicates that the maximum torque that can be developed by the motor is independent of the rotor resistance; it is, in fact, fixed by the design constants of the machine. Another significant point is that it is generally impractical to reduce the full-load speed below 50 per cent of rated value because this involves a considerable increase in secondary resistance (over 30 per cent for the given curves) with a substantial drop in overall efficiency (somewhat more than the percentage drop in speed below synchronism). For the starting of extremely high-inertia loads, however, wound-rotor motors are superior to those with squirrel-cage rotors because rotor losses that normally increase the temperature rise in the latter are dissipated principally in the external resistors of the former.

Referring to Fig. 6·8, it may be desirable to calculate how much external resistance per phase R_E should be used in the rotor circuit if a given

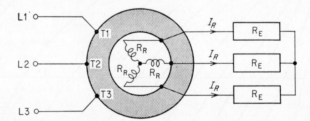

Fig. 6·8   Simplified sketch of wound-rotor motor with external rotor resistors.

motor is to develop, say, rated torque at standstill. For the machine represented by Fig. 6·7 this would be about 60 per cent of the rotor resistance per phase, i.e., $0.6R_R$. (For 50 and 200 per cent starting torques, the resistances are, respectively, $R_E = R_R$ and $R_E = 0.3R_R$.) In general, however, the following relations apply to the three-phase motor:

The power transferred across the air gap to the rotor by electromagnetic induction, i.e., the *rotor power input* RPI, is equal to the power output of the rotor, i.e., the *rotor power output* RPO, divided by (1-s);

thus,

$$\text{RPI} = \frac{\text{RPO}}{1 - s} \quad \text{watts}$$

But the rotor power output per phase is

$$\text{RPO} = \frac{\text{hp} \times 746}{3} \quad \text{watts}$$

Therefore, the rotor power input per phase is

$$\text{RPI} = \frac{\text{hp} \times 746}{3(1 - s)}$$

where s is the slip of the motor for rated horsepower output. The power input to the rotor per phase at rated load is also equal to

$$\text{RPI} = I_{R_{FL}}^2 \frac{R_R + R_E}{s}$$

and at standstill, when $s = 1$,

$$\text{RPI} = I_{R_{FL}}^2 (R_R + R_E)$$

Equating the two relations for RPI and solving for the external resistance,

$$R_E = \frac{\text{hp} \times 746}{3(1 - s) I_{R_{FL}}^2} - R_R \qquad (6\text{·}1)$$

where $I_{R_{FL}}^2$ is the full-load rotor current.

Example 1. A $7\frac{1}{2}$-hp 230-volt 3-phase 60-cycle wound-rotor motor runs at 1,710 rpm when operating at rated load. If the full-load rotor current is 23.6 amp and the rotor resistance per phase is 0.23 ohm, calculate the ohmic value of each of the external rotor resistors if the motor is to develop 100 per cent starting torque, i.e., for $T_{ST} = T_{\text{rated}}$.

Solution

$$s = \frac{1,800 - 1,710}{1,800} = 0.05$$

$$R_E = \frac{7.5 \times 746}{3(1 - 0.05)(23.6)^2} - 0.23 = 3.3 \text{ ohms}$$

Starting methods for squirrel-cage motors. The three general methods for starting squirrel-cage motors can be classified as (1) full-voltage starting, (2) reduced-voltage starting, and (3) part-winding starting. In the first of these, illustrated by the circuit diagram of Fig. 5·13 and a photograph of a small (NEMA size 0) line starter in Fig. 6·9, the

Fig. 6·9   Small line starter for a 3-phase squirrel-cage motor. Note two thermal overload units on sides of START-STOP push-button station.

inrush current is extremely high at the instant the starting equipment connects the motor to the line; reference to Table 6·1 indicates that motor starting currents for NEMA classes A, B, C, and D are about five to seven times rated values. Although there is actually no limitation on the size of the motor that can be started in this way, it is well to understand that objectionable line-voltage dips will generally occur, especially when large motors are started frequently. Moreover, the driven equipment will usually be subjected to severe shock because comparatively high starting torques are applied suddenly. Whether or not full-voltage starting is employed will, therefore, depend upon such factors as (1) the size and design of the motor, (2) the nature of the application, (3) the location of the motor in the distribution system, (4) the capacity of the power system, and (5) the rules established by the public-service company that govern motor installations.

Reduced-voltage starting is, of course, less severe to motor-driven equipment and power lines and is employed where it is necessary to limit inrush currents and/or accelerate a load slowly. Several procedures can be used to accomplish this, the most common of which are (1) the autotransformer method (Fig. 2·23), (2) the line-resistance method (Fig. 2·17), (3) the line-reactor method (Fig. 2·20), (4) the star-delta (Y-delta) method. For all the reduced-voltage schemes indicated, it is well to understand that the inrush current to the motor is lowered in *direct proportion* to the reduced voltage and that the starting torque drops in proportion to the *square of the reduced* voltage. Thus,

$$I_{ST} = \frac{E_{ST}}{E_{\text{rated}}} I_{(\text{inrush at } E_{\text{rated}})} \qquad (6\text{·}2)$$

and

$$T_{ST} = \left(\frac{E_{ST}}{E_{\text{rated}}}\right)^2 T_{(\text{starting at } E_{\text{rated}})} \qquad (6\text{·}3)$$

where the values of I and T at rated voltage are given in Table 6·1.

Table 6·1   Starting currents and torques for 220-volt 60-cycle squirrel-cage induction motors

| Hp | Currents | | | Per cent starting torque (minimum) | | | | | | | |
| | Rated | Starting (maximum) | | Classes A and B | | | | | Class C | | |
		Classes B, C, D	Class F	P* = 2	P = 4	P = 6	P = 8	P = 10	P = 4	P = 6	P = 8
½	2.0	12	.....	...	...	...	150	150			
1	3.5	24	.....	...	275	175	150	150			
1½	5.0	35	.....	175	265	175	150	150			
2	6.5	45	.....	175	250	175	150	145			
3	9	60	.....	175	250	175	150	135	...	250	225
5	15	90	.....	160	185	160	130	130	250	250	225
7½	22	120	.....	150	175	150	125	120	250	225	200
10	27	150	.....	150	175	150	125	120	250	225	200
15	40	220	.....	150	165	140	125	120	225	200	200
20	52	290	.....	150	150	135	125	120	200	200	200
25	64	365	.....	150	150	135	125	120	200	200	200
30	78	435	270	150	150	135	125	120	200	200	200
40	104	580	360	135	150	135	125	120	200	200	200
50	125	725	450	125	150	135	125	120	200	200	200
60	150	870	540	125	150	135	125	120	200	200	200
75	185	1,085	675	110	150	135	125	120	200	200	200
100	246	1,450	900	110	125	125	125	120	200	200	200
125	310	1,815	1,125	100	110	125	125	120	200	200	200
150	360	2,170	1,350	100	110	125	125	120	200	200	200
200	480	2,900	1,800	100	100	125	125	120	200	200	200

* P = poles.

NOTES: For higher voltage motors, reduce current values in proportion to increased voltages.
For class A motors, starting currents may be more than for classes B, C, and D.
For class D motors, starting torques shall be 275 per cent minimum.
For class F motors, starting torques shall be 125 per cent minimum.

The *part-winding* method may be used when the stator winding is divided into two identical parts and the motor is designed to run normally when the two sections are connected in parallel. With this scheme one-half of the winding is energized when the machine is first started; then, after a definite time delay which permits the rotor and its load to come up to speed, the second section is placed in parallel with the first one, under which condition normal operation continues. By this starting procedure, the inrush current is limited to about 60 to 70 per cent of the full-voltage surge and the starting torque is reduced to a value that is somewhat *less* than one-half that obtained if both sections are connected in parallel and to the line simultaneously.

Line starters for squirrel-cage motors. Detailed and elementary wiring diagrams for a line-start squirrel-cage motor are shown in Fig. 6·10. In accordance with NEMA standards, line starters are listed by number and, depending upon the kind of service the motor will perform, the associated horsepower ratings at several supply voltages. As Fig. 6·9 illustrates, most starters are equipped with overload devices and, when desirable and convenient in the smaller sizes, with a START-STOP push-button station mounted on the cover. A complete listing of standard starters for three classifications of motor service and voltage ratings is given in Table 6·2. As specified, they shall not be used with squirrel-cage motors whose full-load currents or horsepower ratings exceed the

Table 6·2   Standard NEMA horsepower ratings of 3-phase
across-the-line magnetic starters

Size of starter	Average operating conditions, nonplugging or nonjogging duty			Plug-stop, plug-reverse, or jogging duty			Multispeed constant-hp motors, plug-stop or jogging duty			Continuous current, amp
	110 volts	220 volts	$440/550$ volts	110 volts	220 volts	$440/550$ volts	110 volts	220 volts	$330/550$ volts	
00	¾	1	1	⅓	½	½	⅓	½	. . .	9
0	2	3	5	¾	1	1	¾	1	1	18
1	3	7½	10	2	3	5	1½	3	5	27
2	7½	15	25	5	10	15	5	10	15	45
3	15	30	50	10	20	30	10	20	30	90
4	25	50	100	15	30	60	15	30	60	135
5	. . . .	100	200	. . . .	75	150	. . . .	75	150	270
6	. . . .	200	400	. . . .	150	300	. . . .	150	300	540
7	. . . .	300	600	. . . .	. . . . .	. . . . .	. . . .	. . . . .	. . .	810
8	. . . .	450	900	. . . .	. . . . .	. . . . .	. . . .	. . . . .	. . .	1,215
9	. . . .	800	1,600	. . . .	. . . . .	. . . . .	. . . .	. . . . .	. . .	2,250

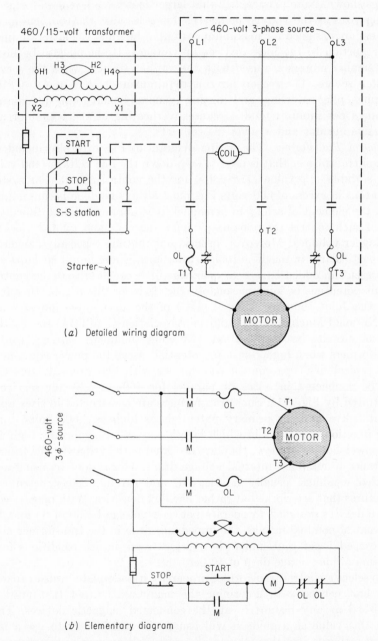

(*a*) Detailed wiring diagram

(*b*) Elementary diagram

Fig. 6·10   Diagrams for a line-start 3-phase squirrel-cage motor.

values shown. Note particularly that larger motors can be used for given starter sizes as the voltage increases; this is because the contacts in the starters are designed for maximum current values and the latter diminish in proportion to the increased motor voltages. It should also be noted that smaller motors are used with the various starters when the service is more severe. The reason for this requirement is that frequent plug-stopping, plug-reversing, and jogging (generally more than five contact openings per minute) tend to cause overheating of the contactor and excessive burning and wear of the contacts.

Control Transformer. The wiring diagram of Fig. 6·10 also includes a control transformer that is used to step down the line voltage (460 volts) to a suitable *safe* value (115 volts) for the control circuit. For system potentials in excess of 600 volts this must always be done because, apart from the element of safety to personnel, it is impractical and difficult to design such control components as operating coils and pilot devices for the higher voltages. Moreover, in locations that are especially dangerous and particularly in machine-tool installations, where liquid coolants and oil present an added hazard to operators, it is essential that the control-circuit potentials be kept reasonably low. It is for this reason, therefore, that the Joint Industry Council (JIC) of the automotive industry and the National Machine Tool Builders' Association (NMTBA) specify that control circuits be energized at 115 volts (nominal) through control transformers when higher system potentials serve the power equipment.

A typical dry-type control transformer with two primary coils that can be connected in series or parallel for 460- or 230-volt service is illustrated by Fig. 6·11. Such transformers are constructed in sizes up to about 5 kva and for primary potentials as high as 4,600 volts. Since they function in circuits in which inrush currents are generally high and the power factors are low, they are designed with extremely low leakage reactance to minimize internal voltage drops. This is done by using well-coupled windings, usually interleaved, and carefully constructed core structures that are worked at rather low flux densities. With magnet coils of a-c devices designed to operate between voltage limits of 85 and 110 per cent of nominal ratings, the impedance drop in the transformer must not exceed 5 per cent under the most extreme inrush conditions for a maximum line-voltage drop of 10 per cent.

To select a control transformer properly with adequate inrush capacity it is first necessary to determine the maximum current that must be furnished at any instant to all the connected magnetic devices. Then with this value as a guide it will usually be satisfactory to use a unit whose rating is less than the total inrush volt-ampere requirements of the individual devices but equal or somewhat greater than the maximum sealed volt-amperes at any time.

Line-resistor reduced-voltage starter. Inserting resistors in the line wires that are connected to a polyphase squirrel-cage motor is equivalent to the practice of using acceleration resistors in the armature circuit of a d-c motor. The purpose is nearly always to limit the inrush current, although reduced accelerating torque is essential in some applications where soft starts must be made. The ohmic values of the resistors not only must be selected to keep the initial surge below an acceptable maximum but must also take into account the ability of the motor to develop sufficient torque to start the load.

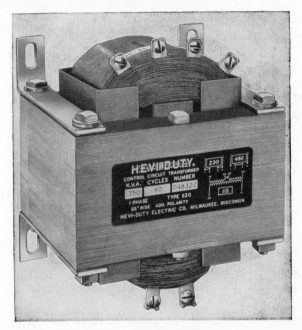

Fig. 6·11   Control transformer.

A typical wiring diagram of a line-resistor starter for a comparatively small motor is given in Fig. 6·12. It is designed to connect the machine initially to the power source through the three S contacts and resistors. When the motor comes up to speed, the resistors are short-circuited by the R contacts, after which the S contacts are opened. Should there be a power failure or a large voltage dip, the undervoltage relay UV will drop out to disconnect all other relays and contactors and stop the motor. Upon the return of power the motor is started in the usual way.

For larger size motors, it is often desirable to use d-c contactors that are connected to the a-c source through a set of selenium, germanium, or silicon rectifiers. This arrangement produces better and more positive action of the magnetic forces and avoids the chattering tendency of a-c contactors. The control-circuit diagram in Fig. 6·13 illustrates how this is done for a motor and resistors similar to that in Fig. 6·12.

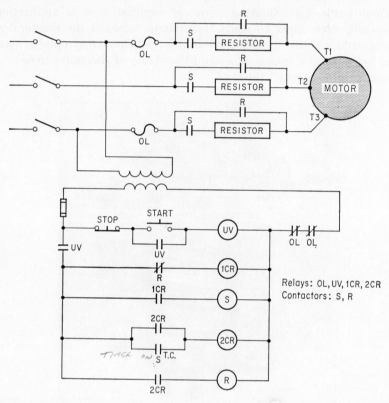

Fig. 6·12 Line-resistance reduced-voltage starter connected to a comparatively small motor.

Several interesting points should be noted in connection with the operation of this circuit. When the start button is pressed, the *UV* relay is energized and is sealed by *its own UV contact.* Another *UV* interlock also closes to permit relay 1*CR* to operate. When the latter is actuated, the *S* contactor picks up through *two* 1*CR* contacts, whereupon the motor starts at reduced voltage since line resistors are inserted. After a timing head on the *S* contactor times out, contact *S*-T.C. closes to permit relay 2*CR* to pick up, the operation of which causes the *R* contactor to close its main contacts and short-circuit the line resistors. At the same time the N.C. *R* contact in the 1*CR* relay circuit opens to disconnect the *S* contactor. When the latter drops out, all associated *S* contacts open but the $2CR$ relay remains energized through its own sealing contact.

It is well to note that a UV interlock is connected across the START button to seal the UV relay. If an S interlock were used to make sure the motor starts *before* the finger releases the START button, a difficulty arises. Under this condition the high-inertia armature of the S contactor

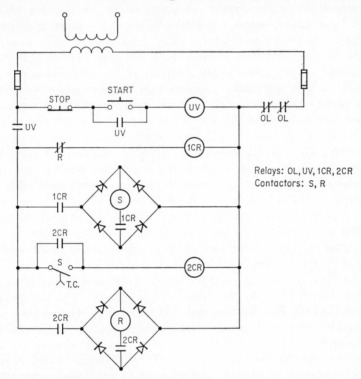

Fig. 6·13 Control circuit of line-resistance reduced-voltage starter for a comparatively large motor (see Fig. 6·12 for motor connections).

goes in rather sluggishly when the $1CR$ contacts close. Should the operator release the START button before the S interlock contact closes, the arm will attempt to open the main contacts. With the latter in the "kiss" position while the current is high, the contacts will tend to weld as they attempt to separate. The possibility indicated is avoided by using a UV contact across the START button as shown; the latter must then remain depressed only long enough to let the UV relay pick up and seal itself. The S contactor will then be operated when the $1CR$ relay is energized.

Observe also that there are control-relay contacts $1CR$ and $2CR$, respectively, in the operating coil circuits of S and R; these are in addition to a similar set of contacts that feed the rectifier bridges. The purpose of this arrangement is to prevent the contactors from discharging slowly through the rectifiers when the control relays $1CR$ or $2CR$ are deenergized. If the contacts in the coil circuits were *not* used, the deenergization of contactors S or R would permit their currents to decay slowly through the rectifiers; this would result in slow dropout and a tendency for the contacts to weld together.

Increment Resistance Starter. In some motor installations that are served by a power source of limited capacity, the starting equipment must not only be designed to minimize "peak" currents but must often restrict the *rate* of current increase to a stated number of amperes per half second or per second. The purpose of the latter requirement is generally to give regulators sufficient time to compensate for excessive voltage drops that are caused by high inrush currents to the motors. Starters that function in this way usually have line resistors that reduce the motor voltage considerably and, in addition, are cut out in two or more steps. They are, in this respect, similar to multistep starters for d-c motors where a set of accelerating resistors in the armature circuit are progressively short-circuited. Since the line resistance used with these starters is somewhat more than those of the two-step type represented by Fig. 6·12, the motor voltage is frequently too low to start the load on the first step. This is not particularly objectionable because sufficient starting torque will be developed on the second step and under conditions that satisfy system stability.

A three-step *increment resistance starter* of the kind described above is given in Fig. 6·14. Note that three sets of contacts $1S$, $2S$, and R are connected to the line resistors and that sections of the latter are short-circuited in a definite time sequence by the respective contactors. When the START button is pressed, the UV relay is energized and is sealed in by its interlock. At the same time, another UV contact closes to permit the $1S$ contactor to operate and start the motor; also, when the $1S$ interlock closes, timing relays TR and $1TR$ pick up. After a delay of 5 sec $1TR$-T.C. closes, and this causes the $2S$ contactor to short-circuit one section of each line resistor. A $2S$ interlock also closes to energize the $2TR$ timing relay which, after another delay of 5 sec, closes its contact in the R contactor circuit. The latter operation results in the complete short-circuiting of line resistors and the connection of the motor to full voltage. Finally, with the opening of the N.C. R contact in the $1S$ contactor circuit, contactors $1S$ and $2S$ drop out as do relays TR and $1TR$. The R contactor, however, remains energized through a connection from the closed UV interlock.

An added feature of this circuit is that timing relay TR is timed to open its TR-T.O. contact in 15 sec. In a normal start the motor is expected to come up to full speed in less than 15 sec, so that the TR-T.O. contact will remain closed until the $1S$ contact opens to deenergize the TR relay. However, in the event the motor does not accelerate properly in 15 sec, the $1S$ contact will remain closed, the TR relay will not drop out, and the TR-T.O. contact will open to deenergize the UV relay. The motor will, therefore, stop.

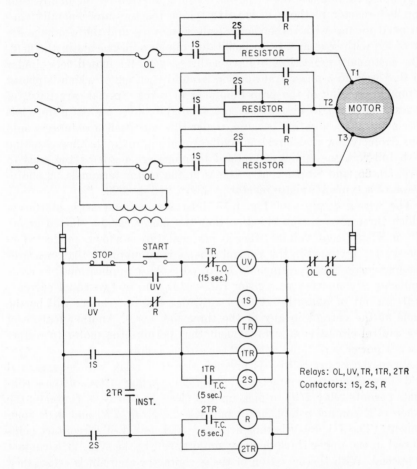

Fig. 6·14 Circuit diagram for an increment resistance reduced-voltage starter.

Line-reactor reduced-voltage starter. Line reactors, like line resistors, will reduce the motor voltage and inrush current during the accelerating period. Although the principles of the two schemes are, in this respect, essentially the same, each possesses certain advantages and disadvantages. For comparatively high-voltage and/or high-current installations where the resistor bank is generally bulky and heat dissipation becomes a problem, line-reactor starters are preferable. The use of reactors, on the other hand, results in exceptionally low starting power-factor conditions and tends to aggravate further the already poor line regulation; this is because line-voltage dips tend to increase as the power factor of the inrush current diminishes. (The phasor diagrams of Figs. 2·19 and 2·21 illustrate how the motor power factor is affected by the line-resistor and line-reactor methods of starting.) Line reactors are generally constructed for use with rather large 3-phase motors, and the windings are provided with several taps to facilitate voltage adjustments in the field. The customary practice is to have a three-legged laminated core similar to that used in 3-phase transformers around each leg of which is placed a tapped winding of the proper number of turns. Special care must, of course, be taken to design the core so that it does not saturate while the motor is accelerating because, under this condition, reactance would be extremely low and inrush currents would approximate those existing with full-voltage starting. Standard tappings are made at the coils to give, 50, 65, and 80 per cent voltages at the motor terminals. A photograph of a typical 3-phase reactor is shown in Fig. 2·22.

The wiring diagram of Fig. 6·15 illustrates a line-reactor starter in which three taps are provided so that the motor can be started at 50, 65, or 80 per cent voltage; the 50 per cent tap is shown connected in the sketch. Included in this control scheme is an ammeter that measures the line current in a presumably balanced 3-phase application. Since the ammeter is connected to register the vector sum of two equal currents that are out of phase by 120 electrical degrees, its reading will be the same as the current in any of the three line wires. Another feature of the control circuit is an arrangement that permits the motor to restart itself if power is restored within 2 sec after a failure.

When the START button is pressed, the $1CR$ relay will be energized and three associated contacts will close. The closing of two of these contacts permits relay $2CR$ to pick up and close its two N.O. contacts; the other $1CR$ contact establishes a path to the S, R, TR, and $3CR$ components. The UV relay is thus actuated, and its UV-T.O. contact helps to seal in the upper three devices; simultaneously the N.C. $2CR$ contact is opened. With the operation of the S contactor a circuit is established through the three line reactors and the motor starts. After timing relay TR times out (it was energized through closed contact $1CR$), relay $3CR$ will pick up, close its contact in the R contactor circuit, and permit the R contactor to operate. The latter action causes the main R contacts to close, to short-circuit the line reactors, and to open the N.C. R interlock. The S contactor, therefore, drops out, completely disconnects the reactors from the system, and the motor runs normally from the full-voltage supply.

Should there be a power interruption, all relays and contactors are deenergized *but* the *UV*-T.O. contact remains closed for the adjusted time of, say, 2 sec. However, if power is reestablished *before* an elapse of 2 sec, the closed *UV*-T.O. and N.C. *2CR* contacts provide a recovery

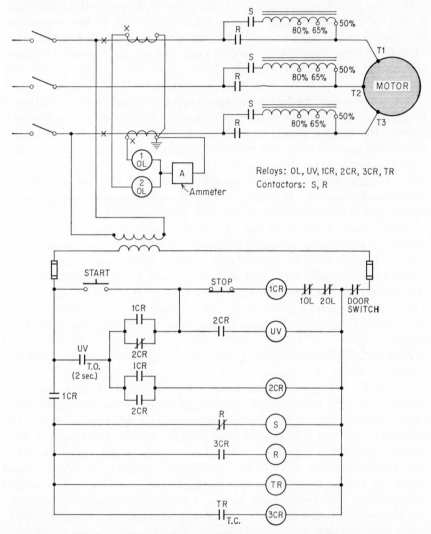

Fig. 6·15   Circuit diagram for a line-reactor reduced-voltage starter.

path to the *1CR* relay. The latter, therefore, is energized automatically, without the necessity for pressing the START button, and the motor accelerates in the usual way.

The motor can be stopped by pressing the STOP button. This action causes the $1CR$ relay to drop out, *but* the $2CR$ relay remains picked up for 2 sec, through the UV-T.O. contact and a $2CR$ interlock. This means that the N.C. $2CR$ contact is open and there is *no* recovery path to the $1CR$ relay. The machine, therefore, performs a normal stop.

In many hazardous installations all control equipment is housed within a steel cabinet whose front door is fitted with an *N.C. door switch in the closed position*. This permits the operator to start the motor only when the door is closed; with the door open, the control circuit is inoperative, as Fig. 6·15 indicates.

Autotransformer reduced-voltage starter—open-circuit transition. When a polyphase motor is started at reduced voltage by the line-resistance or the line-reactance method, the line *and* motor currents are, of course, the same. The inrush current will be less than that existing with full-voltage starting only to the extent that line resistors or line reactors incur voltage drops. The autotransformer method of starting, on the other hand, supplies the motor with reduced voltage by transformer action, and this implies that the *line-side* or *primary* current is the same fractional part of the *motor-side* or *secondary* current as is the ratio of secondary to primary voltage. Thus, neglecting the effect of magnetizing current (which may be appreciable under certain conditions), motor starting voltages of 50, 65, and 80 per cent will lower the line-side currents to 25, 42, and 64 per cent, respectively, of the full-voltage values. The "double-barreled" effect in minimizing the *line-current* inrush is, obviously, a decided improvement over the other "single-acting" schemes.

As Fig. 2·25 illustrates, the usual practice in constructing a 3-phase autotransformer, or *compensator* as it is sometimes called, is to place two tapped windings around the outer legs of a three-legged core and connect the coils in open delta. Standard tappings are at points that yield motor voltages of 50, 65, and 80 per cent of rated value. Since the starting power factor is generally low, usually less than 0.5 lagging, the internal impedance drops act to unbalance the three output motor voltages somewhat, and this has the effect of reducing the starting torque below the balanced-voltage condition. Tests indicate, however, that the unbalance is not particularly serious for most installations, although the three-coil construction is sometimes used where it is desirable that the motor develop maximum starting torque under balanced motor-voltage operation.

A wiring diagram of a *two-coil autotransformer* and its control circuit is shown connected to a 3-phase motor in Fig. 6·16. Note that the START-STOP station and its two relays are operated from the secondary of one control transformer while a d-c RUN contactor R is wired through a bridge rectifier from the secondary of another control transformer. The *S* contactor and the *CR* relay are actuated from the full-voltage source. With the pressing of the START button the *UV* and *TR* relays are picked up and sealed in by their interlocks, and another *UV* contact is closed

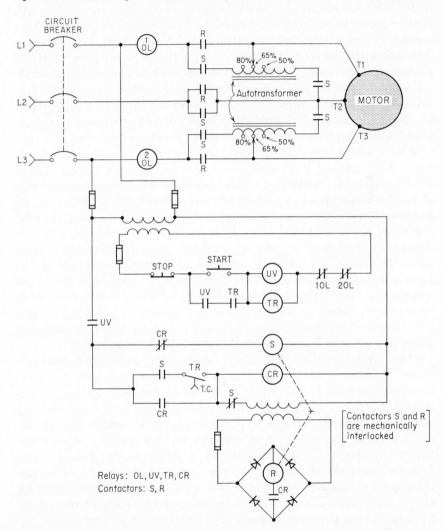

Fig. 6·16   Autotransformer reduced-voltage starter designed for open-circuit transition.

to energize the coil of the *S* contactor. When the latter operates, it immediately opens the N.C. contact in the primary of the lower control transformer (thus electrically preventing the simultaneous operation of contactors *S* and *R*, which are also mechanically interlocked), closes an interlock in the CR relay circuit, and closes five main contacts at the autotransformer. The motor therefore starts on reduced voltage ($0.65E_{rated}$ in the diagram), since the neutral point of the windings is made by two S contacts and power connections are made to winding ends and the motor by the other three S contacts. After relay TR times out, its TR-T.C. contact closes to energize the CR relay. This instantly opens the N.C. contact in series with the S contactor, which drops out, and closes an interlock in the R contactor circuit. With the deenergization of the S contactor, the five main contacts are opened and *the motor is momentarily disconnected from the power source.* The N.C. S contact closes next, permits the R contactor to pick up, and causes the three main R contacts to close. The motor now proceeds to run from the full-voltage source, with the compensator windings completely disconnected.

It is important to understand that there is a small time interval between reduced-voltage and full-voltage operation, i.e., between the opening of the S contacts and the closing of the R contacts, when the motor, running at nearly rated speed, is completely deenergized. Since voltages are, nevertheless, developed in the stator winding during this period, the final closing of the R contacts may result in an exceptionally high inrush current. As explained in Chap. 2, page 50, the peak value will depend upon the phase relation of generated and impressed voltages at the instant the switchover is made and may indeed be higher than the inrush current with full-voltage starting. This so-called *open-transition* switching scheme is, therefore, at a disadvantage where occasional high peak currents may be troublesome. As previously noted, the line-resistance and line-reactance starting methods avoid the difficulty indicated.

✓ *Closed-circuit Transition.* A circuit diagram of a *three-coil autotransformer* wired to a motor and a control circuit that provides for *closed-circuit transition* is given in Fig. 6·17. The arrangement shown is for a 200-hp 208-volt 3-phase squirrel-cage motor used in a comparatively good-sized air-conditioning system. Here again, the start-stop station and the large d-c R contactor with its rectifier unit are shown connected to the 115-volt secondary of a 208/115-volt control transformer. Note particularly that an extra contactor is needed to overcome the objection indicated for open-circuit transition. As will be explained, one of these, $1S$, is used for the transformer neutral, and the others, $2S$ and R, are required, respectively, for transformer and motor line connections.

When the start button is pressed, the UV relay picks up to close three interlocks. One of these is for sealing purposes, a second (on the outside) provides a path to $1S$, $2S$, and TR, and a third (on the inside) handles the circuits associated with normal running, i.e., relay CR and contactor R. With the operation of contactor $1S$, an N.C. interlock opens instantly to prevent electrically the simultaneous closing of contactor R, two $1S$ contacts close to make the transformer neutral, and a $1S$ inter-lock closes to energize the $2S$ contactor which, in turn, causes two $2S$ interlocks to close. Three main $2S$ contacts close next to start the motor at reduced voltage ($0.5E_{\text{rated}}$ in the diagram). After relay TR times out,

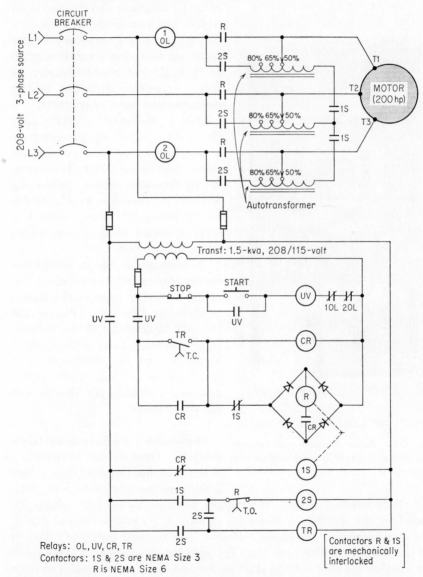

Fig. 6·17 Autotransformer reduced-voltage starter designed for closed-circuit transition connected to a motor in an air-conditioning system.

the TR-T.C. contact closes, control relay CR picks up and immediately opens the N.C. CR contact in the $1S$ contactor circuit. *When this happens, the motor is connected to the source through the three transformer coils, which now act as reactors.* The deenergization of contactor $1S$ is followed by the closing of N.C. contact $1S$ in the rectifier circuit and the opening

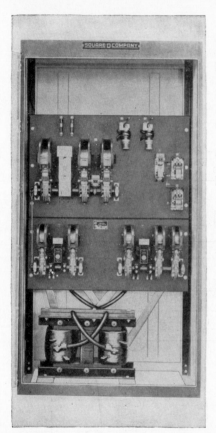

of a $1S$ interlock (to $2S$ and TR) which was previously bypassed by two $2S$ contacts. Contactor R finally picks up to connect the motor directly to the line, after which contact R-T.O. opens to drop out the $2S$ contactor and open its contacts.

It is to be noted especially that the entire switching procedure is accomplished *without opening the power lines to the motor.* The *closed-circuit transition scheme,* called the *Korndorfer connection,* avoids the difficulty previously described for Fig. 6·16, although it does involve an additional contactor.

A photograph of an autotransformer starter and its control components, mounted in a steel cabinet, is given in Fig. 6·18. Observe the two-coil transformer at the bottom, the five-contact S contactor directly above it, and the three-contact R contactor at the top. Three relays and two overloads are also seen in the upper panel.

Fig. 6·18 Autotransformer reduced-voltage starter mounted in steel cabinet with door open.

Star-delta reduced-voltage starter. *Open-circuit Transition.* When a 3-phase squirrel-cage motor is designed to operate *normally* with the stator winding in *delta,* the voltage per phase is the line voltage. If, therefore, the winding is connected in *star* (**Y**) by external means during the starting period, the phase voltage will be reduced to $E_L/\sqrt{3}$, i.e., to 58 per cent of the line potential. The arrangement indicated is the basis for another reduced-voltage starting method but can be employed only when both ends of each winding phase are brought out. A simplified schematic diagram showing how this is accomplished by using three sets of contacts (three contactors) is given in Fig. 6·19. With the pressing

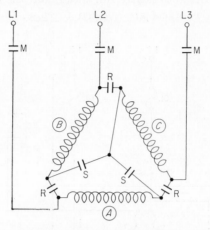

Fig. 6·19 Schematic diagram illustrating the *star-delta* reduced voltage starting method. A, B, and C are the three phases of the stator winding.

of a START button, two S contacts close to connect the three phases A, B, and C in *star*, after which the three line contacts close; the motor thus starts under conditions that are equivalent to a line voltage of $0.58E_L$. After a time interval determined by the adjustment of a timing relay, switching relays first open the S contacts and then close the R contacts. With the three corners of the *delta* now connected to the source, the motor proceeds to run normally at rated voltage. An important advantage of this reduced-voltage starting scheme is that no accessory equipment such as resistors, reactors, or transformers is needed.

The inrush current and starting torque are lowered considerably by this *star-delta* method of motor acceleration. If it is assumed that these are, respectively, $6I_{FL}$ and $1.5T_{FL}$ with full-voltage starting, i.e., when the winding is connected in *delta*, the winding-phase current will be $(6/\sqrt{3})I_{FL}$ under this condition. However, since the voltage across each winding phase is diminished to $0.58E_{\text{rated}}$ when the *star* connection is used, the peak line current, which is the same as the winding current, will be $(0.58 \times 6/\sqrt{3})I_{FL}$ or $2I_{FL}$; also, the starting torque, which varies as the *square* of the winding voltage, will be $(1/\sqrt{3})^2 \times 1.5T_{FL}$ $= 0.5T_{FL}$. Note particularly that both the line inrush current and the starting torque are one-third of their full-voltage values.

A wiring diagram of a *star-delta* control circuit connected to the three winding phases of a motor is illustrated by Fig. 6·20. As explained above, its operation depends upon first closing the S and M contacts, permitting the motor to accelerate to nearly full speed, and the opening the S contacts *before* the R contacts are allowed to close. It is imperative that

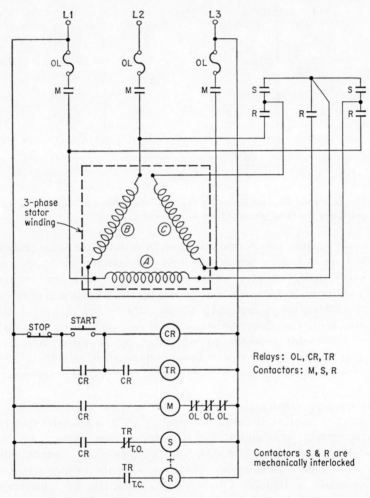

Fig. 6·20   Star-delta reduced-voltage starter designed for open-circuit transition.

the control devices impose the latter restriction on the S and R contactors because their simultaneous operation would involve a serious short circuit. To avoid the latter possibility, both mechanical and electrical interlocking is provided.

The motor is started with the winding in *star* (**Y**) by pressing the

START button. This activates the CR relay which closes four interlocks, one for sealing purposes and the other three to permit the timing relay TR and the M and S contactors to pick up. Note that the two S contacts establish the *star* point in the winding and the M contacts connect the latter to the source. When relay TR times out, contact TR-T.O. *opens first* to drop out the S contactor, after which the TR-T.C. contact closes to energize the R contactor. The opening of the S contacts to open the neutral is, therefore, followed by the closing of the three R contacts to interconnect the winding in *delta* and to the source. Note particularly that this switching scheme involves a short time interval, between the opening of S and the closing of R, when the motor is completely disconnected from the power lines; it therefore represents *open-circuit transition*.

Closed-circuit Transition. To avoid the possibility of high inrush current when the winding is momentarily open-circuited in switching from *star* to *delta*, a special set of resistors and an additional contactor are used in a modified control system to maintain continuity. The scheme is illustrated by Fig. 6·21 which provides closed-circuit transition for a 300-hp 208-volt 3-phase motor that drives a rather large blower. Note that it includes two triple-pole (TP) main contactors $1M$ and $2M$ (both NEMA No. 6 size), a double-pole (DP) neutral-point contactor (NEMA No. 3 size), and a TP transition contactor (NEMA No. 1 size). The operation of the control circuit is as follows:

1. When the START button is pressed, the UV relay is energized and three UV contacts close.
2. The S contactor is actuated to close the neutral at winding points $N1$ and $N2$.
3. The $1CR$ relay picks up, contactor $1M$ operates, and the winding, now in *star*, is connected to the source at terminals $T1$, $T2$, and $T3$.
4. When timing relay TR times out (it is energized through closed contacts S or $1CR$) contact TR-T.C. closes in the T-contactor circuit. This action causes three T contacts to close, which, in turn, connects a resistor in parallel with each of the three winding phases. The winding is still connected in *star*.
5. Contact T-T.O. opens next and drops out the S contactor. The S contacts, therefore, open with the result that each winding phase and a corresponding resistor are connected in series and to full-line potential. The winding is now in *delta*, with a resistor in each phase to limit the current inrush when the switch is made.
6. With the deenergization of the S contactor, the N.C. S interlock closes to permit relay $2CR$ to pick up. The N.C. interlock $2CR$ opens to provide additional safety.

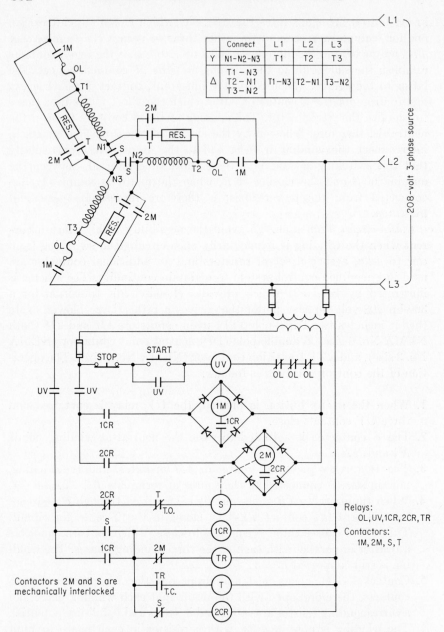

Fig. 6·21 Star-delta reduced-voltage starter designed for closed-circuit transition connected to a 300-hp motor driving a large blower.

7. Contactor $2M$ picks up next, its three contacts close, and each one short-circuits a resistor. The winding is now normally in *delta* and connected to the full-voltage source.
8. The N.C. interlock $2M$ opens, relay TR drops out, interlock TR-T.C. opens, and contactor T is deenergized. The last action open-circuits the three resistor circuits.

Part-winding starter. As explained in Chap. 2, page 30, a dual-voltage motor is sometimes equipped with a starter that initially connects one-half of the stator winding to the source and then, after the machine reaches nearly full speed, parallels the second half of the winding with the section already energized. When started by this so-called part-winding method, such motors must, of course, be operated at the lower of the two voltage ratings, since the higher voltage implies that both halves of each phase be permanently in series. Generally, the windings are connected in *star*, as Fig. 2·7 illustrates, although in special cases *delta* connections have been employed. Two general starter designs are available, designated as two-step and three-step. In the two-step arrangement, briefly described above, the locked-rotor current is about 60 per cent of the value that exists when both winding halves are in parallel and the motor develops about 45 per cent starting torque. Then, after about 3 to 6 sec the parallel connection is made and standard operation proceeds. The three-step type includes a series resistor in each half-winding phase when the motor is first started. After a short time interval (about 2 sec) the resistors are short-circuited and the energized half of the winding is connected to full voltage. Two or more seconds later, usually, normal operation continues as the paralleling action is completed.

As in the star-delta method, part-winding starting requires no auxiliary equipment such as resistors, reactors, or transformers and offers the additional advantage that only two half-sized contactors are necessary. On the other hand, the starting torque cannot be adjusted as in the other schemes (by tap changing, for example) and is often too low to start the load on the first step. Since motor heating is usually a factor under such conditions, it is frequently necessary to resort to increment starting, which reduces the initial torque further while it lessens the heating effect. Another objection to the part-winding starter is that motor acceleration is often accompanied by excessive noise and extremely high transient currents during switching.

A simplified diagram of the stator winding and its terminal markings for a dual-voltage, four-pole motor is given in Fig. 6·22. When arranged for part-winding starting, terminals $T4$, $T5$, and $T6$ are joined as in Fig. 2·7 to form the second star point. Note particularly that one-half of each phase, represented by two diametrically opposite sets of coils that are similarly polarized, is energized on the first step; this occurs when terminals $T1$, $T2$, and $T3$ are connected to the line. For normal running, the controller will connect the second section of the winding in parallel with the first half by joining $T7$ to $T1$, $T8$ to $T2$, and $T9$ to $T3$. A wiring diagram illustrating how this is accomplished by the control circuit is

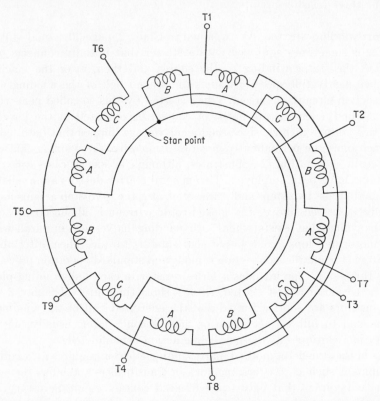

Fig. 6·22 Stator winding diagram and standard terminal markings for a dual-voltage 4-pole motor.

given in Fig. 6·23. Observe that overload protection is provided for both halves of the motor winding.

Two-phase motors connected to 3-phase line starters. Since 2-phase motors have limited use for industrial applications, manufacturers of control apparatus find it uneconomical to build starters that are specially designed for these machines. This is not particularly objectionable because 3-phase starters can usually be adapted to the 2-phase motors which, when employed, generally have low-horsepower ratings and are started by line-voltage equipment. Figure 6·24 shows two wiring schemes for a

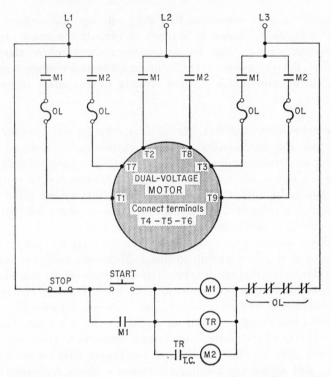

Fig. 6·23   Dual-voltage motor connected to a part-winding starter.

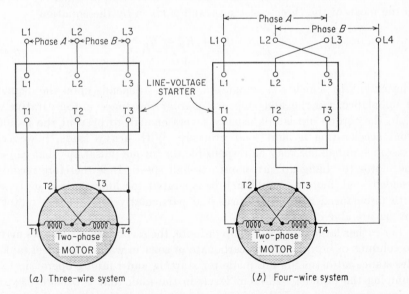

(a) Three-wire system   (b) Four-wire system

Fig. 6·24   Connection methods for a 2-phase motor to a 3-phase line-voltage starter.

2-phase motor that is connected to a *three-wire* and a *four-wire* system. In the first of these, sketch *a*, a junction is made of two ends of the motor winding, i.e., *T*3 and *T*4, and is then connected to the neutral point *L*2 of the three-wire system through the full-voltage starter. In sketch *b*, each motor phase is wired directly to a separate power phase as illustrated.

Wound-rotor motor starters. Magnetic starters for wound-rotor motors are, in some respects, similar to increment-resistance reduced-voltage starters for squirrel-cage machines (Fig. 6·14) in which primary resistances are cut out in steps to control the inrush currents. In contrast to squirrel-cage motor equipment, however, the stator current in the wound-rotor motor is a reflection, by transformer action, of the current in the rotor circuit where resistance adjustments are made.

Two general methods can be employed to cut out the secondary resistors in the motor as it is brought up to speed. More commonly, equal resistances are short-circuited on the three legs simultaneously; this maintains balanced conditions and is theoretically the most desirable arrangement. A more economical scheme is to use single-pole contactors each of which short-circuits one resistor in each leg of the circuit at a time. This takes better advantage of the control equipment, since more accelerating steps are possible, although the unbalanced conditions tend to cause torque pulsations and additional mechanical stresses. Tests indicate that the effective resistance of an unbalanced rotor circuit is approximately equal to the mean of the three resistances and is given by the equation

$$R_{\text{eff}} = \frac{R_A + R_B + R_C}{3} \tag{6·4}$$

The extent to which performance will suffer depends upon the degree of unbalance. On the first step, when rotor resistance is comparatively high, the torque dip is not usually severe enough to prevent the motor from accelerating a light load properly. With heavy loads, however, resistance unbalance may be responsible for torque pulsations that cause the motor to "hang up" at about one-half speed. It is therefore recommended that motors should not be operated for long periods of time with unbalanced rotor resistances and particularly so when the motor ratings are above 50 hp.

For rather large motors, 200 hp or more, the secondary resistances may be columns of liquid, such as carbonate of soda in water, in a steel tank. Resistance adjustments are made for starting and running operation by pumping the liquid to different levels in the tank that is occupied by a unique arrangement of plates. The advantages of this system of control

are (1) smooth, stepless acceleration, (2) high thermal capacity, and
(3) simple construction.

A wiring diagram illustrating a four-step magnetic starter connected
to a small wound-rotor motor is given in Fig. 6·25. As will be noted,

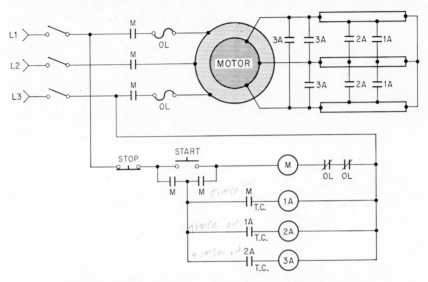

Fig. 6·25   Magnetic starter connected to a wound-rotor motor.

balanced rotor-resistance conditions are maintained by the closing of
successive sets of contacts 1A, 2A, and 3A; double-pole contactors are
used for the first two sets, and a triple-pole unit for the last set. Since
the final step is the full-speed point, short-circuiting three pairs of lines
helps to reduce contact resistance for zero-resistance operation.

Where motors drive intermittent-duty loads such as cranes, ore
unloaders, skip hoists, and other heavy machinery in steel mills and
industrial plants, a-c mechanical timers are generally less rugged than
the contactors to which they are attached and tend to multiply service
problems. The use of *off-delay d-c relays* that are energized through
selenium rectifiers have proved much more satisfactory in such installations and have been widely used. A circuit diagram illustrating a control
circuit that is timed by such components is shown in Fig. 6·26. *Here
the three accelerating relays 1AR, 2AR, and 3AR are picked up and their
N.C. contacts are opened as soon as the main switch is closed;* thus, contactors 1A, 2A, and 3A cannot be energized until the time-closing contacts are permitted to close by the deenergization of the respective relays.
The pressing of the START button causes the three M line contacts to
close to start the motor with maximum secondary resistance; it also

opens an M interlock in the first relay circuit, i.e., $1AR$, which, closing its contact after a time delay, results in the short-circuiting of the first set of resistors. Following this, N.C. interlock $1A$ opens, N.O. interlock $1A$ closes, and relay $2AR$ is deenergized. When the latter times out, its N.C. contact closes and permits contactor $2A$ to pick up and short-circuit the second group of resistors. With the opening of the N.C. $2A$

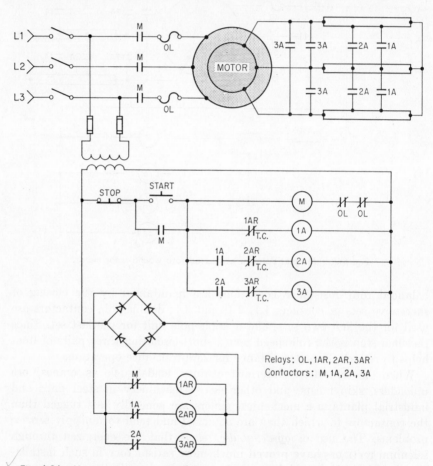

Fig. 6·26 Magnetic starter with off-delay relays connected to a wound-rotor motor.

interlock, relay $3AR$ drops out to close its contact in the $3A$ contactor circuit after a time interval and causes the third contactor, $3A$, to operate. The final set of resistors are thus cut out, and the motor proceeds to run normally.

An important characteristic of all induction motors is that the *rotor voltage and its corresponding frequency* are directly proportional to the rotor slip, having maximum values at standstill and diminishing to zero at synchronous speed. In the wound-rotor motor, these quantities can, of course, be measured at the slip rings and, if desirable, made available to specially designed tuned relay circuits that respond to definite frequencies and act to time the closing of a group of acceleration relays. The scheme has, indeed, been used effectively in a control system illustrated by Fig. 6·27, where each of four acceleration relays is connected

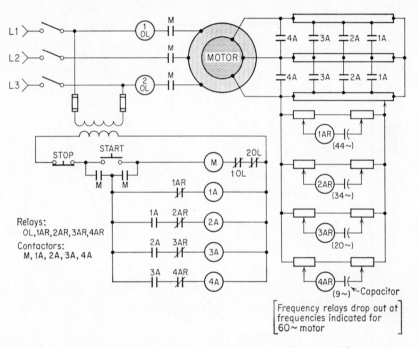

Fig. 6·27 Magnetic starter with frequency relays connected to a wound-rotor motor.

in series with a properly selected capacitor and the combination is energized at the output terminals of a potential divider. The *RLC* circuits represented by similar relays and different values of capacitance are designed to pick up instantly at 60 cps, i.e., at standstill, and drop out at designated frequencies. The latter may be any suitable values as, for example, 44, 34, 20, and 9 cps. For a 4-pole motor whose synchronous speed is 1,800 rpm, these frequencies would, therefore, correspond to 480, 780, 1,200, and 1,530 rpm.

To show further how the series *RLC* circuits function, refer to Fig. 6·28. Note that the position of each *current-versus-frequency curve* depends upon the size of the capacitor and that the current varies considerably with frequency. The relays are adjusted to remain picked up above certain minimum values of current but will drop out in sequence as the motor speeds up and lowers both frequency and current.

The circuit of Fig. 6·27 is designed to operate in essentially the same was as that described for the off-delay relay starter (Fig. 6·26) with the one difference that frequency relays control the acceleration of the motor. At the instant the motor starts with maximum secondary resistance, all four relays pick up (see Fig. 6·28) and the contactor coils are deenergized.

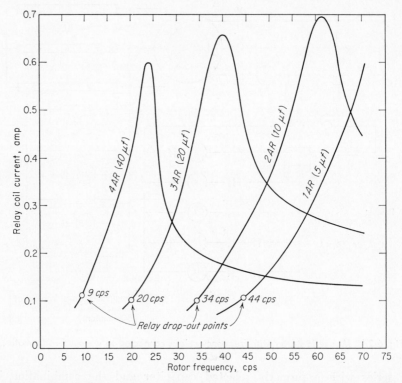

Fig. 6·28 *Current-versus-frequency* relations in a frequency relay connected to a rotor of a wound-rotor motor.

At a rotor frequency of 44 cps (480 rpm for a 4-pole motor), $1AR$ drops out, its N.C. contact closes, and contactor $1A$ pulls in to short-circuit the first set of secondary resistors; also a $1A$ interlock closes to prepare the $2A$ contactor for the next step. When the motor reaches a speed of 780 rpm and the frequency is reduced to 34 cps, relay $2AR$ drops out and the second group of resistors is cut out by contactor $2A$. Similar actions take place at 20 cps (1,200 rpm) and finally at 9 cps (1,530 rpm) when the motor accelerates to its operating speed.

Reversing of a-c motors. Since the rotor of a polyphase (3-phase) induction motor always tends to rotate in the same direction as the revolving magnetic field and the latter depends upon the phase sequence of the impressed voltages, this type of machine can be reversed by interchanging any pair of stator leads. This is generally accomplished by having two mechanically interlocked contactors operate in a control circuit that is quite similar to that used for d-c motors. In addition, the contactors are frequently interlocked electrically to avoid energizing the coil of an opened unit while its mate is in a closed position. With a contactor in a dropped-out position, i.e., with a wide armature air gap, the current is high enough to damage the coil if permitted to flow for an appreciable time. As in reversing d-c circuits, mechanical interlocking is essential to prevent line short circuits, while electrical interlocking is necessary in a-c circuits to guard against overheating and possible burnout of contactor coils. While a line-start induction motor is running in one direction, it can be plugged to perform a quick reversal because, under this condition, the inrush current is not much more than when it is started from rest.

Two circuit arrangements illustrating how a line-start squirrel-cage motor is connected to reversing controllers are given in Fig. 6·29. In diagram a the control circuit is similar to that of Fig. 3·16b, used for armature reversing of d-c motors. Note that the triple-pole forward (F) and reverse (R) contactors are mechanically interlocked and that electrical interlocking is provided by the back contacts of the FOR.-REV. buttons and by N.C. interlocks in the coil circuits. In diagram b three double-pole contactors are employed, one for line-closing and the other two for *forward* and *reverse* running.

A wiring diagram showing a wound-rotor motor connected to a reversing master-type controller (see Chap. 4) is illustrated by Fig. 6·30. Frequency relays are employed in the control circuit as for the nonreversing motor of Fig. 6·27. Of special interest is the use of a plugging relay PR which is tuned to drop out at 61 cps, or just before the plugged motor comes to rest and is ready to rotate in the opposite direction. This provision is felt desirable to make sure that secondary resistance will not be cut out until the motor is about to reverse. With the machine running normally in a *forward* direction, for example, the rotor frequency will be extremely low (the slip frequency, which is about 3 to 5 cps). If the master is then moved quickly to the last point *reverse*, the revolving field will instantly change direction, and with the motor still running forward, the rotor frequency will immediately rise to about 120 cps. Since the PR relay will be picked up, its N.C. interlock in the P contactor circuit will be open and additional external resistance will be connected in the rotor circuit. As the motor slows down in its attempt to reverse, the

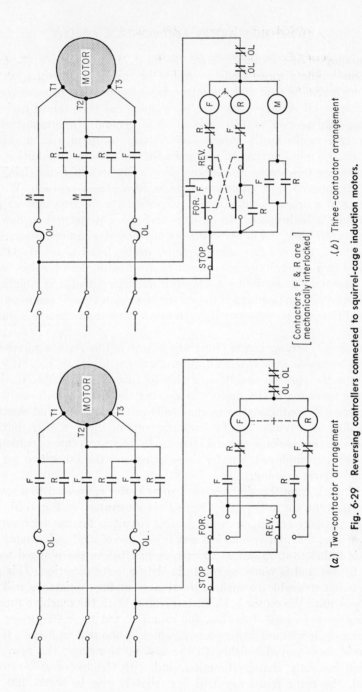

.(b) Three-contactor arrangement

[Contactors F & R are
mechanically interlocked]

(a) Two-contactor arrangement

Fig. 6·29 Reversing controllers connected to squirrel-cage induction motors.

rotor frequency drops and at 61 cps, or just prior to standstill, the *PR* relay drops out, closes its interlock, and normal acceleration proceeds in the opposite direction.

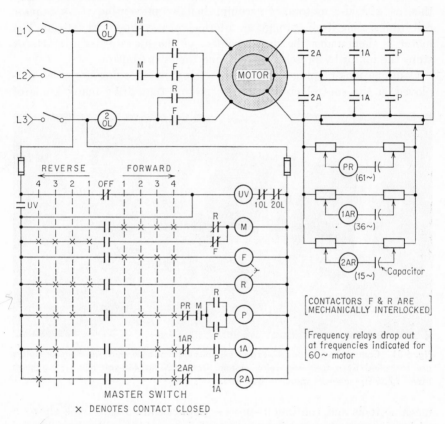

Fig. 6·30   Master-type reversing controller connected to a wound-rotor motor.

A photograph of the control room for a double-cage mine hoist installation is shown in Fig. 6·31. The large drums are driven by wound-rotor motors whose control panel can be seen at the far right. A frequency relay control system like that described for Fig. 6·30 is used.

Plug-stopping and jogging. To plug a 3-phase squirrel-cage motor to a stop it is first necessary to develop a reversing torque by changing the direction of the revolving magnetic field (interchanging any two stator leads) and then, by using a zero-speed plugging switch in the control circuit (Fig. 3·24), disconnecting the power source at the instant the rotor attempts to reverse. As was previously pointed out, the plugging action need not be accompanied by a reduction in motor voltage (as is necessary in d-c motors by the insertion of armature resistance) because the inrush current is not substantially higher than when starting from rest.

Jogging (inching) is frequently required in a-c motor installations, and this, as with d-c motors, is accomplished by preventing the contactor coil from sealing itself through an interlock across the START button. The motor is then under the direct control of the operator, who starts or stops the motor by pressing or releasing the START button.

Combinations of plug-stopping, reversing, and jogging are often employed in the control circuits of a-c motors that drive many kinds of

Fig. 6·31   Control room of double-cage mine hoist installation in which wound-rotor motors are controlled by a frequency-relay system. Data: 400-hp 440-volt motors; maximum travel 2,755 ft; maximum speed = 800 fpm.

machine tools and continuous-process applications. Figure 6·32 shows a wiring diagram of one such arrangement, for a rubber mill, whose control circuit was designed (1) to provide continuous operation in one direction, (2) for plug-stopping, and (3) for jogging in reverse only. Note that there are three relays, a *forward control relay FCR*, a *reverse control relay RCR*, and a *plugging-switch control relay CR*, in addition to a zero-speed plugging switch and mechanically interlocked *forward* and *reverse* contactors. For forward operation the RUN button is pressed. This actuates the *FCR* relay, which in turn opens an interlock in the coil of the *RCR* relay and closes a contact for the energization of the *F* contactor. One *F* interlock also closes across the RUN button, and another does likewise in the plugging-switch circuit. As the motor starts, the plugging-switch contact closes and permits the *CR* relay to pick up. Three *CR* interlocks then close, one to seal in the *CR* relay, another to prepare a circuit for the *R* contactor, and a third to complete the short circuit across the

RUN button. The motor can be plugged to a stop by pressing the STOP button. This deenergizes *FCR*, causes the *F* contactor to drop out, and permits the N.C. *F* contact to close; the *CR* relay remains picked up through its interlock even though a paralleled *F* contact opens. With the closing of the reversing line contacts negative torque is developed

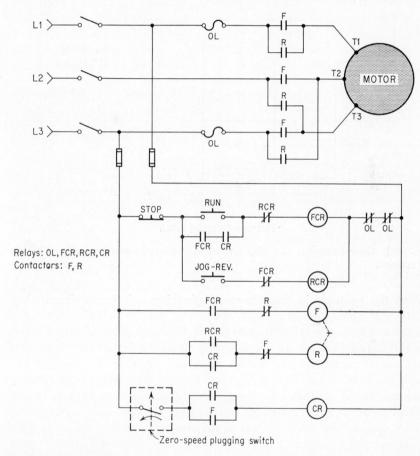

Fig. 6·32 Squirrel-cage motor connected to a controller designed for forward running, jog reversing, and plug stopping.

and the motor plugs. When the motor comes to rest, the plugging-switch contact opens, the *CR* relay is deenergized, and the *R* contactor drops out to disconnect the power source. With the motor at rest it can be jogged in *reverse* by pressing the JOG-REV. button. This merely lets the *R* contactor pick up through closed contact *RCR* without providing a seal for the *RCR* relay coil.

To plug-stop a comparatively small squirrel-cage induction motor from either direction of rotation, the control circuit of Fig. 6·33 can be used.

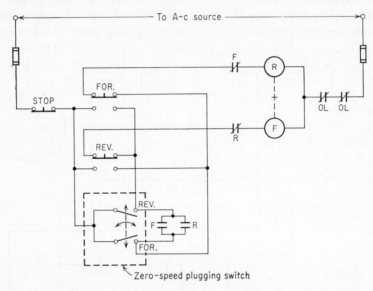

Fig. 6·33 Control circuit to plug-stop a small squirrel-cage motor from either direction of rotation.

When the contacts in the zero-speed plugging switch are closed, they carry the contactor-coil current. As in Fig. 6·32, which also shows the motor and line connections for this circuit, mechanical and electrical interlocking is provided for the contactors. It should be noted that the *forward* and *reverse* contactors are always energized, respectively, through the back contacts of the REV. and FOR. buttons. With the motor running forward, the F contactor is energized through the FOR. contact of the plugging switch, the F interlock, and the back contact of the REV. button. To plug-stop, the STOP button is pressed and quickly released. This drops out the F contactor and permits the N.C. interlock in the R contactor circuit to close. Current now passes to the R contactor coil through the FOR. contact of the plugging switch and back contact of the FOR. button. Since reverse torque is developed, the motor rapidly comes to a stop, at which instant the zero-speed plugging switch opens its contact to deenergize the R contactor coil. Similar operating conditions prevail when the motor is plug-stopped from a reverse direction.

Dynamic braking. An induction motor can be brought to rest quickly by the dynamic-braking principle if, during operation, the power supply to the stator winding is changed from alternating current to direct current. A set of *stationary* magnetic poles are then developed, equal in number to those in the normal revolving field, and these are cut by the rotor conductors. Since voltages are generated in the latter, currents will flow in the squirrel cage (or in the closed-circuit secondary of the wound rotor) and mechanical energy of rotation will be converted to I^2R heating. During the period of energy conversion, determined primarily by the strength of the d-c excitation, the motor continues to slow down until, at zero speed, the interchange is complete. A control circuit that provides for dynamic braking in addition to its other functions is generally equipped with a rectifier unit through which a suitable and adjustable d-c source is made available from the existing a-c power lines. For a quick stop it is customary to adjust the direct current to a value that is about six to eight times the rated motor current.

During the braking period, i.e., between full speed and a complete stop, braking torque varies considerably. Depending upon the magnitude of the d-c excitation, it may be as little as 50 per cent of rated motor torque at the instant the braking torque is initiated, rise to a value of 500 to 600 per cent at about 3 to 6 per cent speed, and then drop rapidly to zero as the motor comes to rest.

Unlike dynamic-braking action in a d-c motor, where the flux is essentially constant and the braking torque diminishes from a high initial value to zero as the armature slows down, the flux in an a-c motor changes greatly during the braking period. Remembering that braking torque depends upon the reaction between *rotor current* and *total flux* (i.e., $T = k\phi I_r$) and that the flux in an a-c motor has a *resultant* value that is represented by the constant d-c flux and a superimposed flux caused by the rotor mmf, it should be clear that the *flux-current product* at any instant will determine the torque at that instant. At the start of the braking period, when the rotor reactance is high, the rotor current has a low lagging power factor. This current, therefore, has a large reactive component that greatly diminishes the d-c field; under this condition the braking torque is low. As the motor slows down and the rotor reactance diminishes (the power factor of the rotor current increases), the demagnetizing mmf is rapidly reduced; the braking torque thus rises even though the motor speed drops.

To illustrate how a set of stationary magnetic poles is produced in the stator core whose winding is excited from a d-c supply, Fig. 6·34 has been drawn. Although the diagram shows the conditions that exist for a 4-pole star-connected winding, similar magnetic relations will prevail in stators with delta-connected windings and for any number of poles. Direct current can be supplied to the 3-phase winding in either of two ways. In one of these, a pair of stator terminals such as $T1$ and $T2$ are connected to the source, in which case two phases are excited and the third phase is left open; the same current then passes through the two energized phases in series. In a more popular arrangement, and one that develops slightly less heating for a given braking torque, two terminals, such as $T2$ and $T3$, are tied together and their junction and $T1$ are connected to the d-c supply.

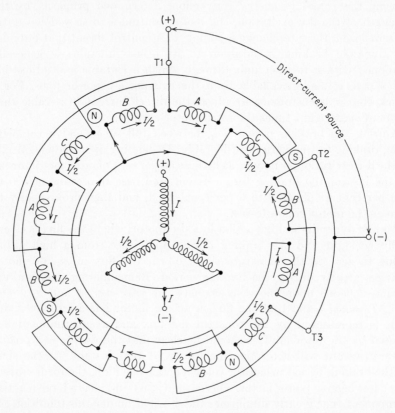

Fig. 6·34 Diagram showing how four stationary magnetic poles are produced when a star-connected winding is energized by a d-c source.

A circuit diagram for a squirrel-cage motor connected to a line-start controller arranged for dynamic braking is given in Fig. 6·35. Note particularly that an extra back contact on the START button opens in the dynamic-braking coil circuit when the motor is started; this is added assurance that direct current and alternating current will not be applied to the stator winding simultaneously. When a normal start is executed by pressing the START button, the N.C. interlock M opens and the N.O. contact TR-T.O. closes. Thus, the main contactor M is energized and the dynamic-braking contactor is made ready for a quick stop. When the STOP button is pressed and contactor M and timing relay TR both drop out, the DB contactor is energized through the *time-opening TR* contact and the closed M contact. An N.C. DB contact is also opened to make doubly sure that alternating current is not applied to the stator

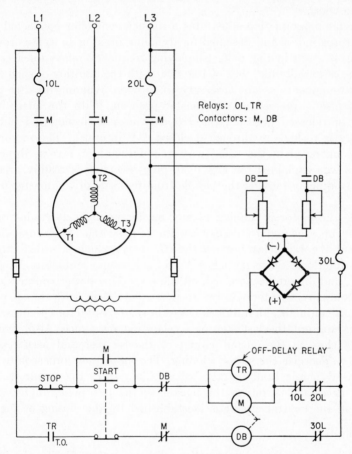

Fig. 6·35 Squirrel-cage motor connected to a line-start controller arranged for dynamic braking.

winding at this time. The motor now proceeds to brake dynamically, and when the speed drops to zero, the TR-T.O. contact opens.

An interesting application of dynamic braking is the lowering operation of a hoist or crane which is driven by a wound-rotor motor. Here, the action is to *retard*, not stop, the downward motion of an overhauling load. Direct current is applied to the stator winding as previously explained, and the machine then behaves like an alternator that "feeds" its energy into the secondary resistors. Under this condition the mechanical energy of the overhauling load is converted to electrical energy that heats the resistors. Master controllers, employed in such installations, adjust the secondary resistance which, in turn, determines the lowering speed; greater retardation is obtained with lower values of resistance, and vice versa.

A circuit diagram that illustrates a master controller connected to a wound-rotor motor and designed for dynamic braking as well as reverse plugging is given in Fig. 6·36. Employing frequency relays for acceleration it is essentially like that of Fig. 6·30 with the addition of such other circuit components as are necessary to provide dynamic braking when a *foot switch* is pressed at any master position. With the latter in its normal or released position, the *DB* contactor is deenergized but the *CR* relay is picked up through a closed *UV* interlock. The motor can then be operated in the usual way in a forward or reverse direction. Also, as was explained for Fig. 6·30, plug-reversing is readily accomplished by quickly moving the handle from one side of the master to the other.

While the motor is running at any master point, a dynamic-braking stop can be executed by depressing the foot switch. This immediately drops out the *UV* relay but *not* the *CR* relay, which is sealed through its own interlock. Since the left *UV* contact opens, contactors *M*, *F*, *R*, 1*A*, 2*A*, and 3*A* are deenergized, all the respective power contacts open, and accompanying interlocks return to normal. Then, with the operation of the *DB* contactor, an interlock closes in the *P* contactor circuit, which, in turn, cuts out the last step of acceleration resistance. Also, two *DB* contacts close in the output circuit of the rectifier and permit direct current to energize the stator winding. The motor now proceeds to slow down as dynamic braking takes place. To restart the motor it is, of course, necessary to return the master to the OFF point, under which condition the control circuit is reestablished by the closing of the *UV* contacts.

Synchronous motors—construction and operation. Unlike the induction motor, which receives electrical power through its stationary stator winding only and is, therefore, *singly fed*, a synchronous motor is *doubly fed* because both stator and rotor are connected to different sources of supply. The stators of the two types of machine are, however, identical, and as was previously explained, when either winding is excited from polyphase power lines, a revolving field is created that rotates at a constant speed, the so-called synchronous speed. The latter is directly proportional to the line frequency and inversely proportional to the number of poles for which the stator is wound, or in equation form, $S = 120f/P$.

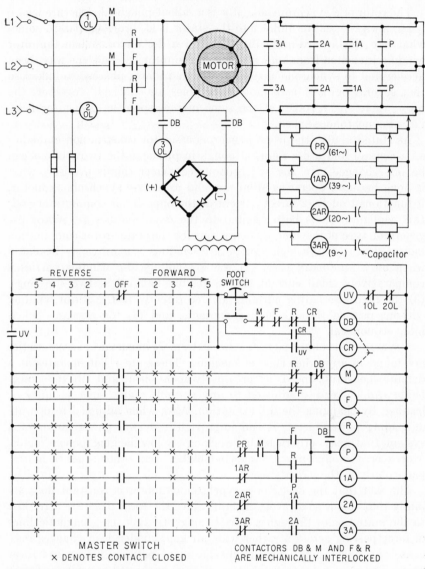

Fig. 6·36 Master-type reversing controller, with dynamic braking, connected to a wound-rotor motor.

The rotor of a synchronous motor is a salient-pole field structure whose poles, always equal in number to the stator field, project outward somewhat like the inward-projecting poles of a d-c motor. When the rotor winding is excited from a *d-c source* and the rotor has, in some way, been accelerated to synchronous speed, rotor and stator phases are linked so that several pairs of *north* and *south* poles are "locked" together; the rotor, therefore, revolves at the same *average* speed as the stator field, i.e., at synchronous speed.

As stated above, the usual synchronous-motor construction embodies an a-c stator and a d-c rotor, although in principle the two sections can be interchanged. (In the synchronous converter this is precisely what is done because, operating from one end as an a-c synchronous motor, it functions to deliver direct current at the opposite or commutator end; this configuration is better suited to the double-acting a-c motor–d-c generator type of machine.) The reasons for the stator-rotor combination indicated are (1) the high-voltage a-c winding is more easily insulated when on a stationary core, (2) the low-voltage d-c winding carries a comparatively small current that is readily handled at two slip rings, (3) the high alternating currents are fed directly to the stator winding without passing through sliding contacts, and (4) the arrangement is more economical.

A synchronous motor possesses two extremely important characteristics not found in any other type of machine. These are (1) it operates at a definite, constant, speed, as previously mentioned, and (2) the power factor can be readily adjusted to any suitable value, such as unity or leading, by changing the d-c excitation. Also, when adjusted to operate at leading power factor, it can improve the overall power factor of a system that may otherwise be working at low lagging power factor. Moreover, the efficiency is somewhat higher than other types of motor, particularly in the larger sizes.

The fact that the speed of a synchronous motor is constant does *not* imply that this is so at every instant. Since the rotor poles are attracted to the stator poles through a *flexible* magnetic link, i.e., magnetic lines of force passing across a rather wide air gap, sudden load changes may cause variations in the angular displacement between stator and rotor poles. The rotor poles therefore tend to shift backward and forward with respect to the uniformly revolving stator poles, so that the instantaneous rotor-pole speed may be more or less than synchronous speed. This so-called "hunting" action then gives rise to changes in motor torque which increase and decrease, respectively, as the angular displacements widen and narrow. Should a heavy load increase the pole separation beyond the stability limit, the rotor will pull out of step and the motor will stall. Note particularly that, unlike an induction-type machine, the rotor of a synchronous motor cannot slip but must always operate at an *average* constant speed, the synchronous speed.

An obvious disadvantage of the synchronous motor, although not a serious one, is that the rotor field must be energized by a separate d-c source. The latter may be an available d-c supply, usually at 115 or 230 volts, or, what is frequently the case, a small d-c generator that is mounted in the shaft extension of the motor. Moreover, the machine, as described, is incapable of developing starting torque but must be brought up to speed in some way before it can continue to operate normally. The method commonly employed to accomplish this is to construct a squirrel cage in the pole faces of the rotor and accelerate the latter an as induction motor. Special precautions must be taken when this is done, but these will be described in the next article, where synchronous-motor starters are discussed in some detail. Also, the very same squirrel cage that provides the machine with its starting torque is instrumental in minimizing hunting, i.e., it tends to damp out oscillations; this is because, by Lenz's law, any change in flux that links with the cage as the latter attempts to oscillate back and forth is accompanied by a current in the moving structure whose direction is such as to oppose a variation in flux that normally links the armature and the rotor. The effectiveness of the *amortisseur* (a term often used for the squirrel cage) depends upon its resistance; the lower the resistance, the stronger is the damping action. But since a high amortisseur resistance is necessary for good starting torque, it is customary to employ a compromise value.

The primary purpose of the squirrel cage is to accelerate the motor, since it carries no current during normal operation except for short periods, when it acts as a damper winding. It is, therefore, designed with low thermal capacity, i.e., for short-duty service, and must be protected against the possibility that the machine will come up to speed too slowly under certain conditions and run as an induction motor for an extended period of time. The kind of protection that suggests itself is a timing relay whose N.C. contact will open the control circuit in the event the motor fails to synchronize properly. As with induction motors, synchronous machines can be started by line-voltage, reduced voltage, or part-winding controllers. The type of equipment used will, of course, depend upon such factors as power-service restrictions, kind of load and frequency of starting, the extent to which voltage dips can be tolerated, and others.

During the accelerating period, the stator revolving field moves across the rotor poles, quickly at first and diminishing to a *relative* speed of zero as synchronism is approached. Since a great many turns are used in the d-c field winding, an extremely high voltage can be induced in the latter because of the rapid change in flux linkages. The condition is particularly serious at the instant of starting, when the relative motion of flux to poles is at a maximum, and voltages of 10,000 or more may be developed if the field winding is left open-circuited. To avoid the difficulty indicated it is generally necessary to close the field winding through a discharge resistor; this will enable a demagnetizing component of the resulting current to reduce the flux linkages and limit the induced emf to a safe value. Since the current in the discharge resistor is almost constant during the starting period, decaying somewhat as the motor approaches synchronous speed, the current capacity of this component must be based on the rms value. Its ohmic value, on the other hand, must depend on the design of the field structure and the required starting torque; the latter does, in general, diminish as the resistance is decreased. In some cases where good starting torque must be developed, the field is left open-circuited but is sectionalized to reduce the voltage across separated portions; this keeps the emf below the puncturing or dielectric strength of the insulation. In other cases where the motor must exert an extremely high starting torque, a special *insulated* pole-face winding is used and is connected to a set of three slip rings. External resistance can then be inserted in this secondary circuit, as in a wound-rotor motor, for the purpose of making starting-torque adjustments. Such a motor is obviously rather costly and has five slip rings, two for the d-c field and three for the pole-face winding.

Synchronous motor starters. Summarizing the foregoing discussion concerning synchronous-motor operation, it should be clear that a controller must perform the following functions: (1) It must bring the machine up to speed as an induction motor, making sure that the d-c field is *not* excited during the accelerating period; (2) it must keep the field-winding circuit closed through a discharge resistor (in special cases it must sectionalize an open-circuited field); (3) it must synchronize the machine by energizing the field *after* the rotor has reached its top speed—somewhat below synchronism—as an induction motor; (4) it must open the control circuit to deenergize the motor should the machine be sluggish in coming up to speed and operate too long as an induction motor; (5) it must permit adjustment of the d-c field excitation after the synchronizing process is completed so that the motor can be operated at a desired power factor.

A wiring diagram of a synchronous motor connected to a starter that automatically fulfills the requirements listed is given in Fig. 6·37. An important aspect of the control circuit is a special *polarized field frequency relay PFR* that functions to apply d-c excitation after the upper induction-motor speed is reached and at the very instant that the rotating poles are moving toward and in the direction of the proper revolving-field poles. The construction of the relay, clearly showing the d-c polarizing coil and the a-c coil, is illustrated by the photograph of Fig. 6·38; its operation will be described after the control circuit is analyzed.

Referring to Fig. 6·37, the following points should be noted in connection with the two contactors M and FC and the three relays PFR,

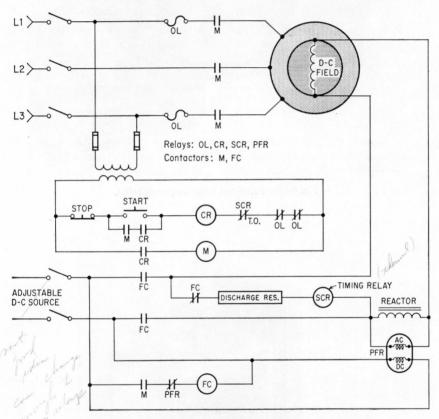

Fig. 6·37   Control circuit for starting a line-start synchronous motor.

SCR, and CR: The main contactor M has three power contacts and two N.O. interlocks; the field contactor FC has one N.C. contact and two N.O. contacts; the polarized frequency relay has one N.C. contact; the squirrel-cage relay SCR has one N.C. *time-opening* contact; the control relay CR has two N.O. contacts. At the instant the motor is started, by pressing the START button, the field winding is closed through the N.C. FC contact, the discharge resistor, the SCR relay, Fig. 6·39, and the reactor, Fig. 6·40; the M contactor is energized through one CR contact, and the CR relay is sealed in by another CR contact and an M

Fig. 6·38   Polarized field-frequency relay.

Fig. 6·39   Squirrel-cage timing relay used in the control circuit of synchronous-motor starter.

interlock. Since a voltage is induced in the field winding (at 60 cps just before rotation begins), an alternating current passes through the a-c coil of the *PFR* relay which is shunted across the reactor; also, the d-c coil of the *PFR* relay is permanently excited from the d-c source. This relay, therefore, picks up very rapidly and opens its N.C. contact in the *FC* contactor-coil circuit *before* the N.O. *M* interlock closes. Thus, the *FC* field contactor remains dropped out. The motor now proceeds to accelerate as an induction motor until, at about 92 to 97 per cent of synchronous speed when the field voltage and frequency have diminished

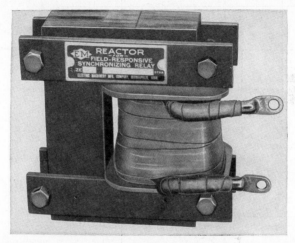

Fig. 6·40 Reactor used with polarized field-frequency relay (Fig. 6·38) in the control circuit of a synchronous-motor starter.

sufficiently and when the rotor position is correct, the *PFR* relay drops out. The *PFR* contact therefore closes and permits the field contactor *FC* to pick up through the *M* interlock. The motor now pulls into synchronism because d-c excitation is applied to the field through the two closed *FC* contacts. *After* the field circuit is established, the N.C. *FC* interlock opens in the discharge-resistor circuit, whereupon the *SCR* relay is deenergized. The *SCR*-T.O. contact, therefore, remains closed, assuming, of course, that the motor is synchronized *before* the relay times out. However, should the machine fail to come up to speed within a normal time period, say 5 to 10 sec depending upon the thermal capacity of the squirrel cage, the *SCR* relay will time out, open its contact in the *CR* relay circuit, and bring the motor to a stop before overheating causes injury to the rotating structure.

Another point that should be noted is that there is a positive overlap between the N.C. and the two N.O. contacts of the *FC* contactor because the field must never be open-circuited, either during or after the switching period.

Another circuit, designed to accomplish essentially the same results as Fig. 6·37, for a 150-hp 2,300-volt line-start synchronous motor, is shown in Fig. 6·41. Note that it includes an *exciter* which must be in operation

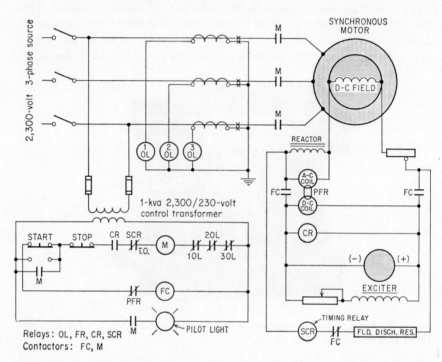

Fig. 6·41   Control circuit for starting a 150-hp 2,300-volt line-start synchronous motor.

before the synchronous motor is started; also, a *CR* relay across the exciter terminals must be energized first to close an N.O. contact in the main contactor circuit to make the START button effective. A pilot light is included in the circuit to indicate, when illuminated through an *M* interlock, that the motor is running.

The polarized frequency relay (Fig. 6·38) performs an extremely important function in the control circuit during the synchronizing process. As Fig. 6·42 shows, the relay is energized by two coils, one of which is excited from a constant d-c source and polarizes the core while the other, connected directly across the reactor, receives a-c excitation of varying magnitude and frequency from the induced emf in the field of the synchronous motor. Two sets of magnetic flux are thus created and superimposed on each other in the core and armature. When the currents in the two coils

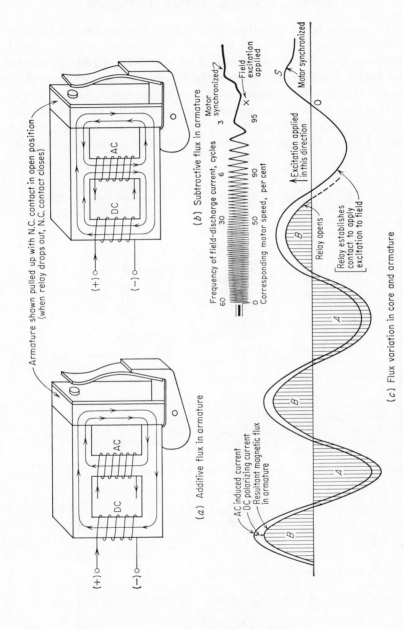

Armature shown pulled up with N.C. contact in open position
(when relay drops out, N.C. contact closes)

(+) o

(−) o

DC

AC

(a) Additive flux in armature

AC induced current
DC polarizing current
Resultant magnetic flux in armature

(+) o

(−) o

DC

AC

(b) Subtractive flux in armature

Motor synchronized

Frequency of field-discharge current, cycles
60   30   6   3

Corresponding motor speed, per cent
0   50   90 95

× Field excitation applied

B   A   B   A   B   S

Excitation applied in this direction

Relay opens

Relay establishes contact to apply excitation to field

O   Motor synchronized

(c) Flux variation in core and armature

Fig. 6·42   Sketches illustrating the operation of a polarized frequency relay.

199

are in the *same* relative directions (Fig. 6·42*a*), *their superimposed fluxes in the armature are additive* and the current variations are represented by the lower portions of the waves, *A* of Fig. 6·42*c*. However, on the reverse half of the a-c cycle the *fluxes in the armature are subtractive* (Fig. 6·42*b*) and the current variations are shown by the upper portions of the wave, *B* of Fig. 6·42*c*. As the motor speeds up and the induced emf and frequency diminish, the upper current waves continue to shrink until the magnetic force is no longer sufficient to keep the armature picked up.

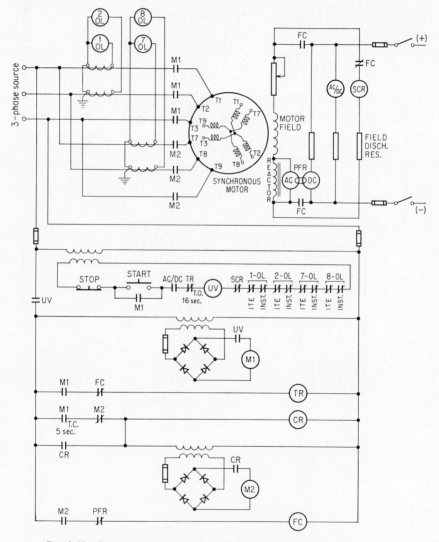

Fig. 6·43   Part-winding starter for a 600-hp 550-volt synchronous motor.

The relay, therefore, drops out, closes its contact, and permits the *FC* contactor to apply d-c excitation at the point indicated on the wave. Excitation is always applied in the direction shown by the large arrow, opposite in polarity to that of the induced field current at the point of application. This is done to compensate for the time required to build up excitation due to the hysteresis effect, i.e., magnetic inertia, of the motor field winding. The inertia is such that the d-c excitation does not become appreciably effective until the induced current has reversed, point *O* on the wave, to the same polarity as the direct current. Excitation then continues to build up until the motor synchronizes at point *S* on the curve.

To illustrate further the application of the various components of Fig. 6·41 for *part-winding starting* (Fig. 6·23) of a 600-hp 550-volt synchronous motor, Fig. 6·43 is given. Combination-type thermal-magnetic overload relays (Fig. 5·12) with inverse-time elements (ITE) and instantaneous-trip elements (INST) are included as are d-c contactors that are energized through bridge-connected rectifiers. Operation of the circuit follows essentially the same procedure as previously described, with the polarized frequency relay acting to time the proper "pull-in" or synchronizing point of the motor. An added measure of protection is provided by timing relay *TR* which, after a time delay, will open its contact in the *UV* relay circuit to stop the motor should the machine pull out of step. When the latter occurs, the *PFR* relay will be energized by the slip-frequency voltage across the reactor in the motor-field circuit and the N.C. *PFR* contact will open. The *FC* relay will then drop out, close its *FC* contact, and energize the timing relay *TR*. If, within a 16-sec period, the motor does not pull into step, the *TR*-T.O. contact will open and disconnect both contactors from the line. Note particularly that, in starting, one-half of the stator winding is energized first through *T*1, *T*2, and *T*3 when contactor *M*1 picks up and about 5 sec later the second half of the winding is connected in parallel with the first half through *T*7, *T*8, and *T*9, when contactor *M*2 operates.

QUESTIONS AND PROBLEMS

1. Referring to Fig. 6·2, discuss open and partially closed slots in stator cores as these designs affect winding construction, motor performance, servicing.
2. Describe the two general types of rotor construction for induction motors. How would the controllers for the two kinds of motor be likely to differ?
3. Discuss the constructional and operational differences of the various classes (A, B, C, D, and F) of squirrel-cage motor. Indicate also several practical applications for each class of motor.
4. Distinguish between die-cast and brazed rotor construction, single- and double-cage rotor construction, skewed rotor and skewed stator construction.

5. What is the effect upon the speed and efficiency of a wound-rotor motor by inserting secondary resistance?

6. Referring to Fig. 6·7, what relative values of external rotor resistance should be used if the motor is to develop (a) 150 per cent starting torque? (b) 250 per cent starting torque? (c) maximum starting torque?

7. A 25-hp 440-volt 3-phase 60-cycle 1,128-rpm wound-rotor induction motor has a rotor resistance of 0.15 ohm per phase and carries a current of 45 amp in its rotor circuit at rated load. What external resistance should be inserted in each line of the rotor circuit if the motor is to develop a starting torque (a) equal to its rated value? (b) equal to 150 per cent of rated value? Assume that the slip and rotor current vary directly with the torque.

8. List three general methods for starting squirrel-cage induction motors. Which of these are not applicable to the starting of moderately sized or large d-c motors?

9. Under what conditions is it permissible to apply full voltage to a squirrel-cage induction motor at the instant of starting?

10. The inrush current and the starting torque to a 75-hp 440-volt 93-amp 60-cycle 865-rpm 3-phase squirrel-cage induction motor are, respectively, $5.2I_{FL}$ and $2.3T_{FL}$ when started at rated voltage. (a) Calculate the inrush current and starting torque of the motor when 65 per cent voltage is applied at the instant of starting. (b) What voltage should be applied if it is desired that the motor develop 150 per cent torque at starting? What will be the inrush current under this condition?

11. What is meant by *part-winding* starting? Make a simple sketch showing the winding of a 3-phase motor designed for part-winding starting. Label the winding terminals and indicate how the latter are connected for starting and running.

12. Why are *across-the-line* starters for a-c motors rated on the basis of horsepower *and* voltage? Why do the horsepower ratings of a given starter increase with increasing values of motor voltage?

13. Check the two diagrams of Fig. 6·10 to verify the equivalence of detailed and elementary connections of a 3-phase motor and its line starter.

14. What purpose is served by a control transformer in a control circuit? How does the design and construction of such a transformer differ from a standard distribution transformer? At what power factor do control transformers generally operate? What voltage range can be expected across the windings of such transformers during normal operation? How does high inrush current affect control transformer operation.

15. Describe the operation of the *line-resistance* reduced-voltage starter circuit of Fig. 6·12.

16. Modify Fig. 6·12 to include a second START-STOP station at some *remote* position with operation as follows: To start the motor it should be necessary to press the *remote* START button first and then the *main* START button; it should be possible to stop the motor by pressing either STOP button.

17. Referring to Fig. 6·13, explain (a) why a *UV* contact rather than an *S* contact is connected across the START button; (b) why 1CR and 2CR contacts are used, respectively, in the *S* and *R* contactor-coil circuits; (c) why d-c contactors and associated rectifiers are frequently used for large motor control systems.

18. Under what operating conditions is it desirable to use an *increment resistance* reduced-voltage starter? Explain the operation of the control circuit of Fig. 6·14, indicating the particular function of the *TR* relay.

19. Discuss the advantages and disadvantages of *line-reactor* reduced-voltage starters as compared with *line-resistor* types of starter.

20. Carefully describe the operation of the *line-reactor* reduced-voltage control circuit of Fig. 6·15 (**a**) for normal starting, (**b**) during a power failure of less than 2 sec, (**c**) for normal stopping.

21. After analyzing the operation of the autotransformer *reduced-voltage* starter circuit of Fig. 6·16, explain why it is essential that the S contactor drop out *before* the R contactor is energized. What timing sequence of contacts is necessary for this mode of operation.

22. Why does the control circuit of Fig. 6·16 provide open-circuit transition?

23. Distinguish between two-coil and three-coil autostarters with regard to construction, degree of voltage unbalance, motor starting torque, cost.

24. Referring to the closed-circuit transition autotransformer diagram of Fig. 6·17, tabulate first the *sequence* of contact closings and openings in the upper part of the diagram and then proceed to describe the operation of the control circuit as a whole.

25. Carefully explain how the compensator in Fig. 6·17 acts as a simple 3-phase *autotransformer* when the motor is started and as a *line-reactor* during the transition period from reduced to full voltage.

26. When a motor is started by the star-delta method, what are the values of the inrush current and starting torque in terms of full-voltage starting?

27. The inrush current and starting torque of a 20-hp 550-volt 21-amp 1,750-rpm squirrel-cage motor are $5.4I_{FL}$ and $2.6T_{FL}$ when the motor is started at rated voltage. What values of inrush current and starting torque can be expected if the motor is started by the star-delta method?

28. Where should *three* overloads be placed in the control circuit of Fig. 6·20 if it is desired to protect the three stator winding phases directly?

29. Carefully list the sequence of contact closings and openings in the *star-delta closed-circuit transition* starter of Fig. 6·21. Why must contactors S and $2M$ be interlocked mechanically?

30. List advantages and disadvantages of *part-winding* starting.

31. Why are four overloads required in the part-winding control circuit of Fig. 6·23?

32. Discuss balanced and unbalanced secondary-resistance starting methods for wound-rotor motors. What advantages do liquid rheostats have for wound-rotor motor starters?

33. Referring to Fig. 6·26, why is it desirable to use a set of N.C. contacts in the accelerating relay-coil circuits $1AR$, $2AR$, and $3AR$? Design a control system that employs N.O. contacts in the accelerating relay-coil circuits.

34. The open-circuit voltage at the slip rings of a 6-pole 60-cycle wound-rotor motor is 270 volts. Calculate the voltage and frequency at rotor speeds of 260, 480, 720, 1,000 rpm.

35. Describe the operation of the frequency-relay control circuit of Fig. 6·27. What adjustments should be made if it is desired to change the dropout frequencies of the accelerating relays to 41, 31, 23, and 12 cps?

36. Why are the reversing controllers represented by Fig. 6·29 interlocked both mechanically and electrically?

37. In the master-type reversing controller of Fig. 6·30, why is it desirable to tune the plugging relay PR to drop out at 61 cps? Modify the circuit so that triple-pole contactors are used.

38. Modify Fig. 6·32 so that the motor can be jogged in *forward* as well as in reverse.

39. In the plug-stop control circuit of Fig. 6·33 the contacts on the zero-speed plugging switch carry the contactor-coil currents. Design a circuit in which relay contacts replace those of the plugging switch.

40. Explain why the dynamic-braking torque of an induction motor varies so greatly during the braking period. Indicate when and state why it is low at the instant braking is initiated and high when the motor approaches standstill.

41. A 50-hp 220-volt 126-amp 6-pole squirrel-cage motor has a stator winding with 72 coils, each of which has six turns. During the dynamic-braking period the d-c excitation is adjusted to $5I$. Calculate the number of ampere-turns per pole developed by the winding if it is connected in star (**Y**) and direct current is applied to: (**a**) terminal $T1$ and the junction of terminals $T2$ and $T3$; (**b**) terminals $T1$ and $T2$, with terminal $T3$ disconnected. Make similar calculations (**c**) and (**d**) for a delta-connected winding.

42. Make a sketch similar to that of Fig. 6·34 showing the winding delta-connected and illustrating how the poles are formed when the d-c source is connected to any pair of motor terminals.

43. What is the purpose of the two rheostats in the dynamic-braking circuit of Fig. 6·35.

44. Carefully explain the operation of the control circuit in Fig. 6·36 (**a**) for normal acceleration, (**b**) for dynamic braking.

45. What is the purpose of the pole-face winding (the squirrel cage) in a synchronous motor?

46. What types of starters are used for synchronous motors?

47. Why must the d-c field of a synchronous motor be protected against insulation breakdown during the starting period? How is this accomplished in the control circuit?

48. List the functions which must be performed by a properly designed starter for a synchronous motor.

49. Carefully describe the operation of the synchronous-motor control circuit of Fig. 6·37, emphasizing particularly the sequence of contact closings and openings of relays *PFR* and *SCR* and contactor *FC*. Why is it desirable to have an overlap between the N.O. and N.C. *FC* contacts?

50. Why is the armature of the polarized frequency relay pulled up to open its N.C. contact during the accelerating period? When does this relay drop out to close its contact? Why is excitation applied to the field in a direction opposite to that of the induced field winding when the motor is synchronized?

51. Explain the operation of the part-winding starter of Fig. 6·43 for a synchronous motor. Why are d-c contactors used?

CHAPTER **7**

Speed Control of D-C and A-C Motors

General aspects of d-c motor speed control. One of the most valuable characteristics of the d-c motor is its ability to provide a wide range of easily adjustable speeds. This benefit is particularly important because a high degree of speed control is often essential to certain motor-driven installations. It is significant that d-c series, shunt, and compound machines can generally be made to serve very effectively for such applications because voltage and flux changes, when properly made, greatly influence the behavior of these motors; this is in contrast to the induction-type motor, whose speed does not change substantially under similar conditions.

As was pointed out in Chap. 2, an emf E_c is generated in the armature winding that acts in opposition to the impressed voltage V. Being a speed voltage, i.e., a voltage that depends upon the speed of rotation, this cemf is directly proportional to the rate at which flux is cut by the armature conductors. Moreover, E_c is always less than the terminal voltage by the value of the $I_A R_A$ drop. Thus

$$E_c = V - I_A R_A = k_1 \phi S$$

where all terms have been previously defined.

Solving for speed,

$$S = k \frac{E_c}{\phi} = k \frac{V - I_A R_A}{\phi} \tag{7.1}$$

This equation indicates that the speed of a d-c motor can be controlled by changing either the cemf or flux. The means for doing this include:

205

1. Introducing a rheostat in series in the armature circuit; this lowers the voltage V_A across the armature terminals and reduces the speed as a function of the added resistance.
2. Varying a rheostat in the shunt-field circuit of a shunt or compound motor; when the resistance is increased, the field flux is diminished and the speed is raised.
3. Varying the voltage across the terminals of a series motor or varying the voltage across the *armature* terminals of a shunt or compound motor while the shunt-field flux is maintained at a constant value by a separate source; the speed is lowered with diminishing values of voltage.
4. Varying a rheostat shunted across the armature terminals of a series motor; decreasing the ohmic value of the shunted resistor increases the series-field flux and lowers the motor speed.
5. Varying a rheostat shunted across the armature terminals of a shunt of compound motor while another resistor, connected in series with the paralleled combination of armature and shunted resistor, is kept at some fixed value; this introduces an artificial IR drop in the armature circuit (across the series resistor) to reduce the motor speed as the ohmic value of the shunted resistor is decreased.

Torque and horsepower relationships. The torque developed by a d-c motor is directly proportional to both the armature current I_A and the flux per pole. Thus

$$T = kI_A\phi \tag{7·2}$$

The horsepower output is, on the other hand, proportional to both the torque and the speed. Thus

$$\text{Hp} = kTS \tag{7·3}$$

These equations indicate that, for a *given* armature current I_A, *rated value* for example,

1. If a speed adjustment is made by changing only the voltage impressed across the armature terminals, torque remains constant while the horsepower output varies directly with the speed. In Table 7·1 *constant-torque–variable-horsepower* speed-control methods are illustrated in I-*A* and I-*D* for the series motor, II-*D* for the shunt motor, and III-*D* for the compound motor.
2. If a speed adjustment is made by changing only the flux, torque varies while the horsepower output remains constant. Thus

$$\text{Hp} = kTS = k(kI_A\phi)\left(k\,\frac{V - I_A R_A}{\phi}\right) = k'$$

Table 7·1 Control circuits and equations for d-c motors

CONTROLLED ELEMENT	I – SERIES	II – SHUNT	III – COMPOUND
A FIELD RESISTOR	$$S \doteq k\,\frac{V-I\,(R_S+R_A+R_{SE})}{\phi}$$ Increasing R_S decreases speed	$$S = k\,\frac{V-I_A R_A}{\phi_{VAR}}$$ where: $$I_A = I - V/(R_f + R_F)$$ Increasing R_f increases speed	$$S = k\,\frac{V-I_A(R_A+R_{SE})}{\phi_{VAR}+\phi_{SE}}$$ where: $$I_A = I - V/(R_f + R_F)$$ Increasing R_f increases speed
B FIELD AND ARMATURE RESISTORS	$$S = k\,\frac{V-IR_{SE}-I_A R_A}{\phi}$$ where: $$I_A = I - \left[\frac{V-IR_{SE}}{R_p}\right]$$ Decreasing R_p decreases speed A current-limiting resistor R is often connected in series with R_{SE}	$$S = k\,\frac{V-I_A\,(R_a+R_A)}{\phi}$$ where: $$I_A = I - V/(R_f + R_F)$$ Increasing R_a decreases speed Increasing R_f increases speed	$$S = k\,\frac{V-I_A(R_a+R_A+R_{SE})}{\phi_F+\phi_{SE}}$$ where: $$I_A = I - V/(R_f + R_F)$$ Increasing R_a decreases speed Increasing R_f increases speed
C SERIES AND SHUNT ARMATURE RESISTORS	$$S = k\,\frac{V-I(R_S+R_{SE})-I_A R_A}{\phi}$$ where: $$I_A = (I-I_p) \text{ and}$$ $$I_p = \frac{V-I(R_S+R_{SE})}{R_p}$$ Increasing R_S decreases speed Decreasing R_p decreases speed	$$S = k\,\frac{V-I_S R_S-I_A R_A}{\phi}$$ where: $$I_S = I - V/R_F \text{ and}$$ $$I_A = I_S - \left[\frac{V-I_S R_S}{R_p}\right]$$ Increasing R_S decreases speed Decreasing R_p decreases speed	$$S = k\,\frac{V-I_S(R_S+R_{SE})-I_A R_A}{\phi_F+\phi_{SE}}$$ where: $I_S = I - V/R_F$ and $$I_A = I_S - \left[\frac{V-I_S(R_S+R_{SE})}{R_p}\right]$$ Increasing R_S decreases speed Decreasing R_p decreases speed
D TERMINAL VOLTAGE	$$S = k\,\frac{V_{adj.}-I\,(R_A+R_{SE})}{\phi_{SE}}$$ Increasing V increases speed	○To separate exciter○ $$S = k\,\frac{V_{adj.}-IR_A}{\phi_F}$$ Increasing V increases speed	○To separate exciter○ $$S = k\,\frac{V_{adj.}-I\,(R_A+R_{SE})}{\phi_F+\phi_{SE}}$$ Increasing V increases speed

where the three k's are merely different constants of proportionality. *Variable-torque–constant-horsepower* speed-control methods are illustrated in Table 7·1 by II-*A* for the shunt motor and III-*A* for the compound motor.

In connection with the variable armature-voltage method of speed control, it should be pointed out that for extremely heavy loads, i.e., those that cause the $I_A R_A$ drop to become an appreciable part of the numerator $(V - I_A R_A)$ of the speed equation, Eq. (7·1), voltage changes have a diminishing effect upon the range through which the speed can be adjusted.

Several of the speed-control schemes shown in composite Table 7·1 are designed to adjust both the flux and the armature-terminal voltage. In I-*C* for the series motor, for example, a decrease in resistance R_p results in an increase in the current through the series field R_{SE} and the series resistance R_s. These changes result in (1) a flux increase, (2) a larger voltage drop represented by $I(R_{SE} + R_s)$, and (3) a lower armature-terminal voltage V_A. This combination of changes produces a marked reduction in speed. Other speed-control schemes can be similarly analyzed.

Shunt-motor control by field weakening. An extremely popular method of increasing the speed of a shunt motor above the base value is to weaken the field , i.e., by inserting resistance in the field-winding circuit. This must, however, be done with some discretion, and depending upon whether the motor is designed for standard or adjustable-speed service, the maximum speed range may be as little as 1.5 to 1 or as much as 6 to 1. Limits within which the speed can be changed are set by both mechanical and electrical factors, although the latter are often more significant. Moreover, the higher the permissible speed range, the more expensive does the motor become because special fields and frame structures must be employed to ensure stability at high speed.

An important restriction that is imposed upon a shunt motor operating with a weak field at high speed is the commutation limit, i.e., the current which can be safely commutated. This is because armature reaction, which is more effective as the main flux is reduced, tends to cause serious arcing and instability. Since torque depends upon both the flux and the armature current and the latter must be low enough to inhibit sparking, the high-speed output of the motor is therefore limited even though cooling is improved. Still another aspect of this method of control is the way in which the speed regulation is affected by field weakening. In the low-speed ranges when the field is strong, a given torque is developed by a comparatively low armature current; under this condition the cemf is high and the high-torque speed is not much less than the no-load value. Regulation is, therefore, good, and the motor operates at nearly constant speed between the two load torques. At high-speed ranges, on the other hand, when the field is weak, a proportionately higher armature current is necessary for the same motor torque; this means, then, that the cemf is reduced below the previous value, with a resulting reduction in speed and an increase in speed regulation.

The foregoing discussion is summarized by the family of speed-regulation curves in Fig. 7·1. Note particularly that the following four general

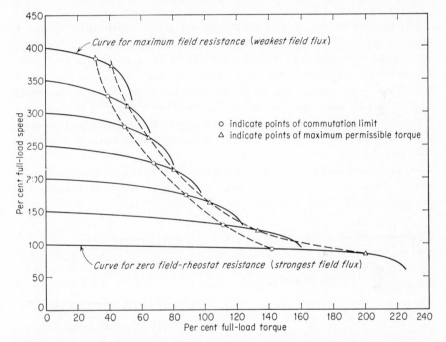

Fig. 7·1   Characteristic *speed-versus-torque* curves for shunt motor with field rheostat control.

changes take place as the field is weakened: (1) the speed-versus-torque curve is shifted up to give higher speeds; (2) the commutation limit is reduced to lessen the permissible armature current; (3) the maximum permissible torque output is reduced; (4) the speed regulation, i.e., speed changes between no-load and equal values of torque, increases to alter the normal tendency of the motor to operate at constant speed.

An important modification of the adjustable-speed shunt motor is the addition of a stabilizing field. This is a few turns of heavy wire around each of the main poles that carry line or armature current. It is, in this respect, a comparatively weak series field whose function it is to nullify the effect of armature reaction. Without this winding and with the shunt field greatly weakened, the motor may attempt to operate at excessive speed under load; it may, in fact, exceed the no-load speed under this condition. The stabilizing field acts to strengthen the otherwise reduced flux and exercises a desirable moderating influence.

Controllers for shunt motors and particularly those for adjustable-speed types should be designed to provide full-field starting (Fig. 5·20). This is essential if reasonably high starting torque is to be developed at minimum values of inrush current.

Series-type motor in adjustable-speed crane hoist application. A crane is an excellent example of an adjustable-speed drive in which a series-type motor is employed. During the hoisting portion of the cycle, the acceleration resistor serves to vary the voltage across the series-connected motor and, in accordance with I-*A* of Table 7·1, the speed is adjusted by moving the master to insert or remove resistance. For the lowering portion of the cycle, the series field, properly protected by a line resistor, is shunted across the armature, under which condition the machine operates essentially as a shunt motor. In addition, the acceleration resistor is connected in the series-field circuit and is varied by the master to adjust the speed. Diagrams showing the connections for the hoisting and lowering operations are given in Fig. 7·2.

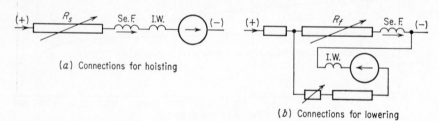

(*a*) Connections for hoisting

(*b*) Connections for lowering

Fig. 7·2   Schematic diagrams for a crane hoist.

Typical speed-torque curves that illustrate the characteristics of the series motor (for hoisting) and the shunt motor (for lowering) with different relative values of the variable resistance are shown in Fig. 7·3. Note particularly that the speed of the series-connected motor fluctuates considerably with load-torque changes and tends to vary more widely with small variations in load as the series resistance is increased. The motor is, therefore, not suitable for overhauling types of load. The shunt-motor connection that incorporates a variable resistance in the field and a fixed resistance in the line (Fig. 7·2*b*) is, on the other hand, quite satisfactory for the lowering operation because, with flatter curves, performance is much more stable.

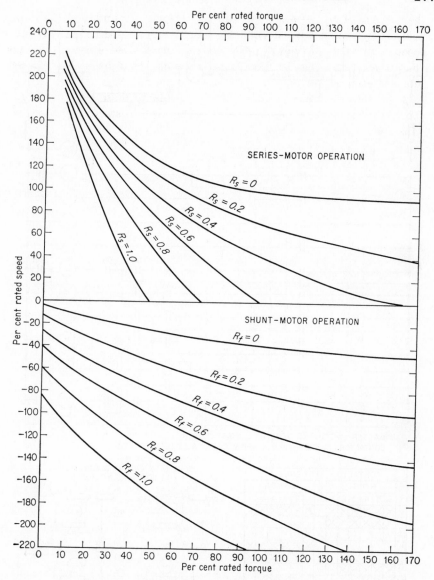

Fig. 7·3 Typical speed-torque curves for a crane hoist.

The foregoing scheme of control is illustrated by the wiring diagrams of Fig. 7·4 in which a six-point reversing master is used. Before analyzing the circuit in some detail it will be desirable to describe its general behavior. For hoisting, the N.C. *DB* contact opens on the first point and the speed of the motor is controlled by short-circuiting series resist-

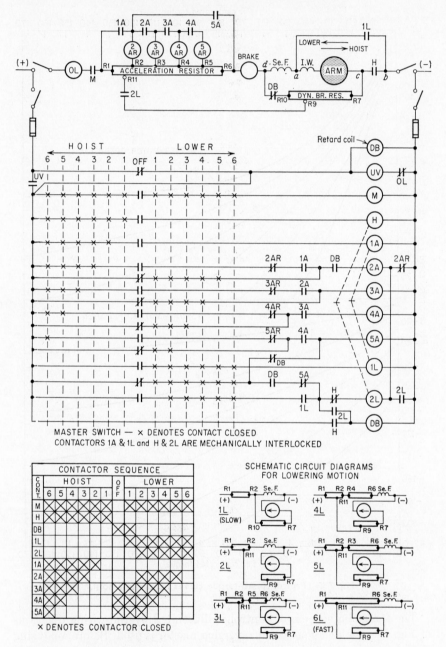

Fig. 7·4 Crane hoist application of a series motor connected to a controller provided with dynamic braking and dynamic lowering.

ance as the master is moved toward the sixth point. Since the first accelerating relay is omitted, there is no time delay in cutting out the first step of resistance when the handle is shifted to position 2. The torque developed on the first point will not move the load but is sufficient only to tighten the cables and set the gears. It should also be understood that the acceleration resistors are not merely used for starting purposes but are designed for continuous heavy-duty service.

The lower set of schematic sketches in Fig. 7·4 indicates how the motor and the two sets of resistance (accelerating and dynamic) are connected during the lowering portion of the cycle. On the first point, $1L$, the series field is at full strength ($R_f = 0$), the line resistance is a maximum, and the resistance in series with the armature is a maximum; the motor, therefore, operates at minimum speed. When the master is moved to the second point, $2L$, the series field is weakened and the line resistance and the armature-circuit resistance are both reduced slightly; these changes cause the motor to speed up. For positions $3L$ to $6L$ the line and armature resistances remain unchanged but the series-field circuit resistance is progressively increased; these changes, therefore, weaken the field and cause an increase in the motor speed. In position six, $6L$, the highest speed is attained.

Consider next the HOIST portion of the main diagram. The retard coil of the DB contactor is energized on the OFF point, and its mmf tends to keep the DB contact closed until the operating coil (at the bottom of the diagram) is energized. Since the mmf of the retard coil exerts a delaying action, the contactor drops out with a snap action when the operating-coil flux rises to a point well up on the magnetization curve. Contactors M and H are energized on all six points of the master, and their main contacts close. Also, an H interlock closes in the DB contactor-coil circuit, so that unit drops out to open the dynamic-braking resistor loop during the entire hoisting period.

1. All the acceleration resistance is in the armature circuit on the first point, the minimum-speed point.
2. The $1A$ contactor picks up *instantly* on point 2, there being no accelerating relay to delay the closing of the main $1A$ contact that short-circuits one step of resistance. This is the second speed point. Also, the $2AR$ relay is energized to open its N.C. contact in the $2A$ contactor-coil circuit before the N.O. $1A$ interlock closes. Note also that a DB interlock was previously closed in the same circuit. After a time delay the $2AR$ relay drops out and permits its contact to close.
3. Similar sequences of action take place when the master is moved to point 3 for the third speed and points 4, 5, and 6 for the fourth, fifth, and sixth speeds. For all these motions the respective portions of the

acceleration resistor are short-circuited with adjusted time delays after the proper accelerating relays drop out.

4. The master can be moved to a lower numbered point at any time to reduce the motor speed.

5. When the master is returned to the OFF point, all contactors drop out and power is removed from the motor.

Attention is directed to the lower *contactor-sequence* table which indicates which contactors are closed in the various positions above referred to.

The following procedure applies to the LOWER portion of the diagram:

Position 1. The M contactor is energized and closes its main contact· The $1A$ contactor will continue open on *all* master points, and this means that the $R1$-$R2$ portion of the acceleration resistor will always remain in the circuit. Since the operating coil of the DB contactor is deenergized, its DB interlock will remain open in the $2A$ contactor circuit and the latter unit will be dropped out. The $3A$ contactor will pick up to close an interlock in the $4A$ contactor-coil circuit. The $4A$ contactor will do likewise to close an interlock in the $5A$ contactor-coil circuit. The closing of the main $5A$ contact then short-circuits the $R2$-$R6$ portion of the accelerating resistor. The $1L$ contactor is energized on *all* master points, and the closing of its contact connects terminals a and b in the motor circuit. With all the connections made as indicated, current passes from $R1$ to $R2$, through relay $2AR$, through closed contact $5A$ to $R6$, through the brake (which releases) and thence to point d where the two parallel paths are formed. One of these is the series field to point a, through closed contact $1L$, and to point b, while the other is through the N.C. DB contact to $R10$, through the dynamic-braking resister to $R7$, and finally through the armature from points c to a to b in a direction *opposite* to that for the hoisting operation. The schematic sketch $1L$ at the bottom shows the circuit described. This is the slow-speed point because there is no resistance in the series-field circuit and the flux is at its maximum.

Position 2. The $2L$ contactor picks up through the closed $1L$ interlock. This closes a $2L$ interlock directly below and causes the DB contactor to operate and open the N.C. DB contact in the motor circuit. A $2L$ contact also closes to connect $R11$ to $R9$. With these connections made, a small portion of the dynamic-braking resistor, $R10$-$R9$, is removed from the armature, and resistance $R11$-$R2$ is transferred from the line to the series field. Since the field is weakened and the armature is strengthened, the motor speed increases. The schematic sketch $2L$ shows the circuit described.

Position 3. Contactor $5A$ drops out to open its contact between terminals $R11$ and $R6$. With contacts $2A$, $3A$, and $4A$ still closed (contactors $2A$, $3A$, and $4A$ remain energized), an additional resistance, $R5$-$R6$, is inserted in the series field. This weakens the field still further, and the motor speeds up. The schematic sketch $3L$ shows the changed arrangement.

Position 4. Contactor $4A$ drops out to open the main contact, and this action adds another section of the acceleration resistor, $R4$-$R5$, to the series field. With the strength of the latter reduced again, the motor speeds up. Refer to schematic $4L$ for the simplified circuit.

Position 5. Contactor $3A$ drops out to open its main contact. Again a part of the acceleration resistor, $R3$-$R4$, is inserted so that now the series-field circuit includes resistors $R11$-$R2$ and $R3$-$R6$. Again the motor speeds up with the smaller field current. See schematic $5L$ for arrangement on this point.

Position 6. Contactor $2A$ drops out, opens its main contact, and permits the final section of the acceleration resistor, $R2$-$R3$, to be connected into the series field. The motor now operates on its weakest field and at the highest speed.

Adjustable-voltage speed-control system. To control the speed of shunt and compound motors steplessly and over a considerable range, the adjustable-voltage system offers many advantages. (See II-*D* and III-*D* in Table 7·1.) Generally referred to as Ward Leonard control after its originator, the method requires two separate d-c sources of supply, one of which is a *main adjustable-voltage generator* that supplies power to the armature of the motor and the other a small *constant-voltage generator* that excites both the generator and motor fields. The main generator and exciter must, of course, be driven by a prime mover, and this is often a constant-speed a-c motor when a polyphase source of power is available or, if not, some other type of mechanical machine . All control is centered in a rather small rheostat in the field circuit of the main generator where a reversing arrangement is frequently installed to reverse the polarity of the generator and, therefore, the direction of rotation of the motor. A diagram illustrating how the machines and other components are interconnected is given in Fig. 7·5. Note particularly that a solid connection forms the power loop comprising the controlling generator and controlled motor. Since the motor field flux usually remains unchanged throughout the entire speed range, the Ward Leonard control is essentially a constant-torque variable-horsepower system.

Although an adjustable-voltage installation of this type is rather expensive, involving as it does three machines to control a motor, it does, nevertheless, have wide application wherever low and high speeds must be accurately made and where the service is severe and exacting. It offers the following advantages:

1. The speed range is much greater than that obtainable with a straight shunt motor with armature and field control. Adjustments can readily be made for "crawling" as well as for high speeds.
2. The control component is generally a small field rheostat whose power requirement is about 1 to 2 per cent of the total input to the motor. Since the rheostat usually has a great many tapped points, the control is practically stepless.
3. All heavy armature contactors are eliminated because the loop circuit where the power current flows is solidly connected. The motor is started, accelerated, speed adjusted, and stopped by merely adjusting the generator voltage.
4. Reversing is simply and effectively accomplished in the field circuit of the generator; reversing contactors are, therefore, unnecessary.
5. Main generators having special characteristics can be employed to match specific motor-load requirements. This is particularly desirable in certain machine tools and for such heavy equipment as excavators.
6. Magnetic and rotating amplifiers, with their outstanding and astonishing performance characteristics, can be readily adapted to this system of control. Where this is done, control power can be greatly reduced and regulation is considerably improved.

The two field rheostats shown in Fig. 7·5 serve to adjust not only the field flux of the generator, and therefore its output voltage, but also

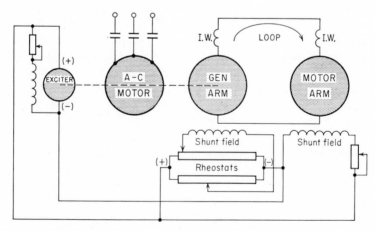

Fig. 7·5   Adjustable-voltage (Ward Leonard) system of speed control for a reversing shunt motor.

the polarity of the latter. With both movable arms at the *same* ends of the rheostats, either left or right, the field voltage is zero; the generator (motor) emf and the motor speed are, therefore, zero. However, if the two arms are at *opposite ends* of the rheostats, the field voltage is at the maximum; the generator (motor) emf is then at the highest value, and the greatest motor speed is reached. Moreover, when the upper arm is at the left and the lower arm is moved to the right, a given generator polarity causes, say, clockwise rotation of the motor. If, on the other hand, the lower arm is at the left and the upper arm is more to the right, the generator polarity is reversed and counterclockwise rotation of the motor results. In any case, the generator voltage and speed are directly proportional to the magnitude of the flux at any position of the movable arm.

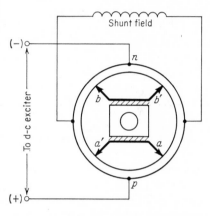

Fig. 7·6 Rheostat with provision to reverse the shunt-field polarity.

The specially constructed single-unit device illustrated by Fig. 7·6 accomplishes the same results, i.e., voltage adjustment and motor reversal, as the two-rheostat arrangement. In the position shown, the shunt-field voltage is zero and, as before, the motor speed is zero. If the inner contacts are rotated clockwise so that a and b touch, respectively, at points p and n, current through the shunt field is from left to right; the resulting polarity produces, say, clockwise rotation of the motor. However, when the inner contacts are rotated counterclockwise and a' and b' touch, respectively, at points p and n, the shunt-field current is from right to left and the motor rotates counterclockwise. The actual motor speed is, of course, determined by the position of the inner contacts on the circular resistor.

A difficulty arises with this control system when the motor operates in one direction only and an attempt is made to stop the motor by reducing the shunt-field current of the generator to zero. Under this condition a residual voltage, resulting from the cutting of residual flux, produces a substantial current in the low-resistance loop circuit. The torque developed by the motor may, therefore, cause the motor to "crawl," especially when the load is light. This situation does not arise in a reversing drive because the residual flux can be reduced to zero by applying a small reversed mmf to the generator field. The action described can be avoided by making the generator neutralize its own residual flux. As shown in Fig. 7·7, the neutralizing action, or "suiciding," as it is called, is accomplished by throwing the DPDT switch to the right, assuming that, with the switch closed to the left, the field current was first reduced to zero. When this is done, the small *residual voltage* sends a comparatively low reversed current through the shunt field to "kill" the residual flux. We assume, of course, that the polarities of the exciter and generator are correct with respect to each other.

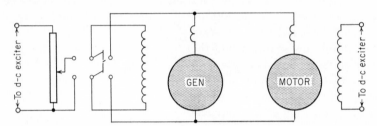

Fig. 7·7 Circuit diagram to neutralize the residual field at zero field current.

Multimotor adjustable-voltage control system. Many complex industrial machines employ several similar motors which must be operated and controlled simultaneously from one central master or console. This generally involves such functions as accelerating, speed adjusting, and stopping all motors in unison so that long lines of materials are moved and retarded smoothly. Good examples of such installations are processing lines in food industries, tinning and annealing lines in steel mills, papermaking, and textile manufacturing.

The adjustable-voltage system functions admirably under these conditions because any or all of the motors can be preselected for operation and, after the drive is running, those at rest cannot be started until the system is shut down. Moreover, rheostats can be preset so that the various motors can be brought up to their proper speeds by the single control unit.

A diagram illustrating a typical three-motor application is given in Fig. 7·8. Here, the motor fields are joined in parallel, each with its individual field rheostat for speed presetting, and are connected to the common exciter; the armatures of the motors are energized by the main generator whose voltage is adjusted by a special type of field rheostat. The simple control circuit permits the operator to close any or all contacts in the armature circuits *only when the system voltage is zero and all motors are at rest.* Once the machinery is set in motion, an idle motor cannot be started until the line is shut down again.

After the main generator and exciter are brought up to speed by the driving motor (or other type of prime mover), the exciter voltage is properly adjusted. The field-control switch is then closed, and with the rheostat in the "all-out" position so that contact m is made, the CR relay picks up to close its contact. At this time the emf at the armature terminals of the main generator is a low residual value because the field voltage and current are zero.

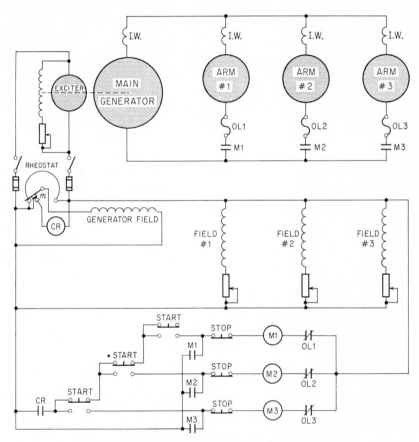

Fig. 7·8 Multimotor adjustable-voltage control system.

To start any or all of the three motors and bring them up to the desired speed simultaneously, the proper START buttons are pressed. This energizes the respective contactors $M1$, $M2$, and $M3$, which, in turn, close their contacts in the armature circuits; also the interlocks $M1$, $M2$, and $M3$ close to seal in the contactor coils. Note particularly that the contactors pick up on the front contacts of the START buttons and become ineffective when the CR relay drops out; this is to prevent the energization of a particular contactor coil that was not picked up when the system was started. Since the armature voltage is still the low residual value, the motors do not start. To bring them up to speed, the rheostat is turned toward the "all-in" position; this has the effect of slowly increasing the field excitation and the armature voltage. Moreover, as the rheostat arm leaves its initial zero-field position, contact m opens to deenergize the CR relay, which, in turn, opens its contact; the START buttons, as indicated above, now have no function, since they are in an open circuit.

Two-exciter adjustable-voltage control system. An interesting departure from the fundamental adjustable-voltage system of Fig. 7·5 employs *two* exciters which operate to regulate simultaneously the field currents in the generator and motor fields. With the voltage of one of the exciters maintained at a constant value, a rheostat in the field of the other unit is varied so that the motor-field excitation is reduced as the armature voltage is increased, and vice versa. Since the two changes indicated always have the effect of altering the speed in the same way, i.e., both up or both down, speed changes respond very rapidly to the single control unit. The upper speed point is, moreover, extended beyond that obtained with constant motor-field current.

A diagram of the modified Ward Leonard system of control is illustrated by Fig. 7·9. Note that the *main and intermediate exciters* EX_M

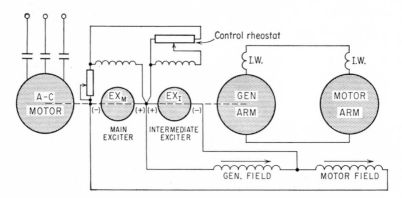

Fig. 7·9   Modified adjustable-voltage system of speed control.

and EX_I are connected in series *bucking* and that all speed adjustments originate at the potential divider in the intermediate-exciter field, which, in turn, is energized by the main exciter. If the potential of EX_M is maintained constant at 230 volts and that of EX_I varies between 0 and 115 volts, (1) the motor speed will be zero when $EX_I = 0$, under which condition the motor field will be excited at 230 volts and will be at full strength, and (2) the motor speed will be a maximum when $EX_I = 115$ volts, under which condition 115 volts will be impressed across the motor field and the potential across the generator field will be 115 volts. In other words, speed changes from zero to a maximum are accompanied, respectively, by voltage variations of $E_{\text{gen field}} = 0$ to 115 volts and $E_{\text{motor field}} = 230$ to 115 volts. This is no longer a constant-torque system, since speed changes result from variations in both armature voltage and field flux.

Adjustable-voltage control system with three-field generator. The electric excavator or shovel is an important application whose motors must operate over considerable speed and torque ranges. Such equipment generally goes through four basic motions each of which is provided by an adjustable-speed reversing motor whose power, in turn, is derived from a specially designed three-field excavator generator. The torque and speed of the several motions are controlled by varying the output of the generators through small master switches that are manipulated by the operator. The *crowd motor* fills the bucket as the latter is thrust into and out of the load material (gravel, coal, rock, etc.), after which the *hoist motor* raises the loaded bucket to clear obstructions as the *swing motor* rotates the boom and its bucket to the unloading position. The fourth or "propel" motion is provided by the hoist motor.

The Ward Leonard adjustable-voltage system of control has proved the most satisfactory for the severe and exacting demands of shovel operation. This is because loads can be moved very rapidly when they are light and motors develop high torques at low speed under heavy load conditions. Moreover, each motion has an entirely independent power source which not only is designed to provide the most suitable speed-torque characteristics but gives inherent overload protection without the use of protective or load-limiting devices. The latter is accomplished by using two shunt fields, one of them separately excited and the other self-excited, and a differentially connected series field. Then by adjustment of the proportion of self- to separate excitation with the field rheostats, the shape of the *voltage-versus-current curve* can be altered to give the desired performance. In addition, the differential series field exercises a load-limiting influence, since its opposing mmf restricts the maximum voltage to a safe low value under *stalled motor* conditions, as, for example, when the bucket strikes an immovable load in a bank.

All control is centered in the rheostat of the separately excited field which raises the speed of the motor when that field is strengthened, and vice versa; the self-excited shunt field always follows by building up in an aiding direction on the basis of its adjusted resistance. The drive is reversed by first reducing the separate excitation to zero, which brings the motor to rest, and then increasing the current in the opposite direction. These actions reverse the generator polarity with the result that the self-excited field builds up in the opposite direction to aid again the separately excited field; with the current reversed in the load circuit, the series-field mmf acts differentially with respect to the shunt fields. Shovel generators are extremely rugged and commutate extremely well when loads are applied suddenly and even at currents that are as high as 240 per cent of rated value.

Figure 7·10 shows the elementary connections for an adjustable-voltage system employing a three-field generator. For the polarities indicated,

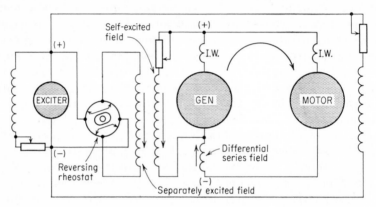

Fig. 7·10   Adjustable-voltage control system with three-field generator.

note particularly the two shunt fields with their aiding mmfs and the differentially acting series field. The actual operating characteristics of the generator will, of course, depend upon the relative strength of the two shunt fields. Although the three-field generators were widely employed in large shovel applications and are indeed presently used in existing installations, they have been largely replaced by shunt generators that are regulated by rotating- and magnetic-type amplifiers. Such control systems will be discussed in a subsequent chapter.

General aspects of a-c motor speed control. Unlike a d-c motor, which is a doubly fed machine in the sense that the stator (field) and the rotor (armature) are both energized by direct current, the a-c induction motor is singly fed. Only one section, usually the stator, receives its power directly from the a-c mains. The d-c motor develops torque by the interaction of two stationary magnetic fields, both *conducted* to the field and armature windings. In contrast, torque is produced in the induction motor when a synchronously revolving magnetic field, created by currents conducted to one winding, induces currents in the conductors of the other winding because the rotor turns in the same direction as the revolving field but slightly slower.

The induction motor is essentially a constant-speed energy converter; under normal operating conditions the speed changes only slightly (about 2 to 8 per cent) between no load and full load. In a sense, operation resembles that of a d-c shunt motor but differs markedly from the performance of compound and series types, which are characterized by rather wide speed changes with normal torque variations. The speed of an induction motor is definitely "tied to" the line frequency, to the number of poles for which the machine is wound, and, for motors with phase-wound rotors, to the resistance that is inserted in the secondary circuit. It follows, therefore, that speed adjustment is fundamentally possible by employing (1) a variable-frequency source, (2) one of several pole-changing schemes, and (3) variable resistors (or their equivalent) in the rotor circuit of the phase-wound-rotor motor. These speed-control procedures differ greatly from those commonly used for d-c machines, which, as the foregoing discussions showed, involve adjustments of armature voltage and field flux.

Pole Changing. When a squirrel-cage induction motor is connected to a constant-frequency source, its speed N in revolutions per minute, given by the equation $N = (120f/P)(1 - s)$, is inversely proportional to the number of poles for which the stator is wound. The slip s is generally a fractional part (0.02 to 0.08) of the synchronous speed N_s ($= 120f/P$) of the revolving field. The foregoing expression for N is the basis for several pole-changing methods of control, which are suited to applications that do not require continuous (stepless) adjustment and may be driven at two or more so-called base speeds. In one arrangement, two independent windings, completely insulated from each other, are placed in the stator. The windings are designed to produce two sets of poles having some relationship other than 2 to 1. When such a motor is in operation, one of the windings is idle while the other is energized. Because the two windings are inductively coupled, it is generally necessary to have the idle winding open-circuited (it will be if it is star-connected, but one corner must be opened in a normally connected delta) to prevent it from being energized by transformer action. The slots in the stator cores of such machines are made deep enough to accommodate the two windings, and the usual practice is to install the high-speed winding first in the bottoms of the slots and then place the low-speed winding directly over it. Terminal leads are brought out for connection, usually to an automatic push-button line-start controller that is arranged for one of several operating procedures, such as starting at either speed and shifting from one to the other at will, low-speed starting first, high-speed starting first, shifting from one speed to another after time delays that are controlled by accelerating and decelerating relays, and others. *Two-winding two-speed motors* of this type are available in standard horse-power ratings to operate at various combinations of 60-cps base (synchronous) speeds, such as 1,800 and 1,200 rpm, 1,200 and 900 rpm, and 1,200 and 720 rpm; the motors delivering these speeds are wound for 4 and 6 poles, 6 and 8 poles, and 6 and 10 poles, respectively.

Another method commonly employed to provide two operating speeds is to use a single winding whose connections may be changed externally for either of two pole combinations, *one of which is twice the other*. This scheme involves the so-called *consequent-pole principle* and yields a high base speed that is always twice the low base speed. The latter speed is obtained when the winding is connected consequent-pole, as will be explained later. *Single-winding two-speed motors* avoid the disadvantages of their *two-winding* counterparts, which are (1) high cost, (2) comparatively large size for a given output, and (3) some sacrifice in performance characteristics. Single-winding motors with 2-to-1 speed combinations are available in standard horsepower ratings for 60-cps operation at 3,600 and 1,800 rpm, 1,800 and 900 rpm, 1,200 and 600 rpm, and other base speeds. Two windings are sometimes installed in a stator to provide four base speeds, as, for example, 1,800 and 900 rpm with one winding and 1,200 and 600 rpm with the other. In rare cases, as many as three windings have been used to yield base speeds of 3,600, 1,800, 1,200, 720, 600, and 360 rpm. Such motors are intended for special applications, where cost, efficiency, and physical size are incidental to the need for wide-range speed control.

Frequency Changing. An excellent way to control the speed of an induction motor is to vary the frequency of the supply voltage. With this method it is, of course, necessary to provide a separate source whose frequency and voltage can be adjusted simultaneously and in direct proportion to each other. The simultaneous variation is desirable because flux densities in the motor, which are proportional to volts per cycle, should be kept essentially constant. An installation of this kind is expensive, since it requires a motor-generator set, which frequently consists of an adjustable-speed d-c unit coupled to an alternator or frequency converter. Speed adjustment is, however, virtually stepless over an extremely wide range. In another arrangement, a slip-ring induction motor can be made to serve as a variable-frequency source for the speed-controlled motor if the rotor is driven by another machine (an adjustable-speed d-c motor) at the same time that its stator is connected to the available constant-frequency supply mains.

Rotor-resistance Changing. In all induction motors the speed-torque characteristic widens with increasing values of rotor resistance. Advantage is taken of this in phase-wound rotors by connecting varying values of resistance to the rotor slip rings for speed-adjustment purposes. When this method is employed, the speed can be adjusted almost steplessly over an extremely wide range. The control technique does, however, involve a power loss in the added resistors that, for a given load torque, is directly proportional to their ohmic value. Hence, overall efficiency suffers increasingly as the speed is reduced. Generally, the speed of a phase-wound rotor motor is not permitted to drop below half of synchronous speed because efficiencies are then less than 50 per cent.

Cascading. The so-called cascade (or concatenation) principle is equivalent to the pole-changing technique. Two induction motors that are mechanically coupled to drive the load are employed. They are wound for different numbers of poles, and one of them has a phase-wound rotor. By an arrangement of this kind, three practical base speeds can be obtained: (1) the operating speed when the phase-wound rotor motor with, say, 4 poles is used alone; (2) the speed when the other machine with, say, 6 poles is used alone; (3) the speed represented by the tandem connection. In the last case, the stator (primary) of the first (4-pole) motor is excited by the a-c source and its rotor (secondary) is connected to the stator (primary) of the second (6-pole) machine. Under this condition the cascaded set behaves like a motor with $P_1 + P_2$ $(4 + 6 = 10)$ poles.

When a phase-wound rotor is used for the second machine also, a finer degree of control *between* the three base speeds and below the lowest base speed is possible. A fourth base speed can be obtained by employing a "differential-cascade" connection, but because of the complex starting procedure and restricted output, it is seldom used. To start the set, which eventually results in a speed represented by the difference of the numbers of poles $(6 - 4 = 2)$, the first motor is brought up to speed initially; then its phase-wound rotor is connected to the stator of the second motor and, at the same time, the direction of the revolving field is reversed by interchanging one pair of primary leads at the first machine.

Injecting a Voltage into Phase-wound Rotor. When the speed of a phase-wound rotor motor is adjusted below synchronous speed by the insertion of resistors in the rotor circuit, there is a drop in motor efficiency caused by additional I^2R losses. A portion of the energy that is transferred across the air gap from stator to rotor is converted into electrical energy that heats the resistors, and the remainder powers the load. If it is assumed that the motor drives a constant-torque load, the rotor current must remain unchanged for all values of external resistance. Because flux and rotor-circuit power factor are essentially independent of load and motor speed, any addition to the rotor-circuit resistance must be matched by an increase in the emf induced in the rotor winding to maintain constant current. Since the induced rotor voltage depends upon the relative motion between revolving field (synchronous speed) and the rotor conductors, the rotor slip and its corresponding speed $N = (120f/P)(1 - s)$ will always depend upon the ohmic values of the resistors in the controller. In other words, the speed must decrease with increasing values of rotor-circuit resistance because the latter incurs a voltage drop that must be compensated for by a rise in rotor emf.

Another way to reduce the speed of a phase-wound rotor motor is to *inject* a voltage in the rotor. This is equivalent to incurring a voltage drop across a resistor, because the net result is a reduction in the voltage across the rotor winding. Assuming constant-torque operation as before, the injection of a voltage is accompanied by a drop in speed and a rise in the induced rotor voltage that is to maintain constant rotor current and torque. If the injected rotor voltage aids the induced emf, the value of the latter will drop automatically and speed will rise. It is, in fact, possible to cause the motor to run above synchronous speed by introducing a sufficiently large aiding voltage. This voltage-injection method is superior to the simpler rotor-resistance scheme because it eliminates the I^2R losses in the external resistors and permits the motor to operate over a much wider speed range, above as well as below synchronous speed.

The frequency of the injected voltage must, of course, be the same, at every instant, as the slip frequency f_r $(= sf)$ of the emf induced in the rotor winding. Several ways have been devised to accomplish this automatically, some of which require auxiliary equipment and machines and special commutator types of motors that employ brush shifting mechanisms or external sources of adjustable voltage.

Still other speed-control methods have been developed involving the use of such equipment as friction and eddy-current clutches, d-c motors that are energized by rectified alternating current from wound-rotor motors to which they are coupled, and systems that employ electronic and solid-state devices as well as rotating and magnetic amplifiers. Some of these will be discussed subsequently.

Two-winding two-speed motors. Squirrel-cage motors that have two stator windings, each of which creates a different number of revolving poles, are generally more expensive than singly rated machines and have several constructional and operating disadvantages. Because the stator core has deep, narrow slots, the leakage flux and the resulting leakage reactance are high, and these contribute to reduced output and low power factor. Cooling is also difficult, since considerable insulation around the copper has the effect of restricting the transfer of heat from buried sections of the winding. A further objection is that the motor requires a special stator lamination that means a compromise in the winding designs.

As was previously pointed out, these motors are generally controlled by line-start equipment and the windings are usually star connected. The latter arrangement permits control-circuit simplification; since the two windings are inductively coupled and only is one energized at a time, no circulating current can flow in the idle winding. A delta-connected winding would, however, be open-circuited by control contacts at one corner during the idle period.

An elementary diagram of a two-winding two-speed motor connected to a line-start controller is given in Fig. 7·11. The control circuit is

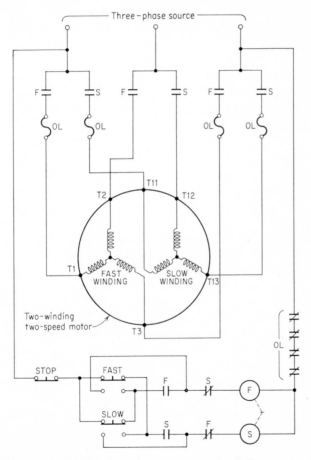

Fig. 7·11 Two-winding two-speed motor connected to a line-start controller.

arranged to start and operate the motor on either of the two windings FAST or SLOW, and shifting from one speed to the other is accomplished at will and without delay.

To reverse the direction of rotation of a single-winding induction motor it is merely necessary to interchange one pair of stator-line terminals. For two-winding motors this, therefore, means that it is possible (1) to reverse the motor in high-speed only, (2) to reverse the motor in low-speed only, and (3) to reverse the motor in both speeds. In all control-circuit arrangements it is, however, necessary to provide proper electrical and mechanical interlocking to avoid short circuits and winding damage. One such scheme is illustrated by Fig. 7·12, which shows a

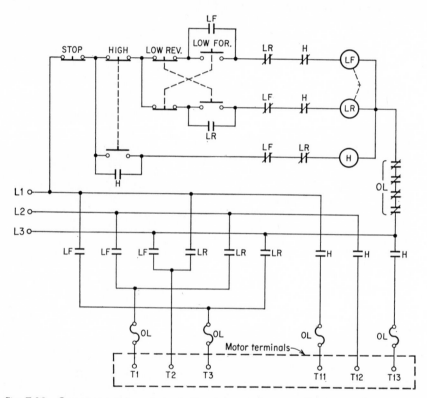

Fig. 7·12 Control circuit for a two-winding two-speed motor with provision for reversing in LOW only.

controller connected to a two-winding two-speed motor that can be reversed in LOW only. Note particularly that the *low-forward LF* and the *low-reverse LR* contactors are mechanically interlocked to prevent line short circuits and that electrical interlocking is provided among the three contactors *LF*, *LR*, and *H*. The operation of this circuit should be clear without further explanation.

The control-circuit diagram of Fig. 7·13 illustrates how a two-winding two-speed motor is connected to provide reversing at both speeds. A selector switch, in conjunction with a pair of reversing contactors F and H, sets the direction of rotation desired, and the LOW and HIGH contactors operate when the proper push button is pressed. Thus, if the

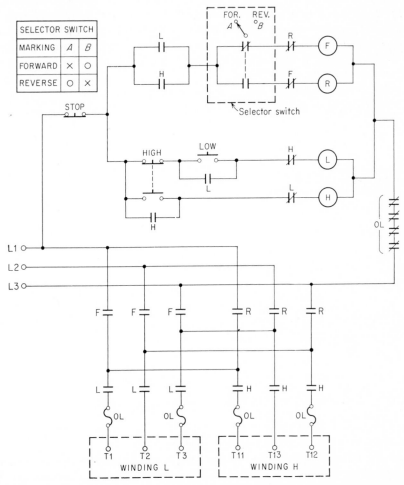

Fig. 7·13 Control circuit for a two-winding two-speed motor with provision for reversing at both speeds.

selector switch is set at FOR. to close contact A and open contact B (as indicated), only the forward contactor F is operative. The pressing of the LOW button will then energize winding L through the F and L contacts. If the HIGH button is pressed, winding H will be energized through the F and H contacts. The other two combinations are obtained by first setting the selector switch to REV. and then pressing the appropriate button.

Single-winding two-speed motors. Practically all 3-phase induction motors have stators with double-layer windings. For a given slotted core, the number of coils in the winding is the same as the number of stator slots, with one-third of them allotted to each phase. Also, all normal pole spans of 180° must contain a similar succession of three groups, with the number of coils in each group being equal to the total number of coils divided by $3P$. Moreover, before the three phases are interconnected to form a *star* or *delta*, it is necessary that the individual coils of each coil group be joined in series. Figure 7·14a is a schematic diagram

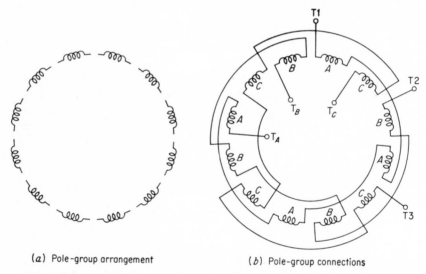

(a) Pole-group arrangement   (b) Pole-group connections

Fig. 7·14 Schematic diagrams representing a 4-pole 36-coil single-speed stator winding.

of a succession of 12 pole groups, each of which contains three series-connected coils for a 36-slot 4-pole, 3-phase winding.

In the *conventional* single-winding single-speed motor, successive pole groups of each phase are connected to produce, at a given instant, poles of opposite polarity; this is frequently done by joining the pole groups in series. Also, to keep the three winding phases properly spaced, it is necessary to bring out similar terminal leads from points on the winding that are displaced 120 electrical degrees (Fig. 7·14b). The final step is to interconnect the three phases to form a *star* or a *delta*. The *star* connection is made by joining T_A, T_B, and T_C and using $T1$, $T2$, and $T3$ as line leads, while the *delta* connection is formed by using junctions of T_A and $T2$, T_B and $T3$, and T_C and $T1$ as line leads.

A popular method for obtaining two operating speeds, *one twice the other*, with a single winding is to connect the latter *conventionally for the high speed and consequent-pole for the low speed*. The technique applies only to 3-phase motors; two-speed 2-phase motors always employ two windings. The change from P poles to $2P$ poles is made with either of two switching schemes. In the series-parallel method, the winding produces P poles (Fig. 7·15b) when all the pole groups of each phase are joined

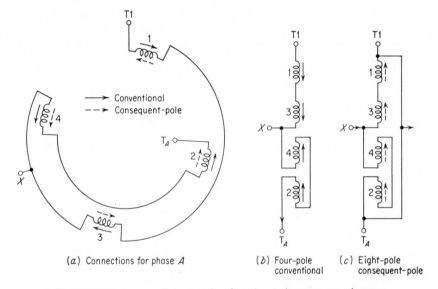

(a) Connections for phase A (b) Four-pole   (c) Eight-pole
                                     conventional   consequent-pole

Fig. 7·15   Series-parallel principle of single-winding two-speed motor.

in series and $2P$ poles (Fig. 7·15c) when the same pole groups are connected in two parallel paths. The reverse is true in the parallel-series method.

In the series-parallel scheme illustrated in Fig. 7·15, four pole groups of one of the three phases of a 4-pole–8-pole winding are connected in series between points $T1$ and T_A and midtap X is brought out. The series connection is made by joining one pair of diametrically opposite pole groups, labeled 1 and 3, to the other pair of diametrically opposite pole groups, labeled 4 and 2, with the X tap emerging from the junction of the two pairs. If current enters terminal $T1$ and leaves at terminal T_A as shown in Fig. 7·15b, four *conventional poles* will be produced. On the other hand, if terminals $T1$ and T_A are joined, the current entering terminal X divides into two parallel paths as it passes through the four pole groups (Fig. 7·15c). Thus, four conventional poles and four *consequent poles*, i.e., 8 poles, are created. This same arrangement applies to phases B and C, and the complete switching system also involves a proper *star* or *delta* connection.

In the second scheme, four conventional poles are developed when the parallel connection of pole groups is employed (Fig. 7·16b); the series

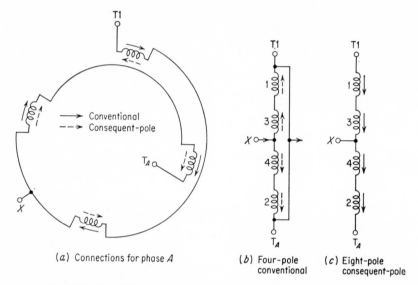

(a) Connections for phase A

(b) Four-pole conventional

(c) Eight-pole consequent-pole

Fig. 7·16   Parallel-series principle of single-winding two-speed motor.

arrangement (Fig. 7·16c) produces four conventional plus four consequent poles.

The windings illustrated by Fig. 7·15 and 7·16 give two speeds that are always in a 2-to-1 ratio. With two such windings in a stator, it is possible to obtain four speeds when one combination of poles differs from the other. Standard squirrel-cage motors are available for such 60-cps speed combinations as 1,800/1,200/900/600 rpm, 1,200/900/600/450 rpm, 3,600/1,800/1,200/600 rpm, and others. Special three-winding six-speed motors have been constructed to yield speeds of 450, 600, 900, 1,200, 1,800, and 3,600 rpm, where windings A, B, and C give, respectively, the first and third speeds, the second and fourth speeds, and the fifth and sixth speeds.

Characteristics of two-speed motors. Three general types of two-speed motors are recognized and standardized by NEMA. These are *constant-torque* motors, which develop approximately the same full-load turning effort with about the same temperature rise at both speeds; *constant-horsepower* motors, whose full-load torques for about the same temperature rise are inversely proportional to the respective speeds; *variable-torque* motors, whose full-load torques are directly proportional to the respective speeds. Each type of 3-phase two-speed motor must obviously receive special design treatment, but as indicated in Fig. 7·17, the winding is always connected consequent-pole for the low speed and conventional-pole for the high speed. The phase connections (series-delta,

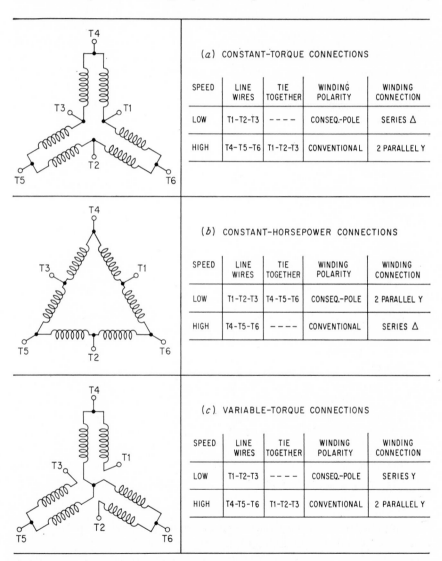

(a) CONSTANT-TORQUE CONNECTIONS

SPEED	LINE WIRES	TIE TOGETHER	WINDING POLARITY	WINDING CONNECTION
LOW	T1-T2-T3	- - - -	CONSEQ.-POLE	SERIES Δ
HIGH	T4-T5-T6	T1-T2-T3	CONVENTIONAL	2 PARALLEL Y

(b) CONSTANT-HORSEPOWER CONNECTIONS

SPEED	LINE WIRES	TIE TOGETHER	WINDING POLARITY	WINDING CONNECTION
LOW	T1-T2-T3	T4-T5-T6	CONSEQ.-POLE	2 PARALLEL Y
HIGH	T4-T5-T6	- - - -	CONVENTIONAL	SERIES Δ

(c) VARIABLE-TORQUE CONNECTIONS

SPEED	LINE WIRES	TIE TOGETHER	WINDING POLARITY	WINDING CONNECTION
LOW	T1-T2-T3	- - - -	CONSEQ.-POLE	SERIES Y
HIGH	T4-T5-T6	T1-T2-T3	CONVENTIONAL	2 PARALLEL Y

Fig. 7·17 Terminal markings and connections for single-winding two-speed induction motors.

two-parallel-star, or series-star) for the two speeds depend upon the motor classification.

Each classification shown in Fig. 7·17 has certain operating characteristics that are particularly suitable for specific applications. Constant-torque motors are, for example, used for compressors, conveyors, stokers, printing presses, many machine tools, and textile, baking, and laundry machines. Common applications of the constant-horsepower motor are lathes, shapers, grinders, and boring mills, where considerable torque must be developed at low speed and much less torque at high speed. In fans, blowers, centrifugal pumps, etc., the required horsepower increases approximately as the cube of the speed; for these applications the variable-torque motor is favored.

The *torque-versus-speed* characteristics of Fig. 7·18 indicate how the three motor classifications are related, assuming that all perform similarly

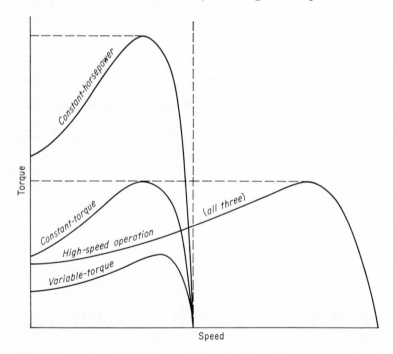

Fig. 7·18 *Torque-versus-speed* characteristics of single-winding two-speed induction motors.

when operated at high speed. Since horsepower output is proportional to the product of torque and speed:

1. For a constant-torque motor, the maximum horsepower output at low speed is one-half the half-speed value.

2. For a constant-horsepower motor, the maximum torque at low speed is twice the high-speed value.
3. For a variable-torque motor, the maximum horsepower output at low speed is one-fourth the high-speed value.

For the purposes of standardization, certain terminal markings have been adopted by manufacturers of single-winding two-speed motors. Figure 7·17 shows the three standard torque and horsepower classifications and presents accompanying tables giving line and terminal connections and the corresponding winding polarities and connections for the low and high speeds.

Figure 7·19 shows the complete internal wiring connections of a *constant-torque motor* designed for 4/8-pole operation. The control circuit permits the motor to start on *fast* or *slow* but requires a *stop* sequence before a shift is made from *fast* to *slow*. By following the *continuous-line* arrows from terminals $T4$, $T5$, and $T6$ it will be noted that the winding is connected conventionally and two-parallel-star for 4 poles when $T1$, $T2$, and $T3$ are tied together; the *dashed-line* arrows from terminals $T1$, $T2$, and $T3$ can be traced to show that the winding is connected consequent-pole and series-delta for 8 poles. Referring to the control section, note that an N.C. $CR1$ interlock opens when the FAST button is pressed and control relay $CR1$ picks up; this, obviously, makes the SLOW button ineffective until the pressing of the STOP button drops out the $CR1$ relay.

Single-winding two-speed reversing motors. The control-circuit diagram of Fig. 7·20 shows the connections for a single-winding two-speed constant-torque motor that can be reversed on the high speed only. Several interlocking features are included to provide the following interesting functions:

1. The motor can be started directly in *low forward LF*, *high forward HF*, or *high reverse HR* by pressing the appropriate button.
2. If the motor is operating in *low forward*, it can be shifted directly to *high forward* by pressing the HIGH FOR. button. Doing this drops out the $1CR$ relay and the LF contactor and permits the HF contactor to pick up through the operation of the $2CR$ relay. Note that the LF contactor and the $2CR$ relay are mechanically interlocked.
3. If the motor is running in *low forward*, a shift *cannot* be made directly to *high reverse*. To do this the STOP button must first be pressed to cause the LF contactor to drop out and permit the LF-T.C. contact to close after a time delay. The HIGH REV. button then becomes effective in energizing the $3CR$ relay and the HR contactor.
4. A shift *cannot* be made directly from *high forward* to *high reverse*. To accomplish this, a stop sequence must first be made to deenergize

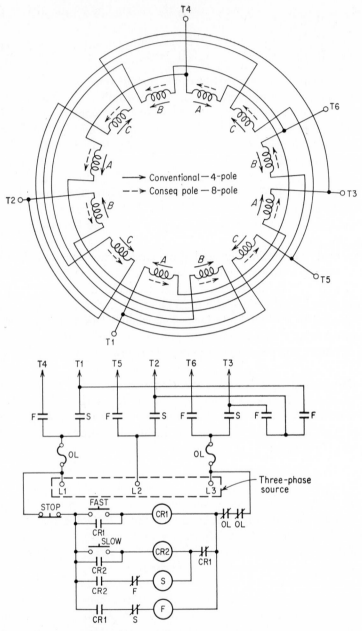

Fig. 7·19 Winding diagram and control circuit for a single-winding two-speed constant-torque induction motor. Motor may be started on *fast* or *slow* but, when operating *fast*, may not be shifted to *slow* until the STOP button is pressed.

the *HF* contactor, after which the *HF*-T.C. contact closes with a time delay to make the HIGH REV. button effective.

5. With the motor operating on *high forward*, a shift can be made directly to *low forward* by pressing the LOW FOR. button. However, the change to the lower speed takes place with a time delay, i.e., after the *HF*-T.C. contact closes.

6. If the motor is running in high reverse, a shift to *low forward* or *high forward* can be made only after pressing the STOP button. Doing this causes the N.C. 3CR contact to close in the *HF* contactor circuit

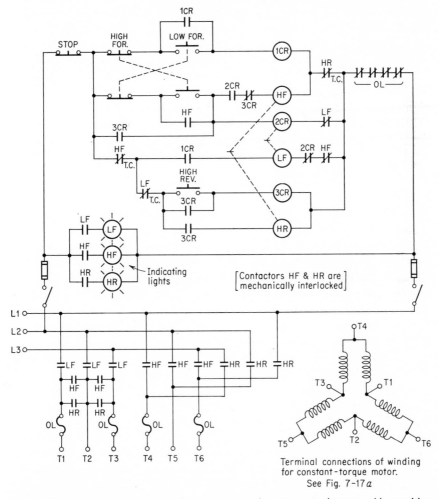

Fig. 7·20 Control-circuit diagram for a single-winding two-speed motor with provision for reversing in HIGH only and with decelerating and reversing timing relays. Motor is designed for constant-torque operation.

and the N.C. *HR*-T.C. interlock to close in the 1*CR-HF* circuit.   The
HIGH FOR. and LOW FOR. buttons then become effective after a time
delay.

Referring to the lower part of the drawing, note that the terminal connections are made in accordance with the sketch and table of Fig. 7·17*a*
for constant-torque characteristics. Terminals $T1$, $T2$, and $T3$ are connected, respectively, to $L1$, $L2$, and $L3$ for low-speed operation in a forward direction. When terminals $T4$, $T5$, and $T6$ are joined, respectively,
to $L1$, $L2$, and $L3$ and $T1$, $T2$, and $T3$ are short-circuited by the two
HF interlocks, the motor will run in *high forward*. Winding terminals
$T4$, $T5$, and $T6$ are switched, respectively, to $L3$, $L2$, and $L1$ for *reverse
forward* running, with $T1$, $T2$, and $T3$ short-circuited by the two *HR*
interlocks.

It should also be noted in Fig. 7·20 that indicating lights, usually
colored and with identifying symbols, have been provided. These are
illuminated when the proper interlock, *LF*, *HF*, or *HR*, closes.

Figure 7·21 represents a control circuit that is designed to permit a
single-winding two-speed constant-horsepower motor to operate in a *forward* or *reverse* direction. Interlocking contacts have been included to
provide the following functions:

1. The motor *must* be started in *low forward* or *low reverse*. It *cannot* be
   started in *high forward* or *high reverse* because the *CR*-T.C. contacts
   are open in the *HF* and *HR* contactor circuits.
2. A shift can be made from *low forward* to *high forward* or from *low
   reverse* to *high reverse* after a time delay set by control relay *CR*.
   Note that the two *CR*-T.O. contacts remain closed until the *CR* relay
   times out; thus, the HIGH FOR. and HIGH REV. buttons remain ineffective until these contacts open. It is, therefore, not possible to shift
   into high speed until the motor has accelerated fully on low speed.
3. To shift from *low forward* to *low reverse* or from *low reverse* to *low
   forward* it is first necessary to go through a *stop* sequence. This is
   because there is an opened *LF* interlock in the *LR* contactor-coil
   circuit when the motor is running in *low forward*, and an opened *LR*
   interlock in the *LF* contactor-coil circuit during *low reverse* operation.
4. To shift from *high forward* to *low forward* or from *high reverse* to *low
   reverse* it is first necessary to press the STOP button. This is because
   the N.C. *HF* interlock or the N.C. *HR* interlock (both in series) is
   open; the LOW FOR. and LOW REV. buttons are, therefore, ineffective
   until the *HF* or *HR* contactors drop out.

Note that the portion of the diagram (upper right) representing the
stator winding and its terminal connections corresponds to Fig. 7·17*b* for

constant-horsepower characteristics. With the motor operating in *low*, line terminals *L*1, *L*2, and *L*3 are connected to *T*1, *T*2, and *T*3 and contactor *LN* joins *T*4, *T*5, and *T*6 to form the neutral point of the two-parallel-star. With the machine running in high, *L*1, *L*2, and *L*3

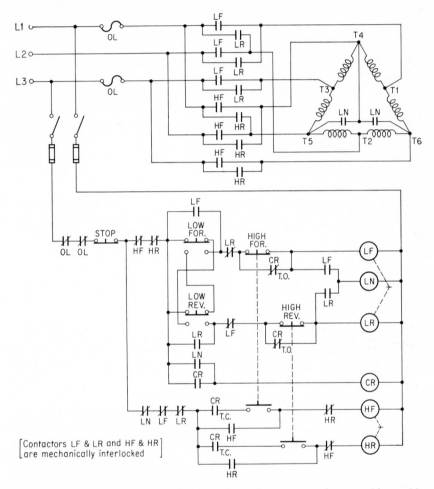

Fig. 7·21   Control-circuit diagram for a single-winding two-speed motor with provision for reversing in both speeds. Motor is designed for constant-horsepower operation (see Fig. 7·17b).

are connected to stator terminals *T*4, *T*5, and *T*6. The two pairs of mechanically interlocked contactors *LF-LR* and *HF-HR* function to interchange the *L*1-*L*2 line terminals to reverse the direction of rotation of the motor.

Two-winding three- and four-speed motors. The winding principles discussed in the foregoing articles in connection with two-speed motors can be extended to include squirrel-cage induction-type machines that have three or four base speeds. This is accomplished by using two windings, and for four-speed motors each one is designed to produce a pair of speeds in the ratio of 2 to 1 that differs from the other pair. A combination of 60-cps base speeds frequently employed is 600/900/1,200/1,800 rpm, with *low* and *third* speeds provided by one 12-pole–6-pole winding and the *second* and *high* speeds developed by another 8-pole–4-pole winding. Since the two windings are inductively coupled and only one of them is energized while the motor is operating at a particular speed, it is essential that the inactive winding be open-circuited during its idle period if circulating currents are to be avoided. The control circuit must, therefore, have N.O. contacts at appropriate points in delta- or two-parallel-star windings, in addition to normal line-terminal and reversing contacts, and only certain contactors must function to close the proper contacts for the active winding.

Figure 7·22 represents a control circuit for a two-winding four-speed nonreversing motor. As is often necessary in such motor applications, interlocking contacts and decelerating relays are included to prevent a direct change from one speed to the next lower speed. This is to avoid possible damage to high-inertia mechanical equipment should the motor attempt to make an abrupt speed reduction. The following points analyze the operating functions of the circuit:

1. The motor may be started in any of the four speeds, i.e., *low, 2nd, 3rd,* or *high.*
2. With the motor running at a given speed, it can be advanced directly to the next higher speed, i.e., from *low* to *2nd, 2nd* to *3rd,* or *3rd* to *high.* An opened *L* contact prevents a change from *low* to *3rd* or *low* to *high;* an opened 2 contact prevents a change from *2nd* to *high.*
3. When the motor is operating at a given speed, there will be a time delay between the pressing of an appropriate button and the change to the *next lower* speed; the actions that follow permit the motor to slow down before the lower speed contactor is actuated. An opened *H* contact prevents a change from *high* to *2nd* or *high* to *low;* an opened 3 contact prevents a change from *3rd* to *low.*
4. Assume that the motor is running on *low* and a shift is to be made to *2nd.* The pressing of the *2nd* button causes the *2CR* relay to pick up, which, in turn, opens the N.C. *2CR* interlock in the *1CR* circuit. The *1CR* relay is thereby deenergized, and its contact opens to drop out the *L* contactor. The closing of the N.C. *L* contact and the N.O. *2CR* contact is then followed by the energization of the *L* contactor.

Similar actions will result if shifts from *2nd* to *3rd* or *3rd* to *high* are attempted.

5. Assume that the motor is running in *high* and a change is to be made to *3rd*. The pressing of the *3rd* button opens its back contact to de-energize the *H* contactor and closes its front contact to pick up the *3CR* relay. Since the *HTR*-T.C. contact in the 3 contactor circuit is still open, the latter does not pick up even though the *3CR* and *H* contacts have closed. This is because the *high timing relay HTR* has not yet timed out by the opening of the *H* contact in that circuit. After an elapse of the timed period, *HTR*-T.C. closes, the 3 contactor is energized and the motor shifts to *3rd* speed. Similar actions will result if shifts from *3rd* to *2nd* or *2nd* to *low* are attempted.

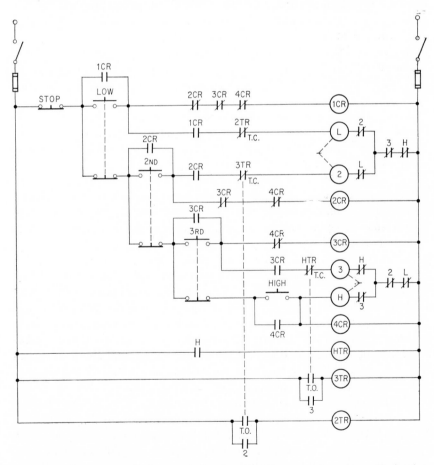

Fig. 7·22 Control circuit for a two-winding four-speed nonreversing motor with decelerating relays.

Control circuits can, of course, be adapted to multispeed motors having any of the characteristics of Fig. 7·17 and, in addition, can be designed to provide reversing service in any or all speeds. Moreover, accelerating and decelerating relays can be used to give time delays between successive speeds in both an increasing and a decreasing direction. The conditions that must be fulfilled will obviously determine the required number of components, the amount of interlocking, and, in general, the degree of complexity of the circuit. These, in turn, will have a direct bearing on the cost of the installation as well as the service problems.

Speed control by frequency variation. A single-winding single-speed squirrel-cage motor can be used in a variable-speed drive if it is provided with an adjustable-frequency source. The latter may take one of two forms: (1) a separate alternator, whose d-c field is excited from a constant-voltage supply and which is driven by a variable-speed prime mover; (2) a frequency changer, driven by a d-c or a-c motor.

The alternator method is generally limited to the use of special variable-frequency laboratory equipment or to large applications such as ship propulsion and wind tunnels, where the adjustable-speed induction motors represent the only loads on the electrical system. When several motors are connected to the alternator, all will change speed simultaneously as the frequency is varied. The control problem thus involves a means of varying the mechanical output of the machine that drives the alternator. Moreover, since the d-c field excitation is maintained at some fixed value, all frequency variations are accompanied by proportionate changes in voltage. This relationship, therefore, results in a constant volts-per-cycle power source and causes induction motors to operate over a wide speed range without flux density changes.

An important aspect of this method of control is that, at elevated frequencies, standard induction motors run at speeds well above their normal base values, under which condition cooling improves. The latter then offers the advantage that higher current densities are permissible even through iron losses (due to higher frequencies), friction, and windage increase. Since torque (at constant flux density) is proportional to current, and power output depends upon both torque and speed, it should be clear that the horsepower developed by the induction motor rises more rapidly than does the speed.

The frequency-changer method is often employed when the speed of a standard squirrel-cage induction motor of moderate size must be greater than that possible with available frequency sources. The changer is usually a wound-rotor induction motor operated as a frequency converter. The latter behaves like an alternator, being driven by another induction motor (when the output frequency must be essentially constant) or an adjustable-speed d-c motor (when the output frequency must be varied).

A phase-wound rotor motor can operate as a frequency changer because a synchronously revolving field of constant magnitude is created when the stator winding is energized by a polyphase source of power. With the rotor locked, the machine acts as a static transformer and line-frequency voltages are developed in the rotor winding. However, if the rotor is *driven at synchronous speed* by a coupled motor in a direction *opposite* to that of the revolving field, the flux cuts the conductors at *twice* synchronous speed and the induced frequency and voltage become *double* the standstill values. Furthermore, at a rotor speed that is *twice the synchronous rpm against the revolving field*, the output frequency and voltage will be *three times* the blocked-rotor values. If, on the other hand, the rotor is driven in the same direction as that of the revolving field, flux is cut at a less rapid rate; the frequency and voltage will, therefore, be reduced proportionately until, at synchronous speed, they become zero. These statements can be expressed in general terms by the equation

$$f_{\text{converter}} = f\left(1 \pm \frac{N_{\text{rotor}}}{N_s}\right) \qquad (7 \cdot 4)$$

where f is the line frequency and N_{rotor} is the speed at which the rotor is driven by the coupled motor. The plus and minus signs indicate rotor rotation that is, respectively, against and with the revolving field.

If the frequency converter must deliver a stepless variable-frequency source to an induction motor, the driving unit is usually an adjustable-speed d-c motor. More commonly, however, the induction frequency converter provides a constant high frequency (about 180 cps) to run a squirrel-cage motor at approximately three times the normal-frequency speed. This can be accomplished with a 4-pole induction motor driving an 8-pole wound-rotor motor. With the stator winding of the frequency converter connected to a 60-cycle source, its revolving field speed is 900 rpm. Thus, when the phase-wound rotor is driven at 1,800 rpm *against* the direction of the revolving field, the frequency and voltage developed at the slip rings are three times the locked-rotor values.

Figure 7·23, which illustrates such an arrangement, includes the control circuit for starting the motor-converter set. Note that the converter winding is not energized until the motor is started; this precaution is necessary because the converter will suffer damage if it is excited while at rest.

Induction-motor speed control by rotor-voltage adjustment. The insertion of external resistors in the secondary circuit of a phase-wound rotor motor to control the speed of the motor is always accompanied by considerable I^2R losses and a loss of efficiency. To overcome the disadvantages indicated, a fundamental practice is to inject voltages in the rotor circuits, avoiding the use of external resistors and the IR drops they normally incur. This accomplishes essentially the same result, since the net voltage available to circulate the necessary rotor current is the

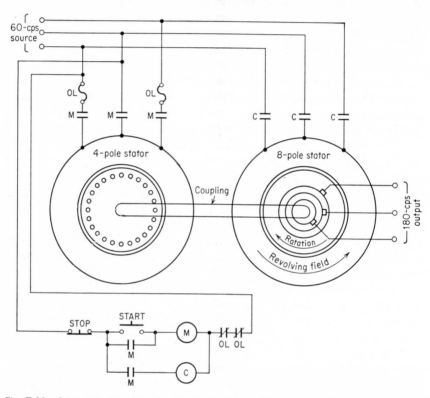

Fig. 7·23 Schematic diagram of an induction frequency converter driven by a squirrel-cage motor to develop 180-cps output.

same whether an IR drop or a countervoltage subtracts from the generated rotor emf. Furthermore, if the injected voltage acts to aid that generated in the rotor, the speed may be made to rise above synchronism. It is apparent that a method such as this has the double advantage: (1) it eliminates wasteful power losses, and (2) it permits the machine to operate above as well as below synchronous speed. The scheme is, however, rather expensive because special types of machines are required to supply the necessary rotor voltages. Hence, it is used only in large installations where the cost of the additional equipment can be justified on the basis of improved overall efficiency and wider speed adjustment. For low-power applications the principle of control has, moreover, been adapted to a unique commutator type of motor in which all essential elements are included in a single structure and a brush-shifting mechanism serves to control the speed. It is discussed in the next article.

Important to this method of control is the need for an injected voltage whose frequency must always be the same as the *slip* frequency, i.e., *sf*. Since the rotor frequency increases progressively as the speed drops below or rises above synchronism, the equipment supplying the injected voltage must automatically adjust itself to generate slip-frequency emf as the main motor changes speed. In one fundamental design, the machine that supplies the adjustable voltage for the phase-wound rotor is somewhat similar to a rotary converter, except that the stator is a laminated core with no winding, acting merely as part of a good magnetic circuit. The rotor contains the usual d-c armature winding that is connected to a commutator at one end and slip rings at the other. As illustrated in Fig. 7·24, the brushes riding on the rings are connected to the 3-phase

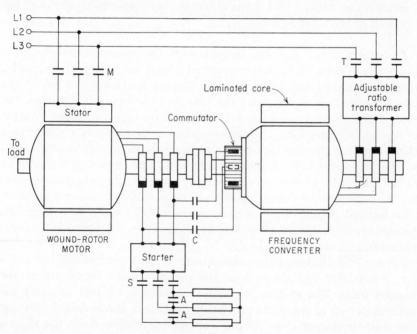

Fig. 7·24 Ring-commutator type of frequency converter and adjustable-ratio transformer connected to a wound-rotor motor for speed-control purposes.

power lines through a set of adjustable-ratio transformers, while another set of brushes, spaced 120 electrical degrees apart, rests on the commutator and is wired to the wound-rotor slip rings. This special machine is coupled to the main motor and therefore always operates at the same speed as the latter.

To understand why the slip-ring frequency of the converter always matches the slip frequency of the wound-rotor motor, it is necessary to recognize the following conditions of operation:

1. Line frequency currents enter the a-c end of the converter to set up a revolving field that rotates at synchronous speed *with respect to the armature-winding conductors.*
2. If the rotor is at rest, the field thus produced revolves in space at synchronous speed.
3. If the rotor is turned at synchronous speed in a direction *opposite* to the motion of the field with respect to the moving conductors, the field is stationary in space. Under this condition, the field is, in effect, rotated backward as rapidly as it is moved forward.
4. At any speed intermediate between rest and synchronism, the field revolves at a speed that is the difference between synchronism and the speed of the rotating machine.

It follows, then, that the frequency at the commutator brushes is always identical with the interconnected wound-rotor rings. Particularly, the commutator-brush frequency and the slip frequency are both f cps when the rotor is at rest, and the two frequencies are 0 cps when the rotor turns at synchronous speed in a direction opposite to the rotation of the field with respect to its own conductors. As in all rotary converters, the voltage at the commutator end always bears a very definite ratio to the impressed emf at the slip-ring end. This means that, with an adjustable-voltage transformer to supply power to the ring end, the correct voltage (for the desired speed) will be available at the commutator. The method permits the operator to control the speed of a constant-torque application over an extremely wide range above and below synchronism and to do so steplessly.

Figure 7·24 illustrates the arrangement of the machines and the auxiliary equipment and shows how the control circuit interconnects the various units. The system is put into operation by first bringing the main motor up to speed. As previously explained, this is done by having the main contactor close its line contacts M, followed by the closing of starting contacts S and accelerating contacts A. The slip-ring brushes are next transferred to the commutator brushes by the closing of the C contacts and the opening of S and A contacts. Finally, the T contacts are closed to connect the adjustable-ratio transformer to the slip rings of the converter, and manual settings are made at the transformer to provide the required injected voltage for the desired speed adjustment.

Although the foregoing scheme is rarely used as described, it is basic to all other systems in which the wound-rotor motor is the main drive and voltage injection in the secondary is the means of adjusting the speed. Its understanding will clarify the operation of some of the more complex systems presently in use as well as the special commutator-type motor discussed in the following article.

Brush-shift method of speed control. The parts of the speed-control system of Fig. 7·24 (wound-rotor motor, frequency converter, and adjustable-ratio transformer) were first combined into a single structure by K. H. Schrage of Sweden to provide a complete, self-contained adjustable-speed motor. The machine consists of a special phase-wound motor, whose *rotor acts as the primary* (it is connected to the 3-phase source through a set of slip rings) and the *stator* of which *performs as the secondary*. This is just the reverse of the functions of the stator and rotor windings in the standard wound-rotor motor. In addition, there is a third winding that functions as an electromagnetic link between primary and secondary and as a source of slip-frequency voltage for the secondary. This so-called *tertiary* winding is placed directly over the primary and is connected to a commutator mounted on the shaft end opposite the slip rings. Instead of being connected to a resistance controller, the secondary winding is wired to *three pairs of brushes* that ride on the commutator. Speed control is effected by moving a lever or turning a handwheel that shifts the three sets of brushes. Because of the simplicity of operation and the ease with which a large stepless speed range can be obtained, above and below synchronsim, the motor is widely used for many moderately small industrial applications.

A schematic diagram of the *Schrage* motor, showing the general arrangement of windings and their connections to the commutator, brushes, and rings, is given in Fig. 7·25. Although the machine would operate with one rotor winding, the two-winding design is preferable in the brush-shifting motor because the voltage between adjacent commutator segments can be minimized to fulfill the requirements for sparkless commutation.

An important detail of construction is the connections of the ends of each stator (secondary) phase to an independent set of brushes. The operation of the lever or handwheel shifts the three sets of brushes simultaneously so that corresponding pairs can be made to move closer together or farther apart. Each pair may come axially into line on the same commutator bar or may be shifted to cross similar electrical positions on the commutator. With the brushes as shown in Fig. 7·25, the motor speed will be below synchronism; when corresponding pairs are lined up axially on the same segment, the speed will rise slightly; with the brushes in crossed positions, it will be above synchronism.

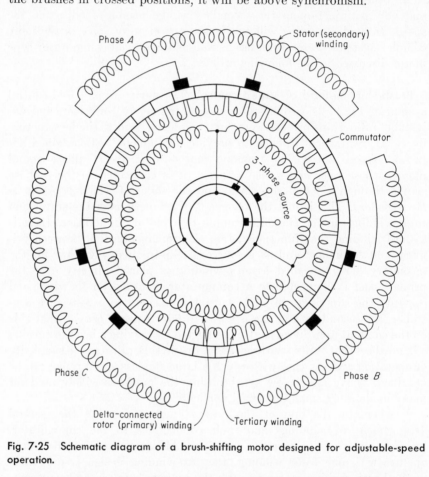

Fig. 7·25 Schematic diagram of a brush-shifting motor designed for adjustable-speed operation.

When the rotor (primary) winding is energized from a 3-phase source, a revolving field is created. The stator *tends* to turn in the same direction, but since the outer structure is fixed and cannot move, the rotor turns instead in the *opposite* direction. Assuming that each set of brushes is lined up axially on the commutator, the three secondary phases will be short-circuited; the motor will then come up to about synchronous speed if the load is light. Under this condition the magnetic field is stationary in space and behaves exactly like the field established by the poles of a d-c motor. The tertiary winding, with its coils connected to the commutator, cuts the stationary field and generates a d-c (zero-frequency) voltage. Note that this is theoretically equivalent to the slip frequency of 0 cps in the stator (secondary) winding because field and winding are stationary with respect to each other.

When the brushes are separated as in Fig. 7·25, an emf from the commutator is connected across each of the stator (secondary) phases; the motor speed, therefore, changes. If the stator-winding connections to the commutator are made so that the injected voltage opposes the emf induced in the secondary, the speed drops. The field then rotates slowly and *at exactly the same rate with respect to both stator winding and brushes.* This means that the frequency of the induced emf in the stator and the frequency of the brush voltage are identical, an important condition if the machine is to function both as a frequency converter and as a variable-speed wound-rotor motor. Obviously, the greater the separation between corresponding sets of brushes, the higher is the injected voltage and the lower the speed. Moreover, if the brushes are moved closer together and then made to cross positions *electrically* on the commutator, the speed rises above synchronism because less voltage is required in the secondary to maintain the same load current for constant-torque operation. Briefly, then, brush shifting merely permits the operator to "pick off" the proper slip-frequency voltage that subtracts from or adds to the secondary emf.

An added benefit of the Schrage motor is that it lends itself to power-factor adjustment. This is accomplished by keeping the same brush separation and moving the entire brush rigging forward or backward. When this is done, the voltage at each pair of brushes remains the same but the phase position with respect to the induced secondary emf is altered. Thus, shifting the brushes one way increases the voltage-current angle to make the power factor more lagging, while shifting them in the other direction decreases the angle for improved or even leading power-factor operation.

Kraemer system of speed control. For variable-speed applications requiring very large motors, from about 500 to several thousand horsepower, the wound-rotor is preferred. When used in such installations, it is, however, associated with auxiliary equipment which, unlike resistance controllers, returns much of the slip-frequency energy of the rotor to the system. One method of doing this was previously described and illustrated by Fig. 7·24.

Another scheme, frequently employed for steel-mill drives, dredges, and other heavy-machinery loads whose speeds must be adjustable over extremely wide ranges, is the *Kraemer* system, two variations of which are in use to convert the slip-frequency energy from the main motor to useful energy. In one, the constant-horsepower system, d-c power from the converter is supplied to a d-c motor coupled to the shaft of the main motor. In the other, the constant-torque system, the d-c output of the converter ultimately drives a synchronous or induction generator that feeds back into the a-c supply lines.

Figure 7·26 shows the wiring connections for the constant-horsepower arrangement. With converter contacts C open, the unloaded main motor is

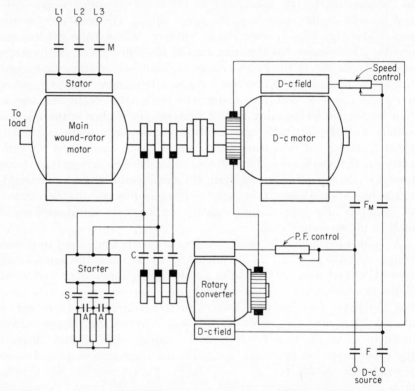

Fig. 7·26   Kraemer system of speed control designed for constant-horsepower operation.

brought up to speed in the usual way with the rotor-resistance starter. Contacts S and A are then opened, normal excitation is applied to the converter, and contacts C are quickly closed to impress slip-frequency voltage across its rings to permit the machine to rotate slowly. The field of the d-c motor is excited next, whereupon that unit, which is coupled to the main motor and running at nearly synchronous speed, generates a d-c voltage that is impressed across the commutator end of the converter. Because a definite voltage ratio always exists between the a-c and d-c ends of the converter, a certain slip-frequency potential appears at the rings and is injected into the rotor of the main motor. The latter therefore slows down, automatically adjusting its own input current for the particular load that was applied after starting. With the speed of the main motor reduced and its slip frequency and voltage increased, the converter speeds up to deliver electrical power to the armature of the d-c motor which, in turn, converts it to mechanical power that aids the main motor in driving the load. The set now stablizes at the adjusted speed for the particular load, delivering a horsepower output equivalent to the power input minus the losses.

To decrease the speed of the main drive, the excitation of the d-c motor must be increased. This raises the cemf of that machine, causes the voltage at the slip ring to increase, and results in a higher injected slip-frequency potential that brings about the desired drop in speed. The reverse is true when the excitation is reduced. In fact, if the excitation to the d-c motor is reduced to zero (when the main motor operates with small slip) and is then reversed with slightly increased excitation, the drive will run at synchronous speed. Under this condition, the converter stops (slip frequency is zero) and the main motor becomes a synchronous motor with d-c excitation. Further increase of the reverse excitation causes the main motor to operate above synchronous speed, and the direction of rotation of the converter reverses. For speeds above synchronism the d-c machine becomes a generator, imposes an additional load on the main motor, and feeds its electrical power through the converter into the rotor of the main motor. A useful feature of the system is that power-factor control is possible by adjusting the excitation of the rotary converter; correction up to about 0.95 is generally practical by overexcitation.

Compared with other systems of speed control for heavy power drives, notably the Scherbius method, which uses such special types of equipment as ohmic drop exciters, induction machines, transformers, and commutator-type motors, the Kraemer system is economical. This is because the d-c motors and rotary converters are, for the most part, similar to those of standard design.

Special five-speed squirrel-cage motor for portable hoist. A characteristic of the polyphase squirrel-cage motor that is used to advantage occasionally for speed-control purposes is its behavior under load when the terminal voltages are unbalanced. Such voltage unbalance is equivalent to two *balanced* systems of voltage whose magnitudes are generally unequal to each other. The larger, *positive-sequence* system of voltages energizes the winding to create a positive revolving field and positive torque, while the smaller, *negative-sequence* system of voltages develops a negative revolving field and negative torque. Since the resultant torque is the difference between the opposing torques, being less than would be developed under normal balanced-voltage conditions, the speed is reduced. In practice, the voltages are unbalanced artificially in one of several ways. But a scheme that has proved quite satisfactory for a portable-hoist application employs an autotransformer whose ratio of transformation can be changed.

Referring to Fig. 7·27a, assume that the autotransformer has the same number of turns between $H1$ and $H4$ as between $X1$ and $X2$. With the entire transformer, between terminals $H1$ and $X2$, connected to lines $L2$ and $L3$ (neglecting resistors which are used for protective purposes), the ratio of transformation will be 2 to 1; the voltage across winding phases B and C will therefore be $0.5E$. Its phasor diagram thus indicates that the three motor voltages are E, $0.5E$, and $0.866E$. With this unbalanced system of voltages the motor will then run at one speed.

Consider Fig. 7·27b next, which shows the $H1$-$H2$ and $H3$-$H4$ portions of the autotransformer connected in parallel. With this arrangement the ratio of transformation will be 1.5 to 1 and the voltage across winding phases B and C will be $E/1.5 = 0.667E$. Calculations then show that the voltage between $L3$ and $L1$ is $0.882E$. Thus, with this unbalanced system of voltages, i.e., E, $0.667E$ and $0.882E$, another motor speed, somewhat higher than before, will result.

Finally, as in Fig. 7·27c with the three voltages nearly balanced, a third, still higher speed is obtained.

If additional speeds are desired by further degrees of voltage unbalance, these could be provided by using other arrangements, tappings and switching. However, since the complications would, in general, be greatly multiplied, other methods are sometimes used if more than three speeds are necessary. Such a system of control is illustrated by Fig. 7·28 in which five speeds are possible. The first three speeds, 100, 225, and 450 rpm, are obtained from one winding whose voltages are unbalanced as described above. The other two speeds, 900 and 1,800 rpm, are given by a second constant-torque winding. Used in a portable-hoist application the five-speed motor and its system of relays and contactors would be controlled by a special pendant-mounted push-button switch shown schematically to the right of the wiring diagram. Two mechanically interlocked push buttons are operated independently for *hoisting* and *lowering*, and only one button can be pushed down at a time. Its mechanism is so constructed that the pressing of a button closes in succession lower contacts as it maintains those above; thus, in the fully depressed position all five contacts are made. Figure 7·27d and e also shows the connections for the second winding, where the fourth speed connects the three winding phases in delta and consequent pole while the fifth speed connects them two-parallel-star and conventionally. It is suggested that Fig. 7·28 be carefully studied and traced for the various *hoist* and *lower*

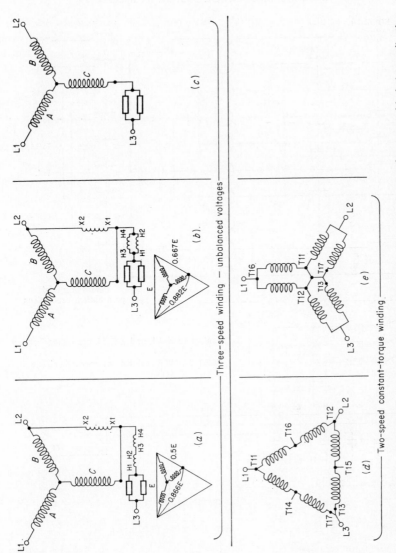

(c)

——— Three-speed winding — unbalanced voltages ———

(b)

(a)

(e)

(d)

——— Two-speed constant-torque winding ———

Fig. 7·27   Sketches showing the connections for a two-winding five-speed motor for a portable-hoist application.

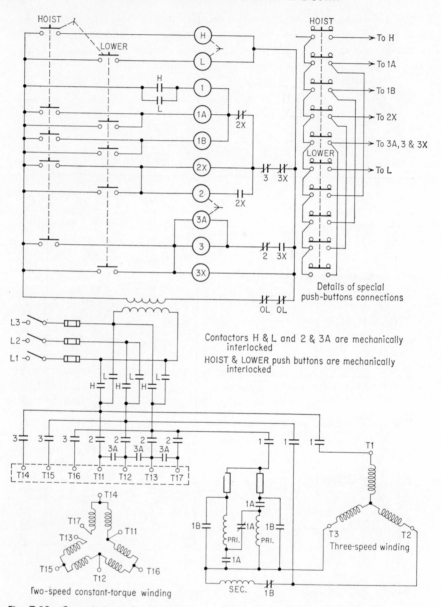

Details of special push-buttons connections

Contactors H & L and 2 & 3A are mechanically interlocked

HOIST & LOWER push buttons are mechanically interlocked

Two-speed constant-torque winding

Three-speed winding

Fig. 7·28 Control circuit for a two-winding five-speed squirrel-cage motor to serve a portable hoist.

switch positions and that these be compared with the resulting circuits of those of Fig. 7·27. Particular attention should be paid to the sequence of the operation of the relays and contactors and to how they close and open the proper contacts in this unique system of control.

Special hoist control for wound-rotor motor. The principle of unbalanced-voltage speed control discussed in the foregoing article for a squirrel-cage motor can be applied equally well to the operation of a wound-rotor motor. In addition, the external rotor resistances can be varied and unbalanced to further control the speed of this type of motor. And when both schemes are employed in a particular system, therefore, it is possible to obtain an extremely wide range of speed control. This was done in a special hoist application in which it was desired to have about eight *hoisting* and *lowering* speeds. As actually worked out, the design provides for eight *hoist* speeds by rotor-resistance change and unbalance and seven *lower* speeds by unbalancing the stator voltages. A drum master is used by the operator to actuate the various contactors and relays, and limit switches are included to avoid overtravel in either direction. A simplified wiring diagram of the system of control, without the drum master, is given in Fig. 7·29, and sketches showing the arrangement of the resistance combinations for the various *hoist* and *lower* positions are represented in Fig. 7·30. Before analyzing these diagrams it should be pointed out that the numbers within the rotor-resistance rectangles indicate the master points at which they are cut out and the letters within the stator-resistance rectangles conveniently identify similar units in both diagrams.

Hoisting. During the hoisting operation, the stator voltages are balanced on *all* master points, since the three winding phases are connected *directly to the three line terminals* through the three closed *M* contacts and the two closed *H* contacts; note that the *M* contactor is energized when the master is moved from the OFF position into the *hoist* direction. The lowest speed is obtained on the first point, since all rotor resistors are in the circuit. The secondary is, moreover, unbalanced to the extent that there are four units in one phase and three units in each of the other two. (It should be recalled that the equivalent resistance per phase, Eq. (6·4), page 176, is the *mean* of the ohmic values in the three phases.) With unit 2 removed on point 2, the secondary is balanced at a lower equivalent resistance; a higher speed therefore results. Continuing through successive points 3 to 8, observe in Figs. 7·29 and 7·30 that resistors are continually being removed until, on point 8, all external resistance is short-circuited. On the latter point the motor will, of course, operate at its highest hoisting speed.

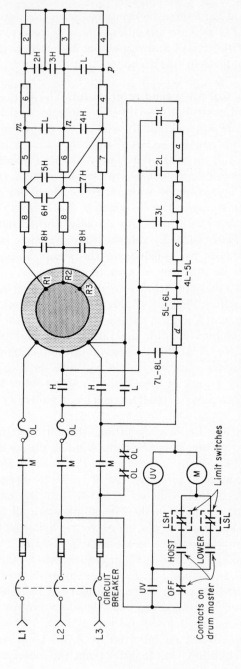

Fig. 7·29   Control circuit for a hoist driven by a multispeed wound-rotor motor.

Lowering. During the entire lowering operation the two *L* contacts in the secondary circuit are closed. This short-circuits the three secondary lines at points *m*, *n*, and *p*, cuts out all resistors to the right of the jumpers, and balances the rotor circuit with two external resistors in each phase. On the primary side, the stator windings are connected to a single-phase source on master points 1 through 4, with varying values of resistance in line *L2*; on master points 5 through 7 to 8, unbalanced 3-phase is

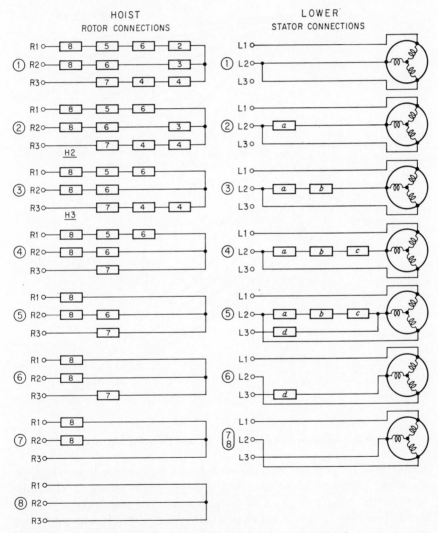

Fig. 7·30 Schematic sketches showing stator and rotor connections for various master switch points of hoist controller of Fig. 7·29.

applied. These successive changes result in varying degrees of positive and negative sequence torques and have the effect of *retarding the motion of the overhauling load* by different amounts. On point 1, for example, maximum retarding torque, i.e., dynamic braking, is exerted and the lowering speed is a minimum. For the higher speed points the dynamic braking action diminishes to permit higher speed operation. For all portions of the speed-torque curves of Fig. 7·31 in the fourth quadrant, the

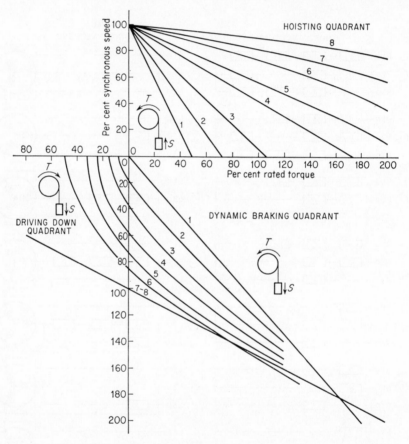

Fig. 7·31   *Speed-versus-torque curves for hoist control system of Fig. 7·29.*

torque will be positive and the motion (downward) will be negative; therefore, dynamic lowering exists. Power torque exists in the third quadrant on master points 2 to 7 and 8 where the empty hook is *driven* down; here, the torque is negative with respect to the hoisting direction and the hook moves down.

QUESTIONS AND PROBLEMS

1. List the various methods that can be used to adjust the speed of a d-c motor. For each method indicate how the speed changes with the adjusted quantity.
2. Distinguish between *constant-torque* and *constant-horsepower* motors.
3. Referring to diagram III-C of Table 7·1, show R_p connected across $R_A + R_{SE}$ and derive equations for the speed S and the armature current I_A.
4. Indicate several difficulties that are frequently encountered when a shunt motor is operated with a weak field at high speed.
5. Why does the regulation of a shunt motor increase when it is operated at high speed with a weak field?
6. What is meant by *commutation limit* and *maximum permissible torque* when referring to the operation of a shunt motor?
7. What is a stabilizing field? When is such a field necessary?
8. In the crane hoist application illustrated by the control circuit of Fig. 7·4, what advantage is gained by operating the series motor with a parallel-connected field during the lowering period?
9. Analyze the control circuit of Fig. 7·4 and carefully check the simple set of schematic sketches given for the lowering operation. Proceed next to draw a similar set of schematic sketches that show the motor connections for hoist positions 1 to 6.
10. List advantages of the Ward Leonard adjustable-voltage speed-control system. Why is it called a constant-torque system?
11. Include field-loss protection in the multimotor adjustable-voltage control system of Fig. 7·8.
12. What important advantage does the two-exciter adjustable-voltage system of Fig. 7·9 have over that of the single-exciter arrangement of Fig. 7·5?
13. Design a control circuit for a 3-phase line-start induction motor that will operate under the following conditions:
    a. There is to be a three-position *selector switch* marked FOR. (forward), FOR.-REV. (forward or reverse), and PLUG-STOP FOR. (plug-stop forward).
    b. With the selector set in FOR. the motor can be operated in a forward direction only. The motor will coast to stop if the STOP button is pressed.
    c. With the selector set in FOR.-REV., the motor can be operated in forward or reverse directions. For either direction of rotation the motor will coast to stop if the STOP button is pressed.
    d. With the selector set in PLUG-STOP FOR., the motor can be operated in a forward direction only. The motor will plug to a stop if the REV. button is pressed; it will coast to a stop if the STOP button is pressed.
14. Distinguish between singly fed and doubly fed motors. Give several examples of each.
15. What are the three *general* speed-control methods for a-c motors?
16. What two *pole-changing* speed-control methods are used for a-c motors? List advantages and disadvantages of each method.
17. In a large machine tool it is desired to operate a constant-torque 50-hp 60-cycle 2-pole 3-phase squirrel-cage induction motor over a speed range of 2,400 to 7,200 rpm. To do this it is decided to employ a variable-frequency power source supplied by an alternator that is driven by an adjustable-speed d-c motor. Make a sketch showing the lineup of required rotating machinery and their ratings for this installation.

18. Distinguish between *conventional* and *consequent-pole* winding connections as they refer to polyphase two-speed single-winding motors. What two general schemes of connection are employed for such windings?

19. Discuss constant-torque, constant-horsepower, and variable-torque two-speed single-winding motors with special reference to winding connections, torque and horsepower relations, and applications.

20. Make a complete winding diagram similar to that given in Fig. 7·19 for constant-horsepower operation. Design the control circuit so that (a) the motor can be started on *fast* or *slow* and (b) when shifting from *slow* to *fast* it is necessary to press the STOP button first and, after a given time delay, a shift can be made to *fast* by pressing the FAST button.

21. Referring to Fig. 7·20, list the various operating conditions under which the motor can be switched to and from *LF*, *HF*, and *HR*.

22. Referring to Fig. 7·21, list the various operating conditions under which the four-speed motor can be switched to and from *L*, *2nd*, *3rd*, and *H*.

23. It is desired to convert a 60-cycle source to 160 cps by using an induction frequency converter. What should be the number of poles in the squirrel-cage driving motor and the frequency converter? What must be the direction of rotation of the rotor of the frequency converter with respect to its revolving field? What frequency will be developed if the relative motion of field and rotor is opposite to that given for the last question?

24. Referring to Fig. 7·24, explain why the frequency of the voltage injected into the rotor of the driving motor by the frequency converter is always the same as that in the rotor conductors of the wound-rotor motor. How is the magnitude of the injected voltage adjusted and how does that voltage affect the speed of the driving motor?

25. Explain why the *Schrage* motor (Fig. 7·25) is, in principle, similar to the control scheme illustrated by Fig. 7·24. List several important advantages of the Schrage motor.

26. Describe the operation of the Kraemer system of control illustrated by Fig. 7·26.

27. Why does the speed of a loaded 3-phase squirrel-cage motor drop when the voltages impressed across the stator terminals are artificially unbalanced. Explain how this is accomplished in Fig. 7·27a, b, and c.

28. Carefully describe the five-speed control circuit of Fig. 7·28.

29. Referring to Figs. 7·29 and 7·30, which represent the control scheme for a wound-rotor motor, carefully check the schematic sketches of Fig. 7·30, hoisting and lowering, with Fig. 7·29 for equivalent master positions 1 to 8.

CHAPTER 8

Control of Single-phase Motors

Types of single-phase motor. Great numbers of motors of comparatively low output ratings are used in single-phase circuits. Most of them, constructed in fractional-horsepower sizes, are technically termed *small motors*, a small motor being one built in a frame smaller than that having a continuous rating of 1 hp, open type, when operating at 1,700 to 1,800 rpm. Many types of such machines have been developed for applications that differ widely, each kind having operating characteristics that meet definite demands. For example, one type performs equally well on direct current or any frequency up to 60 cps; others are designed to develop considerable starting torque and operate at nearly constant speed; another has a commutator and can be readily adjusted over a wide speed range and even reversed by shifting brushes; still another, although not capable of exerting much starting torque, lends itself to speed control, is quiet as well as rugged, and is quite inexpensive.

A special design of small series motor runs with about equal satisfaction when connected to any 115-volt (or 230-volt) d-c or a-c power circuit. Called *universal motors*, they generally operate at high speed and include special design features so that commutation and armature-reactance difficulties, especially on alternating current, are minimized.

The induction principle has been applied advantageously to several types of single-phase motor. Although these machines inherently have no starting torque, several unique ways have been developed to overcome the difficulty; also, a revolving field is created which, it will be recalled, is responsible for the rotation of a squirrel-cage rotor. One scheme is emphasized in the *shaded-pole* motor, an extremely popular machine used in low-starting-torque applications. In another, the so-called *split-*

phase type, the machines are generally larger and are designed in several ways to develop different values of starting torque; mostly, they operate at essentially constant speed, although special constructions are used when speed adjustment is essential.

Repulsion, repulsion induction, and *repulsion-start* motors are other types of single-phase machine that were widely applied until recent years. These machines have been largely replaced by split-phase motors with special capacitors because they can be designed to perform as well as the repulsion types, offering in addition such advantages as lower cost and trouble-free service.

Synchronous motors, as the name implies, operate at synchronous speed for all values of load. There are several constructions of such single-phase machines, although they are generally manufactured in very small ratings. Depending upon the way they are made or the principle of operation, they have special names such as *reluctance* motors, *hysteresis* motors, and *subsynchronous* motors.

Across-the-line starters for single-phase motors. Practically all single-phase motors are designed for line-voltage starting and take inrush currents that may be little more than the rated values in some types and six or more times as much in others. Also, like their larger polyphase counterparts, they are frequently jogged, plugged, reversed, dynamically braked, and plug-reversed. Since the control circuits are not generally burdened by a multiplicity of contactors, relays, and much interlocking, they are, for the most part, simpler in design.

As with polyphase motors, NEMA standards have been established that specify starter sizes and accompanying horsepower ratings of motors. Two important listings have been adopted for different kinds of service, and these are given in Table 8·1. It should be noted that larger motors can be used with given size starters as the voltage ratings increase; this

Table 8·1 Standard NEMA horsepower ratings of single-phase across-the-line magnetic starters

Size of starter	Nonplugging or nonjogging duty			Plug-stop, plug-reversing, or jogging duty		
	115 volts	230 volts	$\frac{440}{550}$ volts	115 volts	230 volts	$\frac{440}{550}$ volts
00	$\frac{1}{2}$	$\frac{3}{4}$	$\cdots$	$\frac{1}{4}$	$\frac{1}{3}$	
0	1	2	3	$\frac{1}{2}$	$\frac{3}{4}$	$\frac{3}{4}$
1	2	3	5	1	2	3
2	3	$7\frac{1}{2}$	10	2	5	$7\frac{1}{2}$
3	$7\frac{1}{2}$	15	25	5	10	15

is because full-load currents diminish proportionately as the voltages are raised and starter contacts are designed for maximum current values. Moreover, larger than normal size starters must be used when the motors are subjected to the more severe service of plug-stopping, plug-reversing, and jogging, since the latter operations tend to cause overheating of contactors and excessive burning and wear of contacts.

The universal (series) motor. The d-c series motor will operate if connected to an a-c source of power because the direction of the torque is determined by *both* the field polarity and the direction of the current through the armature. Since the same current passes through the field coils and the armature winding, it should be clear that the a-c reversals from positive to negative, and vice versa, will affect flux and armature-conductor directions. Note particularly that this type of motor, unlike the induction machine, which depends upon a revolving field to develop torque by induced currents, produces torque by the interaction of a stationary field and currents that are conducted directly to the armature winding. The latter is connected to a commutator to maintain a required stationary current pattern on the armature so that its field will always be in quadrature with the one produced by the field winding.

Special consideration must be given to the construction of the series motor if it is to operate satisfactorily on alternating current as well as direct current. One important requirement is that the field structure be completely laminated to avoid losses due to eddy currents. Another involves a means of combating the bad effects of armature reaction and the resulting poor commutation, and this is usually accomplished by designing the armature so that the volts between adjacent commutator segments are somewhat lower than in equivalent d-c motors. The reason for the more difficult commutation is that a coil undergoing commutation, i.e., when it is short-circuited, is linked by the total alternating flux of one pole and this linking has the effect of inducing a substantial emf in the coil. Higher resistance brushes generally help to improve operation as does the more expensive method of employing special distributed field and compensating windings that are placed in a slotted stator core. A further improvement is made by *skewing* the armature slots and teeth so that the armature conductors are not parallel with the shaft; this helps to keep the motor quiet and eliminates the tendency on the part of the armature to lock when started.

Although the performance characteristics of this so-called *universal* motor are generally similar in shape whether operated on direct or alternating current, the d-c and a-c *speed-versus-torque* curves do not ordinarily coincide. In some motors, the a-c speeds will be higher than the d-c speeds, while in others, the reverse is true; in still other designs, the curves cross between no-load and full load, so that the a-c speeds are higher than the d-c speeds in one load range and lower in another load range. These dissimilar operating conditions result because the speed of the motor depends on several conflicting factors. On direct or alternating current, *armature- and field-resistance drops* tend to reduce speed when load is applied, and *armature reaction* acts to reduce the flux and raise the speed with increasing load. However, when alternating current is used, a third factor, namely, *armature reactance drop*, exerts a speed-lowering effect with increased loading. Whether or not the actual speed for a given load is higher on alternating current than on direct current depends upon which of the foregoing factors change more. If the reactance voltage drop increases more than the air-gap flux decreases, the motor will tend to operate at a lower speed when alternating current is used; otherwise the reverse is true. Also, at the higher loads the iron is saturated, so that the effective flux per a-c ampere is less than that produced per d-c ampere. The result of this condition is a tendency for the motor to run faster on alternating current. Thus, if it were not for the fact that the reactance voltage drop acts oppositely with respect to the effect of armature reaction, the a-c speed would actually be greater than the d-c speed when the motor is loaded. Thus, the factor which predominates determines, in general, whether or not the a-c speed will exceed the d-c speed. In a good design, the reactance voltage drop is kept fairly low, and the difference in flux produced by direct and alternating current is reduced to a minimum by operating at low flux densities. The latter condition, i.e., a weak field motor, is gained at the expense of poorer commutation, correction for which must be made by choosing a proper grade of brush or by using compensating windings.

As with all series-type motors, the speed varies widely with load, and starting and overload torques are good. Also, the fact that the speed of the machine adjusts itself automatically to the magnitude of the load is an advantage in some applications.

Reversing and speed control of universal motors. The direction of rotation of this type of motor is changed (1) by interchanging the field terminals with respect to those of the armature or (2) by using two field windings, wound on the core in opposite directions so that one of them connected to the armature gives clockwise rotation while the other in series with the armature gives counterclockwise rotation. Schematic diagrams showing how this is done are given in Fig. 8·1. Since the two-field method simplifies wiring connections somewhat, it is frequently used in control circuits for such applications as motor-operated rheostats and servo systems. A diagram for such an arrangement is illustrated by Fig. 8·2. Its operation should be clear from previous studies of similar circuits.

Several schemes are employed to adjust the speed of a universal motor. One of these, the *line rheostat* method, merely incurs a voltage drop that results in a reduced voltage across the motor terminals. Table 7·1, I-*A*, p. 207, indicates how speed adjustment is made and that a rather wide speed range is possible. Another less common procedure is to provide

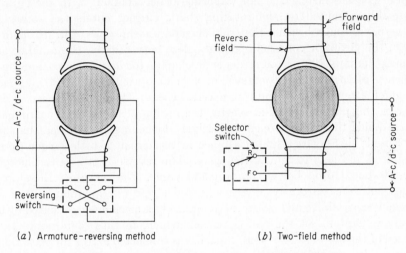

(*a*) Armature-reversing method   (*b*) Two-field method

Fig. 8·1   Two methods for reversing a universal motor.

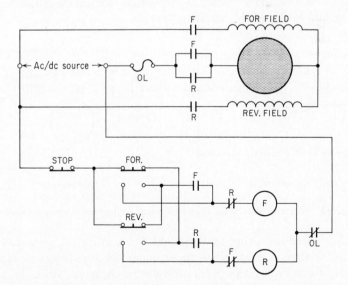

Fig. 8·2   Control circuit connected to a two-field reversing universal motor.

a brush-shifting mechanism, usually hand operated, that raises the speed of the motor as the brushes are moved backward, i.e., against the direction of rotation. Limited speed range is possible by this means because commutation worsens as the brushes are moved farther from the magnetic neutral.

A popular method is to equip the series motor with a specially constructed governor. It consists essentially of a disc of phenolic material upon which is fastened a pair of spring-loaded contacts with the entire assembly mounted on the rotating shaft; current enters the rotating governor through brushes that ride on separate slip rings. During operation, the governor contacts open and close very rapidly, the number of times, perhaps 100 or more, depending upon the natural resonant frequency of the moving contact. For a given spring-tension setting, the contact attempts to vibrate at a certain rate. Then if the speed rises above the particular value set by the spring tension, the centrifugal force holds the contact open a relatively longer period of time than it is closed; this keeps a line resistance in the circuit a little longer than required and acts to reduce the speed. The reverse is true if the motor speed should drop below the adjusted value. The machine, therefore, does not run constantly at the adjusted speed but operates within a narrow range above and below this speed. A link mechanism actuated by a lever is used to set the spring tension which, in turn, determines the speed of the motor; also a small capacitor is connected across the vibrating contact in parallel with the resistor to prevent excessive arcing. Figure 8·3 illustrates how the governor is connected to a series motor for non-reversing and for two-field reversing service.

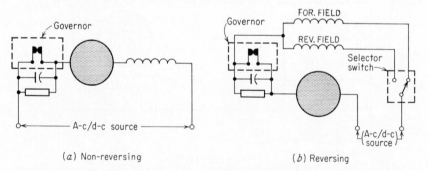

(a) Non-reversing   (b) Reversing

Fig. 8·3 Governor-controlled speed-adjusting method for a universal motor.

Another speed-adjusting arrangement for a universal motor involves the use of a tapped field. Since these machines are always bipolar, the number of turns on the two poles need not be the same, since the air-gap flux is created by any series combination of mmfs. As Fig. 8·4 shows, the field coil that has the larger number of turns is tapped at three points, so that four operating speeds are possible. For a given value of load current, therefore, minimum speed will be obtained when the entire winding is used (this gives maximum mmf and flux) and maximum speed will result on point *H* under which condition minimum mmf and flux are produced. The scheme also offers the possibility of tapping the field at appropriate points to permit the motor to run at the same speed on direct current and alternating current for a particular input current. This does not, however, imply that the d-c and a-c speeds will be the same at some other current.

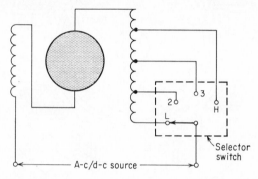

Fig. 8·4   Tapped-field winding speed-adjustment method for a universal motor.

The shaded-pole motor. This type of motor, one of the simplest and cheapest to manufacture, is an induction machine provided with an auxiliary short-circuited winding or windings displaced in magnetic position from the main winding. In its most usual construction it consists of a set of salient poles with a main coil surrounding the whole of each one and a short-circuited copper strap around a portion of the pole. In some designs two sets of shading coils can be used for the purpose of reversing the motor or, when two or three shading coils surround different percentages of the main pole, to obtain desired output characteristics.

Being an induction motor with a squirrel-cage rotor, the latter receives power in much the same way as does the rotor of a polyphase induction motor. There is, however, one extremely important difference between the two types, and this concerns the motion of the magnetic field. Whereas the polyphase induction motor with its slotted stator and distributed winding creates a revolving field that is constant in magnitude and rotates at synchronous speed *completely around the entire core*, the field of the shaded-pole motor varies in magnetic strength and merely *shifts from one side of the salient pole to the other*. The torque of such motors, generally made in very small sizes, from tiny midgets up to about 0.15 hp, is therefore not uniform but varies from instant to instant.

Figure 8·5 illustrates the construction of the shaded-pole motor where each of the laminated poles of the stator iron has a slot cut crosswise

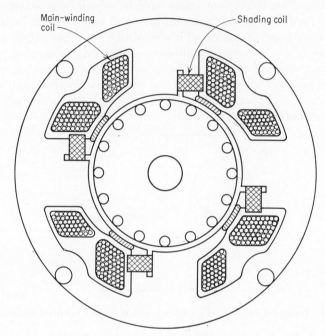

Fig. 8·5   Typical 4-pole shaded-pole motor.

about one-third from one edge. Around the smaller of the two areas formed by this slot is placed a heavy copper shading coil, and an insulated exciting coil surrounds the entire pole. Note that the two windings, the main coil and the short-circuited shading coil, are displaced in space; also, since the induced shading-coil current lags behind the main-winding current, their fluxes are displaced in time. The conditions for a moving field are therefore established, although the ideal 90-electrical-degree relationships of space and time displacements are not attained.

Motors with a single set of main and shading coils are not reversible but always rotate in a direction from the unshaded to the shaded part of the pole. For reversing service the construction of the stator must be changed to include (1) two sets of shading coils that are normally open-circuited and occupy different positions with respect to the main winding, (2) two main windings that occupy different positions with respect to the shading coils, or (3) a special distributed continuous winding that is placed in a slotted stator core and is tapped at appropriate points with respect to the shading coils. In the first of these, Fig. 8·6a, shading coils are placed on both sides of each salient pole; those on one side are then connected in

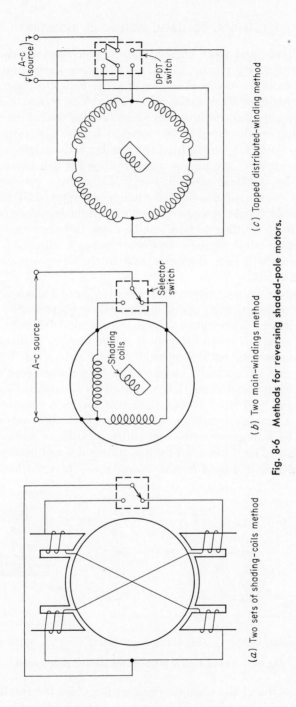

(a) Two sets of shading-coils method

(b) Two main-windings method

(c) Tapped distributed-winding method

Fig. 8·6   Methods for reversing shaded-pole motors.

269

one series circuit, and those on the opposite pole sides are connected in series in another. By means of a short-circuiting switch, either set of shading coils can be made effective, and the coils that carry induced currents determine the direction of rotation. The second scheme, Fig. 8·6b, requires a slotted stator core in which two distributed main windings are placed that are displaced by 90 electrical degrees with respect to each other and 45 electrical degrees from the shading coils. A selector switch is then used to energize the proper main winding for the desired direction of rotation. By the third method, Fig. 8·6c, the stator core must also be slotted and a continuous distributed winding is employed. With the latter tapped at 90-electrical-degree points and the terminals connected to a dpdt switch, the main winding sections can be made to have either of two 90-electrical-degree relationships with respect to the shading coils. In one of these the rotor will turn clockwise, and in the other counterclockwise rotation will result.

The impressed voltage across the main winding of a shaded-pole motor greatly affects its operating speed under load; in other words, the slip of the rotor increases as the motor voltage is lowered. Advantage is taken of this characteristic to control the speed of the motors, particularly when they drive air-moving equipment such as fans or blowers. Three fundamental schemes are generally employed to vary the *winding voltage:* (1) by using an autotransformer that is tapped for several voltages, (2) by using a tapped reactance (choke) coil that incurs a line-voltage drop, (3) by using a tapped exciting winding, so that the constant-voltage source can be impressed across the entire winding or some part of it. Methods 1 and 2 are illustrated by Fig. 8·7a and b and need no explanation. By method 3, Fig. 8·7c, the lowest speed occurs when the entire

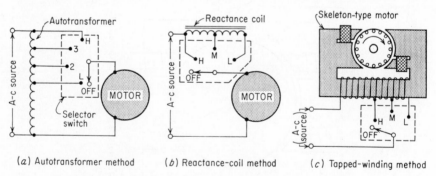

(a) Autotransformer method   (b) Reactance-coil method   (c) Tapped-winding method

Fig. 8·7   Speed-control methods for shaded-pole motors.

winding is used and the highest speed results when the smallest portion of the winding is energized. The reason for this relationship is that the *volts per turn,* which is a measure of the flux and torque, increase as fewer turns are employed. Thus, for example, a higher value of volts per turn is accompanied by more developed flux and torque and therefore increased speed.

Split-phase motors. One of the most widely used types of single-phase motor, made in larger sizes than are those of the shaded-pole construction, is the *split-phase* motor. Fundamentally an induction machine and having operating characteristics that are similar to the usually larger polyphase motor, its field of service includes an extremely wide variety of applications, several of which are refrigerators, washing machines, portable hoists, many kinds of small machine tools, grinders, blowers and fans, woodworking equipment, centrifugal pumps, and many others. When these motors are in operation, a revolving magnetic field that is almost constant in magnitude acts upon a squirrel-cage rotor in much the same way as in a 2-phase motor. The stator core resembles its polyphase counterpart but carries two windings (sometimes three or more for multi-speed operation), one of which, the main winding, is energized continuously and the other, the auxiliary winding, is excited only during the starting period in most types and continuously in some.

A single winding energized by a single-phase source creates a pulsating magnetic field, not a revolving field, when the squirrel-cage rotor is at rest; the motor, therefore, develops no starting torque under this condition. To do so, it is necessary that there be at least two windings, and these must be displaced physically on the stator core and carry currents that are mutually out of phase. Since an attempt is generally made to have the split-phase motor develop good starting torque, the two windings are separated on the core by 90 electrical degrees and the time displacement of the currents is made as much as possible, up to 90 electrical degrees. In the basic design, the *main* and *auxiliary* windings, as they are called, have different ratios of inductive reactance to resistance and, being connected in parallel to magnetize the core, the line current is *split* into two parts so that a time displacement of 30 or more electrical degrees exists between them. It is from this *current-splitting* action that the name *split-phase* motor is derived.

To provide different starting-torque and mechanical (noise) performance characteristics, two general classifications of split-phase motor are available. These are (1) the *standard* split-phase motor and (2) the *capacitor* split-phase motor. Moreover, three special types of capacitor motor are used to extend further the range of usefulness of this excellent machine; they are classified as *capacitor-start*, *permanent-split capacitor*, and *two-value capacitor* motors. All of them are wound similarly, with so-called spiral windings, and differ only by whether or not a capacitor is connected in series in the auxiliary-winding circuit and the way in which the latter is treated after the accelerating period when the motor is running normally.

As previously mentioned, starting torque is developed only when the main and auxiliary windings, connected in parallel, are energized by the single-phase source. In the *standard* construction (with no capacitor), the motor develops nearly as much running torque on the main winding alone at a rotor speed of about 75 per cent of rated value as with both windings acting together. And, when normal operating speed is reached, the two-winding torque drops below that produced by the energization of the main winding only. This is because the rotor mmf exerts a strong influence, together with the main winding, to maintain a revolving magnetic field of nearly constant magnitude. It is logical, therefore, to provide a switching mechanism, actuated by a centrifugal or electromagnetic device, that open-circuits the auxiliary-winding circuit at the proper time. Many kinds of mechanism as well as timing relays and thermal devices are used for this purpose. A sketch illustrating how a standard split-phase motor is connected, with a centrifugally operated switch in the auxiliary-winding circuit, is given in Fig. 8·8. The fact that the auxiliary

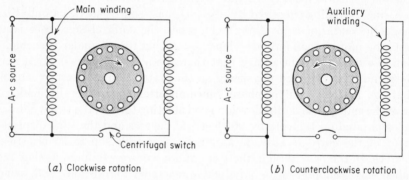

(*a*) Clockwise rotation   (*b*) Counterclockwise rotation

Fig. 8·8 Sketches showing a split-phase motor connected for two directions of rotation.

winding is energized only during the short accelerating period makes it possible to reduce its wire size considerably without causing an excessive temperature rise; this means that the saved space gained by reducing the volume of auxiliary winding copper can be utilized to advantage by increasing the amount of *full-time* main-winding copper. Note also in Fig. 8·8 that the direction of rotation of this type of machine is reversed by interchanging the main- and auxiliary-winding terminals. This is because the field revolves in a direction from a given auxiliary-winding pole to a main-winding pole of the same polarity. Obviously, the motor must be brought to rest or the centrifugal switch must be closed before reversal is attempted.

Split-phase motors are generally started at line voltage, under which condition the inrush current is four to six times rated value. Advantage is taken of this high starting current in connection with the use of an electromagnetic type of relay in hermetic motors. Centrifugal switches cannot be employed in the latter machines. The relay, Fig. 8·9, has a coil that is connected in the main-winding circuit and a pair of contacts, normally open, that is in series with the auxiliary winding. When the motor is started and the main-winding current is high, the contacts close to permit energization of the auxiliary winding. After the motor reaches the proper speed, the main current drops to a value that is low enough to cause the contacts to open. Such relays are mounted outside the sealed motor units, where they can be serviced or replaced if necessary.

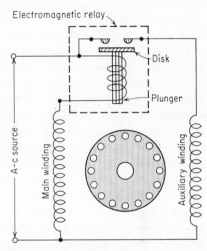

Fig. 8·9 Electromagnetic relay connected in a split-phase motor circuit.

Speed control of these machines is accomplished with some difficulty. The usual methods are to use two or more windings designed to produce different numbers of poles and switching schemes that involve the consequent-pole and conventional techniques common to multispeed polyphase motors. The fact that auxiliary as well as main windings must be arranged for pole-changing adds to the complications. Better adjustable-speed results are obtained with capacitor-type split-phase motors.

Sketches of the three variations of capacitor motors are illustrated by Fig. 8·10. Except for the addition of capacitors and a slightly modified

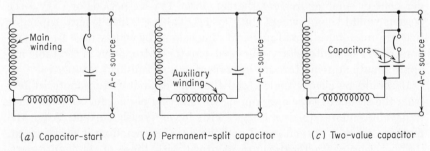

(a) Capacitor-start   (b) Permanent-split capacitor   (c) Two-value capacitor

Fig. 8·10   Three types of capacitor split-phase motor.

switching arrangement for the two-value capacitor type, they are generally similar to the standard construction. It should not be inferred from this, however, that a particular motor can be converted from one type to the other. The winding design of each kind receives special treatment, and this makes it impractical to modify a given machine for a different service.

Capacitor-start split-phase motors. The starting torque developed by a split-phase motor, given by the equation $T_{ST} = kI_M I_A \sin \alpha$, depends upon the inrush currents to the main and auxiliary windings I_M and I_A, respectively, and a function of the angular displacement between these currents, i.e., $\sin \alpha$. Since the angle α is usually between 30 and 40° in the standard type of motor (Fig. 8·8), it is the last term in the formula, in addition to permissible values of I_M and I_A, that limits the maximum starting torque to about $1.5T_{FL}$. However, if a capacitor of the proper size is inserted in the auxiliary-winding phase, its current can be made to lead the main-winding current by as much as 90°, in which case improved starting conditions will result. Moreover, the line current, which is the vector sum of I_M and I_A, diminishes as the angle is widened to 90°, so another advantage results from this change. It is, in fact, possible to more than double the starting torque this way (it may be as much as or more than $3.5T_{FL}$) with a reduction in inrush current of about 40 to 50 per cent. To accomplish this improvement, capacitors of rather large microfarad values must be used. Fortunately, electrolytic types of these units are available in comparatively small canlike structures when they serve for short-time starting duty only. An important aspect of the method of starting is that the voltage across the capacitor increases very rapidly above the normal operating speed of the centrifugal switch. This means that the capacitor and/or the auxiliary winding will be damaged if the switch fails to operate after the motor comes up to speed and remains in a closed position for an appreciable time. Standard recommendations for electrolytic-capacitor service are on the basis of twenty 3-sec periods per hour or some equivalent thereof; forty 1½-sec periods or sixty 1-sec periods would be equivalent duty cycles. Table 8·2 lists typical capacitor sizes (in microfarads) and physical dimensions for widely used fractional-horsepower capacitor-start split-phase motors; also given in the table are approximate starting torques in percentages of full-load torques.

It should be clearly understood that the capacitor-start motor differs from the standard type only during the starting period, for it is then that it develops more starting torque with lower inrush current. After the machine reaches normal speed and the auxiliary winding is deenergized, its operating characteristics are identical with those of its counterpart. Such motors are therefore used in applications which require high starting torque but have starting periods of short duration. The latter point is significant, since electrolytic capacitors are, as previously stated, designed for short-duty service only and will be destroyed if the starting load does not permit the motor to come up to speed rapidly. (Where the starting service is both severe and sustained, it is generally necessary to use a repulsion-start motor.) Centrifugal switches and relays that function in auxiliary-winding circuits must, moreover, be positive in their action, i.e., contacts must not be permitted to flutter; should the latter happen, it is possible for the capacitor voltage to be twice rated value for an

Table 8·2   Typical capacitance values for 60-cycle
capacitor-type split-phase motors

Hp	Poles	Rpm	Approximate capacitance values, μf	Approximate per cent starting torque	Dimensions of capacitors	
					Diam, in.	Length, in.
1/8	2	3,450	80	350–400	1⅜	2¾
	4	1,750		400–450		
	6	1,140		275–400		
1/6	2	3,450	100	350–400	1⅜	3¼
	4	1,750		400–450		
	6	1,140		275–400		
1/4	2	3,450	135	350–400	1⅜	3¾
	4	1,750		400–450		
	6	1,140		275–400		
1/3	2	3,450	175	350–400	1⅜	4
	4	1,750		400–450		
	6	1,140		275–400		
1/2	2	3,450	250	350–400	2	3⅛
	4	1,750		400–450		
	6	1,140		275–400		
3/4	2	3,450	350	350–400	2	4
	4	1,750		400–450		
	6	1,140		275–400		

instant, in which event the capacitor will fail. The double voltage may result if the capacitor is fully charged at the instant the switch opens and then closes on the a-c voltage wave when it is a maximum in the opposite direction.

Permanent-split capacitor motors. These motors are used principally for low-starting-torque loads where they are generally shaft-mounted. Their particular fields of application are air-moving equipment such as fans and blowers, oil burners where quiet operation is desirable, and electrical installations like furnace and arc-welding controls, valves, rheostats, and induction regulators where speed-adjustment and plug-reversing must be simple. As Fig. 8·10*b* shows, no switching device is needed in the auxiliary-winding circuit, since the winding and its accompanying oil-filled capacitor are both designed for continuous-duty operation.

A unique feature of this type of motor is that it can be easily reversed *if main and auxiliary windings are identical.* It was previously pointed out that the direction of rotation of all split-phase motors is always from a given auxiliary-winding pole to an adjacent main-winding pole of the same polarity. If, therefore, an spdt switch is provided as in Fig. 8·11

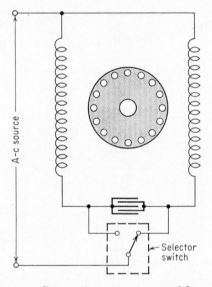

Fig. 8·11   Permanent-split capacitor motor connected for reversing service.

so that the capacitor can be connected in series with one or the other of the *similar* windings, either one can be made to serve as the auxiliary or main winding. The direction of rotation is thus determined by their relative positions.

The speed-torque characteristic of this motor, frequently called a *single-value capacitor* motor, shifts considerably with changes in applied voltage. These relationships are illustrated by the three curves of Fig. 8·12, which are drawn for different values of voltage. Advantage is taken of this basic variation in applying the motor to the operation of multispeed unit heaters. If a normal fan is used, its load curve intersects the motor characteristics at *l, m,* and *h,* which means that these points indicate the *low, medium,* and *high* speeds at the three voltages. Likewise, if a light fan or a heavy fan is used, the operating points will be, respectively, l', m', h', and l'', m'', h''. Voltage-adjusting schemes to control the speed of this type of motor are discussed in a subsequent article.

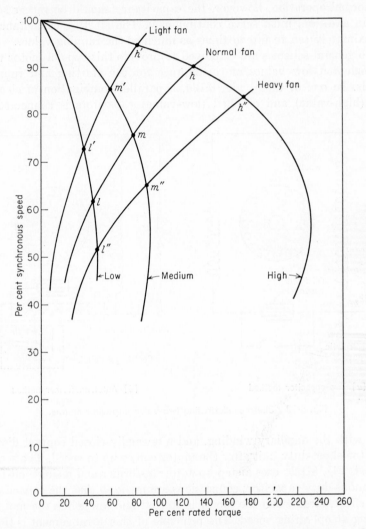

Fig. 8·12   Curves showing how the speed of a permanent-split capacitor motor is adjusted.

Two-value capacitor motors. This type of capacitor motor has the advantage that it develops extremely high locked-rotor torque with one value of capacitance in the auxiliary-winding circuit and approximates the quiet running performance of the 2-phase motor with another capacitance. If the latter condition is not particularly important to the application, the less expensive capacitor-start motor is generally used. As Table 8·2 indicates, comparatively large electrolytic capacitors are needed for high-starting-torque service and these are, moreover, short-duty devices. For normal operation, however, the capacitance should be rather small and, in addition, must serve continuously. The starting capacitance is approximately ten to fifteen times as much as the running value.

Two general schemes are employed to provide this type of motor with the high and low values of capacitance for the starting and running periods. In one of these, Fig. 8·13a, a parallel combination of electrolytic (high-value) and oil-filled (low-value) capacitors is connected in

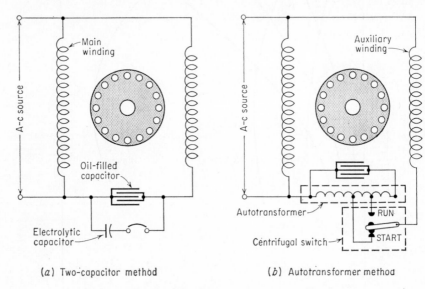

(a) Two-capacitor method   (b) Autotransformer method

Fig. 8·13   Sketches illustrating two-value capacitor motors.

series with the auxiliary winding, and a normally closed contact disconnects the short-duty unit after the motor comes up to speed. The second method, Fig. 8·13b, uses a step-up autotransformer and a single oil-filled continuous-duty capacitor; connections are made so that the capacitance is, in effect, large when the motor is started and low while the motor is running at operating speed. The principle of this arrangement is that a capacitor of value C, connected to the secondary terminals of a step-up transformer, appears to the primary as having a value of a^2C, where a is the ratio of transformation. For example, a 5-μf capacitor connected to the high-voltage secondary of a transformer with a ratio of 6 to 1 would appear to the primary as a $5 \times 36 = 180$-μf unit.

The continuous use of the auxiliary winding with its capacitor during the running period, in contrast to its short-duty service in the capacitor-start motor, improves the performance of the machine somewhat. One effect is to increase the breakdown torque by about 10 to 25 per cent. In addition, double-frequency torque pulsations that tend to cause wrong operation, common to all split-phase motors, are reduced. The motor thus attempts to simulate 2-phase operation, although best results are obtained at a particular load with the given value of capacitance. Design analyses have, indeed, been made that indicate how to proportion a winding and its capacitor to give 2-phase performance at some desired load point.

Starters for capacitor-type split-phase motors. The starting of a small split-phase motor is generally a simple matter when the contacts in the auxiliary-winding circuit are actuated by internally mounted centrifugal switches, electromagnetic relays, or other devices that are placed outside the motor. Proper timing for the contacts is then provided by spring-tension adjustment, thermostats, or current changes during the accelerating period.

Figure 8·14 illustrates one control-circuit arrangement for a two-value capacitor motor. Here, the unit that operates to disconnect the elec-

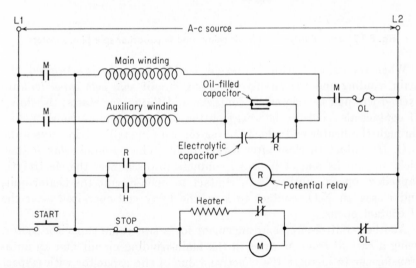

Fig. 8·14   Two-value capacitor motor connected to a starter using a bimetallic timer and a potential relay.

trolytic capacitor is a *heater timer*, a comparatively simple and inexpensive device. Similar to any common thermostat, it consists of a bimetallic strip around which a heater coil is wrapped; in some designs the current is made to pass directly through the bimetal itself. Since the action of such a timer to close and open a contact is sluggish, it is customary to include an electromagnetic relay like that of Fig. 8·15 to provide quick response.

Fig. 8·15   Potential-type magnetic relay used in capacitor split-phase motors.

When the START button is pressed, the M contactor is energized, the main winding and the auxiliary-winding circuit *with both capacitors* are excited through the closed M contacts, and the motor starts; the upper M contact also seals in the START button. Current simultaneously passes through the heater coil which, acting on the bimetallic strip, causes the N.O. H contact to close after a time delay. The *potential relay R* then picks up, opens one of its N.C. contacts to disconnect the electrolytic capacitor, opens another N.C. contact to open-circuit the heater unit, and closes an N.O. contact to keep the relay coil energized after the H contact opens.

Another control-circuit arrangement for a two-value capacitor motor, using a *current relay* in series in the main-winding circuit and an autotransformer to increase the effective value of the capacitor with respect to the auxiliary-winding circuit, is given in Fig. 8·16. The current relay is designed to pick up at about three times the full-load current rating of the winding and to drop out at approximately twice that value. Thus, when the main winding is energized by the closing of the M contacts, the high inrush current actuates the current relay, which, in turn, opens its N.C. contact and closes its N.O. contact. The auxiliary winding is therefore excited through the "stepped-up" capacitor and, developing high locked-rotor torque, the motor starts. As the machine accelerates, the current falls, and at the dropout point of the relay, the two contacts assume their normal positions. Under this condition the autotransformer is open-circuited and the auxiliary winding is established through the capacitor only. Note that the relay remains in the circuit continuously and can pick up again should the load increase sufficiently to raise the current to the pickup point, assuming, of course, that the overload

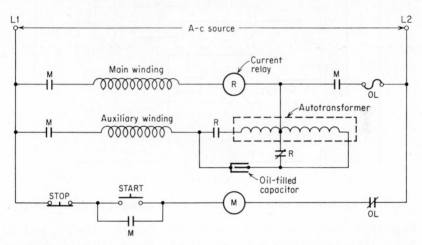

Fig. 8·16   Two-value capacitor motor connected to a starter using a current relay and an autotransformer.

does not trip in the meantime. This is not especially undesirable, since the increased capacitance permits the motor to develop considerably more torque and prevents stalling at the heavier load conditions.

Dynamic braking of split-phase motors. The principle of dynamic braking previously discussed in connection with polyphase motors (Chap. 6 and Fig. 6·35) applies equally well to split-phase motors. As explained, the procedure is to disconnect the stator winding from the a-c lines while the motor is running and substitute a suitable d-c power source, usually through a full-wave rectifier. Proper precautions must, of course, be taken by electrical and/or mechanical interlocking to avoid the possibility that direct and alternating current will simultaneously excite the winding. Also, rheostatic control is usually provided in the d-c circuit to adjust the timing of the stop; this merely determines how rapidly the stored energy in the rotating system is converted to electrical energy and is dissipated.

The control-circuit diagram of Fig. 8·17 illustrates how dynamic braking can be applied to a split-phase motor. In addition to the outside

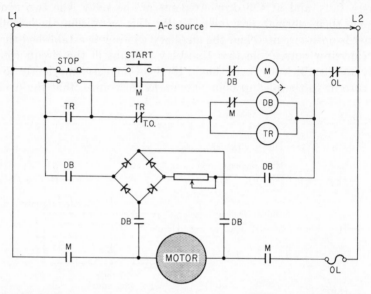

Fig. 8·17   Single-phase motor connected to a starter arranged for dynamic braking.

loop consisting of the START-STOP station, the M contactor, and the motor, there is an inner section, comprising a dynamic-braking contactor DB, a timing relay TR, and a bridge rectifier. The motor will coast to rest if, while running normally, the STOP button is touched lightly, i.e., only enough to open the back contact and without making the front contact. However, when the STOP is pushed down to bridge the front terminals, the dropping out of the M contactor is immediately followed by the energization of the DB contactor and the TR timing relay. The N.O. TR contact then closes to seal in the actuated units, and four DB contacts close to permit direct current to excite the stator winding through the rectifier. After the motor comes to a stop, the rapidity of which is adjusted by the rheostat, the TR-T.O. contact opens to deenergize the DB and TR coils. The N.C. electrical interlocks DB and M and the mechanical interlock between M and DB are included in the circuit to prevent direct current and alternating current from exciting the winding at the same time.

Plug-reversing of capacitor-start motors. Although the conventional capacitor-start motor develops high locked-rotor torque, it suffers from the disadvantage that it is not electrically reversible. For most general-purpose applications which operate in one direction only, this is not particularly objectionable, but for such service as small portable hoists the motor must be made *instantly reversible* by properly designing the control circuit. It will be recalled that the direction of rotation of a split-phase motor is determined by the relative positions of the main- and auxiliary-winding circuits with respect to one another and that, with the centrifugal switch contact open, the motor runs on the main winding only. This means, of course, that a standard motor cannot be plugged to a stop and reversed until the auxiliary-winding circuit is reestablished electrically, to accomplish which an externally operated relay must be used; the centrifugal-switch contact does not close until the motor slows down almost to a stop.

One circuit arrangement, Fig. 8·18, that overcomes the limitation indicated and provides for reverse-plugging requires a special two-contact

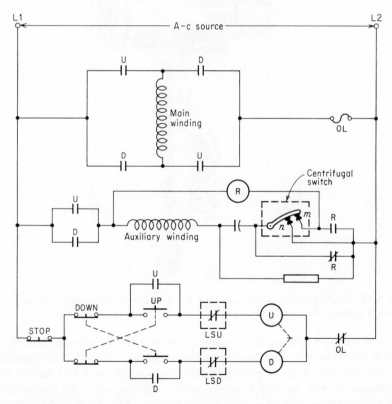

Fig. 8·18 Capacitor-start motor for a portable hoist connected to a controller arranged for plug-reversing.

centrifugal switch and an electromagnetic relay. Included also are mechanically operated *limit switches* each of whose N.C. contacts open when the hoist reaches the upper or lower limit of travel and disconnects the control circuit. For hoisting, the UP button is pressed. This energizes the main winding in one direction with respect to the auxiliary winding, and the latter with its capacitor is connected to the line through contact n (or N.C. contact R); also relay R picks up through centrifugal switch contacts m and n to open an N.C. contact and close an N.O. contact

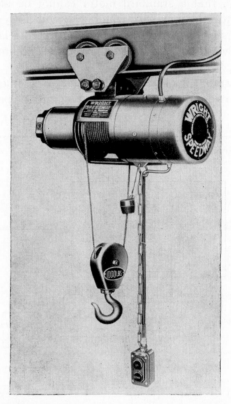

Fig. 8·19 Small portable hoist operated by a capacitor-start motor whose control circuit is designed for plug-reversing.

for sealing purposes. After the motor speeds up sufficiently and the m and n contacts open the auxiliary-winding circuit, the latter is connected to the line through a comparatively high resistance; the motor thus continues to run basically on the main winding only. To plug-reverse the motor the DOWN button is pressed. When this is done, the U contactor is deenergized first, its contacts open, and, at the same time, the relay drops out. Since the N.C. R contact closes, the auxiliary-winding circuit is excited again when the D contactor picks up and closes its contacts. The motor now plugs to a stop because the terminal connections of the two windings are interchanged with respect to each other. As the rotor slows down toward zero, the centrifugal-switch contacts reclose, relay R picks up again, and the motor proceeds to speed up in the opposite direction. A photograph illustrating a portable hoist like the one described is shown in Fig. 8·19. Note the pendant-mounted push-button station at a convenient level to the floor operator.

Another control-circuit arrangement for a portable-hoist application in which a reversing master is used with a capacitor-start motor is given in Fig. 8·20. Here a three-position (H-OFF-L) 3-pole master is operated

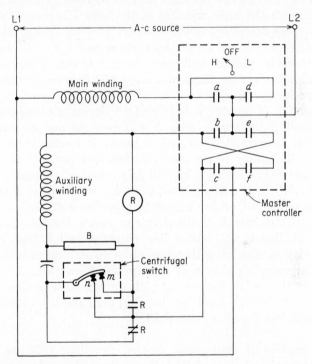

Fig. 8·20 Capacitor-start motor for a portable hoist connected to a master controller arranged for plug-reversing.

by a set of HOIST and LOWER cords, and plug-reversing is effected by merely pulling the LOWER cord if the motor is hoisting or the HOIST cord if the motor is lowering. As in Fig. 8·18, a special two-contact centrifugal switch and a potential relay are employed. For hoisting, the HOIST cord is pulled and a lever closes contacts a, b, and c. The closing of the a contact connects the source to the main winding, and current passes from left to right through the latter from $L1$ to $L2$. Two other circuits are also established by the closing of contacts b and c. These are (1) through the auxiliary winding and its capacitor in an *upward* direction and contact n (or N.C. contact R) and (2) through relay R and contacts n and m. When relay R picks up to open its N.C. contact and close its N.O. contact, it remains energized but is independent of the centrifugal-switch contacts m and n. The motor now proceeds to accelerate until, at a preset speed, contacts m and n open.

To plug-reverse the motor for lowering, the LOWER cord is pulled. This opens contacts a, b, and c *first* and *then* closes contacts d, e, and f. Relay R therefore drops out during the transition from HOIST to LOWER and causes its contacts to assume their normal positions. Current continues to pass through the main winding in the *same* direction (from left to right) but flows downward in the auxiliary winding through the N.C. R contact. (The centrifugal switch contacts m and n are still open.) The motor is therefore plugged and slows down. Eventually contacts m and n close, relay R picks up, the motor stops and then accelerates properly in the opposite direction.

The function of resistor B is important during a momentary power failure. Should this happen, relay R will drop out and its contacts will again revert to normal. Then, when power is reestablished while the motor is still running *in the same direction* and contacts m and n are open, the auxiliary winding is immediately energized through the N.C. R contact. If resistor B were omitted, the relay could not pick up and the auxiliary winding and/or the electrolytic capacitor would fail very quickly. However, with resistor B, in the circuit as shown, the voltage across the auxiliary winding is well above line potential and is high enough to energize relay R. The latter therefore picks up and instantly disconnects the auxiliary-winding circuit at the opened R contact. (It should be pointed out that relay R does *not* pick up when the auxiliary-winding connections are interchanged for plugging; the auxiliary-winding voltage is then much less than line potential.)

Speed control of split-phase motors. Standard split-phase motors (without capacitors) do not lend themselves to speed adjustment by voltage control and are, in this respect, similar to polyphase squirrel-cage types. To alter the speed of these motors it is, therefore, necessary to employ pole-changing methods or use special winding arrangements. Either of these schemes has, however, only been applied to a limited extent because they are generally expensive and require machines that are physically larger than single-speed motors.

In a two-speed motor, for example, there will be two main and two auxiliary windings (for particular applications and by special design, a single auxiliary winding can be used), and each set will be wound to develop a different number of poles. Also, the centrifugally operated switch must be set to open at the lower of the two possible speeds, the one that is somewhat below the value that is best for the higher speed winding. In a 4-pole–6-pole 60-cycle motor, the centrifugal switch will be adjusted to open at about 1,100 rpm, i.e., about 350 rpm less than is proper for the 4-pole winding.

Figure 8·21 shows a two-speed split-phase motor with two main and two auxiliary windings connected to a control circuit that is designed to

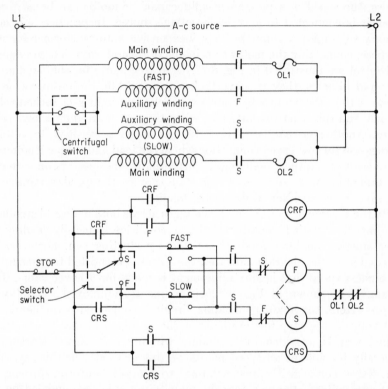

Fig. 8·21 Two-speed split-phase motor connected to a control circuit designed to permit starting at the speed set by the selector switch. After the motor is running at either speed, a transfer can be made to the other speed and back again.

permit starting at the speed set by the selector switch, after which a transfer can be made to the other speed. Also, with the motor in operation at either speed, a change to the other speed is readily possible. For the selector-switch setting shown, note that the motor will not start in *fast* because the *CRS* (*control relay slow*) contact is open with the motor at rest. The pressing of the SLOW button, however, starts the motor on *low* speed, under which condition the *CRS* relay picks up through the *S* interlock and is sealed in by its own *CRS* interlock. After that, the motor speed can be shifted to *fast* and back again to *slow*. When a start is made in *fast*, the centrifugal switch disconnects the auxiliary-winding circuit at a speed that is somewhat below its normal opening point in a single-speed motor. This means that the rotor must accelerate to full speed on the main winding only, but this it can do when the load is moderate.

Speed control of permanent-split capacitor motors. As Fig. 8·12 shows, the speed of a permanent-split capacitor motor can be adjusted when it is connected to a variable-voltage source. In practice it is customary to employ a comparatively inexpensive autotransformer for this purpose, connecting the motor to it in one of several ways. Three systems of control are illustrated in Fig. 8·22. Used frequently for shaft-mounted fans and blowers, they do nevertheless have certain limitations. One of these is that the motors develop extremely low values of locked rotor torque, and this is particularly true when they are started on a low-speed point. Another is that (see Fig. 8·12) the speed is sensitive to voltage changes on the low-speed connection where the load and torque characteristics cross at a rather small angle. Furthermore, the speed points are not the same for low, medium, and high loads, nor are the speeds separated in the same way for different degrees of loading.

By the scheme of Fig. 8·22a, the voltages across both main and auxiliary windings are varied simultaneously, a condition generally leading to poor starting on low speed. Unsatisfactory starting can, however, be avoided if, as the diagram indicates, the motor is permitted to come up to its highest speed first before adjustment is made for a lower value. The arrangement shown in Fig. 8·22b is frequently employed for 230-volt motors. Here, the auxiliary-winding circuit is connected permanently to 230 volts at the step-up terminals of the autotransformer and voltage adjustment is provided for the main winding only. Good starting is generally possible at all speed points by this method. The third system, Fig. 8·22c, connects the two windings to a tapped autotransformer, so that any voltage change across the main winding is accompanied by an inverse change across the auxiliary winding. This scheme, using a properly designed motor, has been applied, with good results, to many blower installations whose speed must not be too sensitive to normal line-voltage variations.

Since two-speed permanent-split capacitor motors are widely used for unit-heater applications, it was found desirable to design a special winding arrangement that eliminated the need for autotransformers. Two general schemes of connection are commonly employed, but the motor for such service must be equipped with a so-called *intermediate* winding in addition to the usual main and auxiliary windings. The same principle of voltage variation previously discussed applies to these methods, although it is

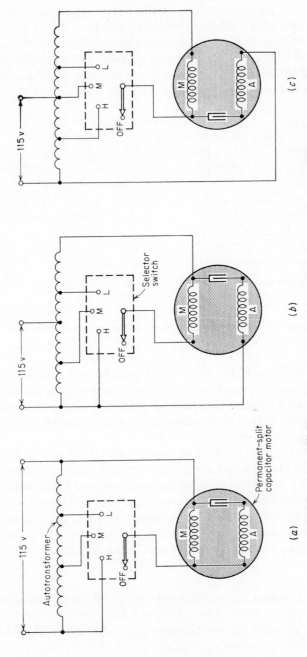

Fig. 8-22 Diagrams showing three systems of speed control for a permanent-split capacitor motor.

accomplished by taking advantage of the phasor relationship of the voltages across the main, auxiliary, and intermediate windings. What is particularly important in this unique motor is that, in addition to the 90-electrical-degree space displacement of main and auxiliary windings, the intermediate winding is in *space phase* with the main winding. The relationships indicated are represented on the diagrams of Fig. 8·23 by

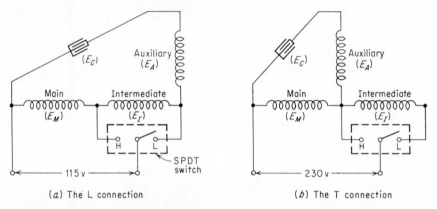

(a) The L connection   (b) The T connection

Fig. 8·23 Diagrams illustrating tapped-winding methods of speed control for a permanent-split capacitor motor.

the relative positions of the three windings. The main and intermediate windings do, in fact, occupy the same slots, with the latter placed directly over the former. They need not, however, have the same numbers of turns, and the wire size of the intermediate is invariably smaller than that used for the main.

The L Connection. This scheme of connections, shown in Fig. 8·23a, is generally used for 115-volt motors. Note that the main and intermediate windings are in *series-aiding*, with the auxiliary and its capacitor across the combination. The outer terminals of the spdt switch, marked H and L, are wired to the intermediate winding, and the pivot point of the switch serves as one line wire. With a power source of 115 volts, the following voltages were measured for a particular motor:

High-speed
$\qquad E_M = 115 \text{ volts} \quad E_I = 110 \text{ volts} \quad E_A = 200 \text{ volts} \quad E_C = 300 \text{ volts}$
Low-speed
$\qquad E_M = 59 \text{ volts} \quad E_I = 56 \text{ volts} \quad E_A = 102 \text{ volts} \quad E_C = 153 \text{ volts}$

Note that the voltages across the main and auxiliary windings, i.e., E_M and E_A, are in about the ratio of 2 to 1 for the high and low speeds.

Thus, the motor behaves as though the impressed voltage is changed by the values given.

The T Connection. In this arrangement (Fig. 8·23b) the auxiliary winding and its capacitor are connected across the main winding only, primarily because the capacitor voltage E_C is less than in the L connection. This makes it possible to use the *T connection* for 230-volt motors without exceeding nominal voltage ratings of capacitors. With a power source of 230 volts, the following voltage values indicate that good two-speed operation will result:

High-speed

$E_M = 230$ volts $E_I = 220$ volts $E_A = 205$ volts $E_C = 308$ volts

Low-speed

$E_M = 118$ volts $E_I = 112$ volts $E_A = 105$ volts $E_C = 157$ volts

QUESTIONS AND PROBLEMS

1. What is a universal motor? List several general types of single-phase induction motors.
2. Modify the reversing control circuit for the universal motor of Fig. 8·2 so that the speed of the machine will be reduced when, and only when, a SLOW button is pressed.
3. Describe several methods for reversing a shaded-pole motor. Indicate also how the speed of this type of motor is adjusted.
4. Why does a capacitor of the proper size, inserted in the auxiliary-winding circuit of a split-phase motor, improve the starting torque of this type of machine?
5. Referring to Fig. 8·13b, show that a 4-μf capacitor behaves like a 121-μf unit during the starting period when the autotransformer ratio is 5.5 to 1 and like a 9-μf unit during running when the ratio is changed to 1.5 to 1.
6. Referring to the control circuit of Fig. 8·14 for a two-value capacitor motor, how would the motor behave during starting if the H contact points were welded together?
7. Discuss the relative merits of potential-relay and current-relay schemes (Figs. 8·14 and 8·16) for two-value capacitor-motor starters.
8. Describe the plug-reversing control circuit for a capacitor-start motor (Fig. 8·18) when going from DOWN to UP.
9. Referring to Fig. 8·21, redesign the control circuit, substituting relays for the selector switch, so that the following operating procedure takes place: The motor must be started in *slow* after which a shift can be made to *fast* and back again to *slow* by merely pressing the appropriate button.
10. Describe the speed-control method for the permanent-split capacitor motor in which an intermediate winding is used in addition to the main and auxiliary windings. Also discuss the L and T connections and their applications.

CHAPTER 9

Rotating and Magnetic Amplifiers
in Control Circuits

The amplidyne—principles of operation. Some types of regulator must frequently be introduced into electrical control systems to help the power equipment maintain certain standards of performance over wide ranges of load. Several devices and machines have been developed to accomplish this, and these generally operate in closed-cycle systems where the output furnishes a signal to the regulator which, in turn, determines the required operating quantity. Examples of the numerous functions performed by regulators are maintaining the motor speed or the generator voltage constant, varying the motor speed as a function of some operating quantity, setting a limit on the maximum current, torque, or power output, and making position adjustments in mechanical systems. A regulator is, in this respect, part of a feedback control loop, since it links the output quantity with factors that tend to cause a deviation from the desired standard. Moreover, it must accomplish its mission with an extremely small power input, i.e., it must have a high amplification factor, and response to output changes must be very rapid.

A widely used rotating type of regulator is the *amplidyne*. Essentially a generator or exciter and usually driven by a constant-speed a-c motor, it belongs to the class of machines known as *armature-reaction* generators, first developed by Rosenberg for train-lighting service. Like its predecessors, whose polarity is independent of the direction of rotation, it has two sets of brushes; one short-circuited set occupies the same relative positions as do the brushes in the conventional generator, while a second set is located along an axis that is displaced 90 electrical degrees with respect to the former and is connected to the load. The field structure is rather complex, consisting of a stack of slotted, one-piece laminations of high-grade silicon steel, somewhat similar to an induction-motor stator, into which is placed one or more *control windings* and three additional windings that serve to create compensating and corrective mmfs. Always used as an auxiliary to the main power equipment, a comparatively low value of excitation watts, derived from the power system and applied to the control winding, is capable of controlling a considerable output with extreme rapidity. The unique feature of the amplidyne is that its armature with the two sets of brushes serves as a two-stage amplifier whose amplification factor may be more than 10,000 to 1; i.e., 1 watt of control-winding power will handle an output of 10 kw or higher.

The principle of operation of the amplidyne can be explained by considering the two basic fields, the control field and the compensating field. Referring to Fig. 9·1, assume that the control field (also called the voltage-controlled field) is energized by some function of the final output and in

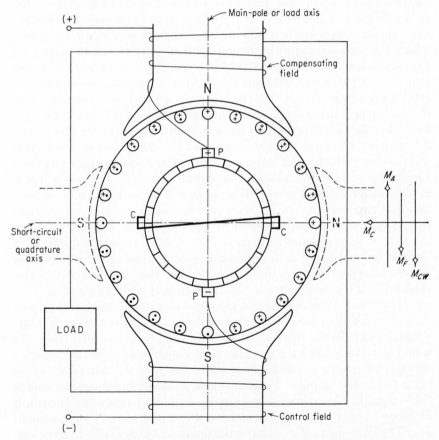

Fig. 9·1 Amplidyne generator with the basic control and compensating fields.

such a manner as to respond to voltage conditions in the controlled unit. For the current direction shown, the main field M_F will be directed *downward* because the upper and lower poles will be polarized, respectively, *north* and *south*. With the armature rotating clockwise, emfs will be generated in the conductors, the directions of which are indicated by the *outer set* of crosses and dots. Moreover, if the C brushes, located along the quadrature axis (the dividing line between the upper and lower halves of armature conductors), are short-circuited, current will pass from brush to brush in the external connector and in the armature conductors in the same directions as the emfs; an extremely small polar excitation is necessary to establish full-load current because the resistance of the short-circuited armature is very low. Thus far the machine behaves exactly like a conventional separately excited shunt generator under short-circuit conditions. With current in the armature conductors, the armature will create a cross-magnetizing mmf directed from right to left, represented in the diagram of phasors by M_C. Next, if a set of brushes P is located along the main-pole (direct) axis to "feed" current into a load, and a low-reluctance magnetic circuit is provided so that the armature-reaction mmf can establish a strong field, the machine begins to develop characteristics that are quite different from the usual. This is especially so because the armature-reaction flux is encouraged in the amplidyne. Assuming that the cross-magnetizing mmf creates an independent field that is directed to the left (horizontally located poles as shown to emphasize this point), the same conductors that cut the main control field (vertically downward) will develop emfs that are indicated by the *inner set* of crosses and dots. Furthermore, if the P brushes are connected to a load, currents will flow in the same directions as the emfs. It should be understood, of course, that the two currents do not exist independently but are regarded as components of an effective current for the purposes of analysis. Carrying the idea of superposition further, the inner set of crosses and dots may likewise be considered as developing a separate mmf M_A *that is directed upward to oppose the main control field*. If the machine were left in this condition, M_A would create a third component of flux, which in turn would establish a new set of emfs, the result of which would be that a new current would flow in the quadrature axis and attempt to nullify the original short-circuit current. Under this condition, the main field M_F would lose control and the machine would operate as a constant-current generator. To alter the situation described so that the generator can be made to display amplifier characteristics, a second field—a *compensating field*—must be placed on the main poles so that, *carrying load current*, it develops an mmf M_{CW} that is equal and opposite to the armature mmf M_A. Thus, the *net mmf* existing in the direct axis is M_F, which, therefore, assumes complete control over the output voltage.

Note particularly that the high power amplification as represented by *the ratio of watts output to excitation watts input* takes place in two stages, one stage from the control field to the quadrature field and the second stage from quadrature field back to the direct axis.

To improve both the operating performance and the degree of amplification, amplidynes are equipped with commutating poles whose windings are energized by the load current, and a series field that is connected in the short-circuit path. With such additions commutation is exceptionally good and amplification factors are elevated. Moreover, response is extremely rapid under transient conditions because, as previously mentioned, the windings are distributed in a slotted, laminated core of high-grade magnetic steel.

In many amplidyne applications that require a number of control functions, a multiplicity of control fields, independently excited by signal devices, are superimposed on one another in the slots of the stator core. The machine will then respond to their resultant mmf and act in a feedback control system as though there were a single control field. Since each of the control windings occupies an extremely small space, it is often possible to employ as many as four such fields, thus permitting several independent signals to control the amplidyne output. Each of these fields is readily adjusted or controlled by a small rheostat, and because the current requirements are low, fast-acting devices can be used. Also, rectifiers can be utilized to block these fields until specified operating conditions or limits are reached. Figure 9·2 illustrates a disassembled

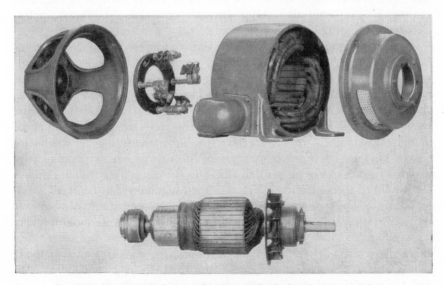

Fig. 9·2 Disassembled view of a 5-kw 250-volt 1,750-rpm amplidyne.

amplidyne having an output rating of 5 kw at 1,750 rpm and 250 volts; the stator clearly shows the slotted core and the distributed windings.

When amplidynes are employed in industrial applications to maintain certain standards of performance, they are generally incorporated in closed-loop control systems where the operating quantity supplies a signal that is balanced against a *reference* signal. A deviation from the desired performance is then accompanied by actions which cause the amplidyne, through the resultant signal, to exercise a corrective function. If an error does exist in a regulating system, and one usually does, the actual performance departs somewhat from the standard. The error should, of course, be as small as practicable because the deviation is thereby minimized; in most amplifier regulating systems the errors are kept within ±1 or ±2 per cent.

After the resultant signal is initiated, a short time will elapse before the system completes its readjustment. How long this will take will depend upon the electrical time constant of the regulating circuit and the mechanical time constant (inertia) of moving machinery whose speed must change. The amplification factor of the amplidyne will, in large measure, determine the response time, which will be somewhat less than ¼ sec. If, for example, the feedback signal is small and proper regulation will result only if amplification is high, the response time will be comparatively long. In systems whose changes take place frequently and rapidly or where the error is low, response time is usually shortened by over-exciting the control field considerably at the instant the correction is initiated. This is called *forcing* and may be as much as three to ten times the steady-state excitation. After correct operation is established, excitation drops to normal.

Another phenomenon that tends to impair proper function of a regulating system is *hunting*. This is a form of instability in which a regulator repeatedly overcorrects and undercorrects a deviation as the system attempts to reach a standard rapidly. The frequency of the oscillations that are thus set up depends upon the constants of the system, and the regulator design will determine whether the latter is unstable, damped, or critically damped. In the unstable regulator where the oscillations are sustained, the system never "homes in" on its standard; this is usually avoided by the use of an antihunt field in the amplidyne which opposes the action of the control field. The damped regulator, on the other hand, stabilizes itself after a few oscillations because the amplitudes of the latter diminish rapidly; this condition is also corrected as a rule by employing an antihunt field. When the regulator is critically damped, correction takes place without oscillation, since the system stabilizes itself without overshooting. The curves of Fig. 9·3 illustrate two degrees of instability and the condition that exists when hunting is critically damped.

Figure 9·4 illustrates two antihunt arrangements that are frequently used with amplidyne regulators. In the first of these, Fig. 9·4*a*, the output voltage of the amplidyne energizes a series *RLC* circuit consisting of an adjustable resistor, an inductive antihunt field, and a rather large (600- to 1,200-μf) electrolytic capacitor. Under steady-state conditions, the field is inactive, since no current passes through the winding. However, when there is a *voltage change* across the amplidyne terminals that

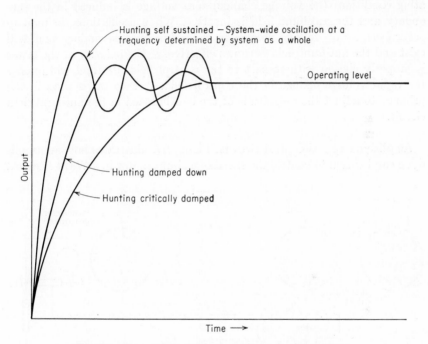

Fig. 9·3   Curves illustrating several degrees of instability.

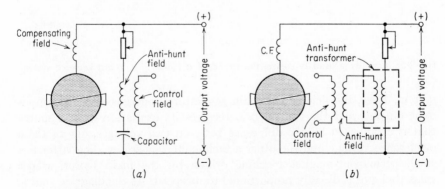

Fig. 9·4   Schematic diagrams of an amplidyne with antihunt fields.

tends to set up oscillations, a current equal to $i = C \, de/dt$ does energize the antihunt field and always opposes the action of the control field which tries to overcorrect the deviation. The rheostat is used to adjust the current and the time constant of the antihunt field circuit. In Fig. 9·4b, the antihunt field is excited through a special type of antihunt transformer, the primary of which is connected to the output terminals of the amplidyne. No capacitor is used in this arrangement. Under stable operating conditions (no voltage change) zero voltage is induced in the secondary and the antihunt field is inactive. When oscillations do tend to occur because of voltage change in the primary, a secondary emf will exist and the antihunt field becomes effective. If the polarity of the latter is properly chosen with respect to the control field, its mmf will oppose the rapid voltage change in the control field. A rheostat is used in the primary to adjust the magnitude of the emf induced in the antihunt field circuit.

Amplidyne speed-control circuits. Figure 9·5 illustrates how an amplidyne can be used to control the speed of a shunt motor in a Ward Leonard

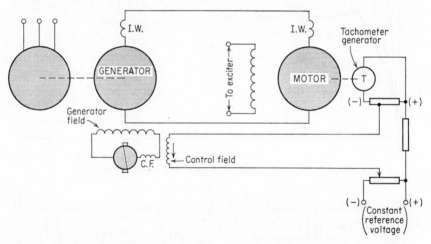

Fig. 9·5 Amplidyne used to control the speed of a motor in a Ward Leonard system.

system. Here, the field of the main generator is energized by the amplidyne output, and the control field receives its excitation from a source that is established by the difference between two voltages, one of them a reference voltage and the other a tachometer generator. The latter is a small permanent-magnet machine driven by the main motor, whose generated emf is directly proportional to its speed. As the diagram shows, the tachometer loads a potentiometer, a portion of whose voltage is a speed signal, while the constant-potential reference is impressed across another potentiometer, a part of which is used as a standard against which the speed is balanced. The magnitude of control-field current is then proportional to the difference between speed and standard voltage signals, while the current direction is determined by relative values of the two emfs. After the motor is accelerated by adjusting the reference rheostat which, in turn, controls the main-field excitation and the voltage of the main generator, the speed will remain constant for all load changes. This is because the reference signal is higher than the speed signal and the direction of the control-field current is as indicated. Thus, a momentary increase in speed is accompanied by a larger speed signal, a reduced voltage across the control field, a lower amplidyne output, and a smaller main generator emf. The reverse actions take place when the speed of the motor tends to drop.

Another amplidyne control scheme is shown in Fig. 9·6, where two motors are "tied together" *electrically* and maintain a definite speed

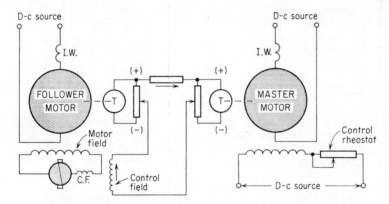

Fig. 9·6 Amplidyne used for a two-motor speed-control system.

relationship with respect to each other. The control unit is a rheostat in the shunt-field circuit of the *master* motor, and the difference between the voltages supplied by two tachometer generators is impressed across the control field of the amplidyne. Note particularly that the polarity of the *follower* tachometer determines the current direction in the control field and the output of the amplidyne energizes the field of the *follower* motor. Since the control field responds to the difference in tachometer emfs, any deviation in the follower-motor speed is immediately corrected by a suitable excitation change. Should the speed of the follower attempt to decrease, its tachometer voltage will drop, the resultant control-field signal will decrease, the amplidyne output will weaken the follower-motor field, and the speed will be brought back to its preset value.

Amplidyne voltage-control circuits. As previously indicated, an amplidyne can be used to good advantage in an adjustable-voltage system to control the speed of a motor; in such an arrangement the output voltage of the generator is varied by adjusting the control-field excitation. Also, for a given setting of a control rheostat, the terminal voltage will remain constant under all load conditions, although small inherent motor-speed changes will occur as the load varies. This is unlike the performance characteristics of the motor in Fig. 9·5, where the tachometer feedback acts directly on the amplidyne control field to prevent the motor speed from changing.

Two circuit diagrams that illustrate how the generator voltage can be controlled and kept constant at a particular adjustment are given in Fig. 9·7. In the first of these (Fig. 9·7a), employed with low-voltage generators,

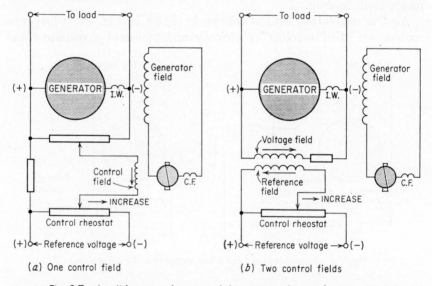

(a) One control field   *(b)* Two control fields

Fig. 9·7 Amplidynes used to control the output voltage of generators.

the control field of the amplidyne is excited by an emf that is represented by the difference between two voltage signals, one of which is derived from the armature terminals of the generator and the other from an adjustable reference source. Note that the *direction* of the current in the control field depends upon the polarity of the reference voltage because it is larger than the armature potential. Thus, for a given control-rheostat setting, the output of the amplidyne will determine the strength of the generator-field excitation and, in turn, the armature voltage. Then, should the generator emf drop, the reduced voltage signal will cause an increase in control-field excitation, a rise in the amplidyne output, strengthening of generator field, and a return to the original voltage.

It is interesting to note that systems such as this respond very quickly to control-rheostat adjustments. This is because the *initial increase* in control-field current is usually several times its normal value at the instant a change is made from one rheostat setting to another. The output voltage of the amplidyne is therefore forced initially to a higher value than normal, and this has the effect of accelerating the voltage change.

For generators having voltage ratings greater than 250, the circuit of Fig. 9·7a has the disadvantage that the operator is subjected to an unsafe potential while manipulating the control rheostat; the potential to ground is then equal to the emf of the generator. The difficulty indicated can be avoided by using two control fields in the amplidyne and connecting them so that their mmfs oppose each other. As illustrated by Fig. 9·7b, one of the control fields, as before, is energized by the generator armature and the other by a comparatively low-voltage reference source through a control rheostat. With this arrangement, the output voltage of the ampli-dyne is proportional to the difference in mmfs between the voltage field and reference field. After a setting is made, the amplidyne will maintain a constant output voltage for all load changes. For example, should the voltage rise momentarily, the resultant mmf will decrease, and this will reduce the amplidyne output; with the generator field weakened thereby, the voltage will be returned to normal.

A circuit diagram illustrating how an amplifier can be used to control the output voltage of a 3-phase alternator is given in Fig. 9·8. Before

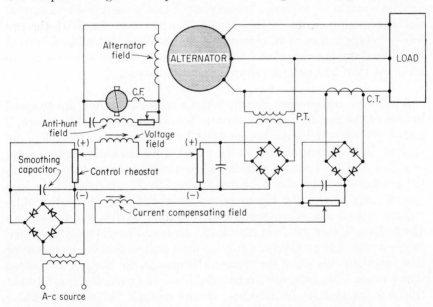

Fig. 9·8 Amplidyne used to control the voltage of an alternator.

analyzing the operation of the circuit it should be noted that there are two control fields and these are energized through full-wave rectifiers because the amplidyne is a d-c exciter and all originating signals are alternating currents; also, smoothing capacitors are connected across the output terminals of the rectifiers to level off the signal pulsations. As in systems previously discussed (Fig. 9·7), the d-c alternator field receives its excitation from the amplidyne, although in this case the latter is stabilized by an antihunt field like that of Fig. 9·4a.

The mmf developed by the voltage field results from two opposing signals, one of which originates at the output terminals of the alternator and is changed in magnitude by a potential transformer P.T. and the other a reference voltage that is adjusted by a control rheostat. The direction of the current in the voltage field is determined by the polarity of the reference, which is always larger than the alternator signal. During operation, the output voltage of the alternator remains substantially constant for any control-rheostat setting and all load conditions. Thus, should the emf rise (or fall) momentarily, the resultant mmf will change inversely, i.e., decrease (or increase). The amplidyne output will, therefore, be lower (or higher), the alternator field excitation will be weakened (or strengthened), and the voltage will return to normal.

Since alternator loads are frequently located some distance from their source of power, it is desirable to compensate for line drop and attempt to maintain the *load voltage* at a constant level. This is accomplished by using a *current-compensating field* in the amplidyne that is energized by a current signal through a current transformer C.T. and a rectifier. With the two fields, voltage and current, aiding, it should be clear that all load-current variations are accompanied by correspondingly larger changes in the resultant mmf and output voltage of the amplidyne.

Amplidyne current-limit circuit. When a motor is brought up to speed by one of the several types of starters previously discussed in Chaps. 2 and 3, the current inrush is controlled by the insertion of acceleration resistors in the armature circuit. This kind of protection is, however, not provided in Ward Leonard systems, where the field rheostat that controls the generator voltage may be moved too rapidly, in which event the loop current may exceed the commutation limit. Also, in such adjustable-voltage systems, acceleration and deceleration of the motor between speed changes may result in high loop currents. The difficulty indicated can be avoided by employing a current-limit control field in an amplidyne that functions only when the current attempts to rise above a designated upper value. Used primarily in adjustable-speed drives, the control-limit field remains inactive for all loop currents that are below the allowable maximum but serves to oppose an increase or decrease in amplidyne output voltage when the current increases above this value.

Referring to Fig. 9·9, note that a signal that measures the loop current is provided by a dropping resistor mn, and a constant-voltage reference

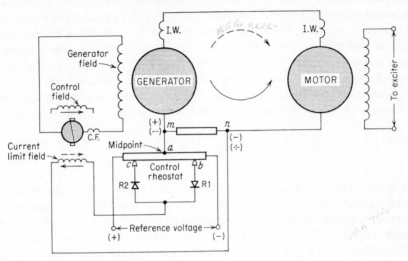

Fig. 9·9 Amplidyne used for current-limit control.

loads a control rheostat whose midpoint is connected to junction m at an armature terminal. Sliders b and c are moved along the control rheostat and adjust the potentials ab and ca to set, respectively, the acceleration and deceleration current-limit values. Assume that the loop current is below the permissible maximum and in the direction of the solid arrow. Since potential ab is greater than mn, current tries to flow through the current-limit field from a to m to n to b but is blocked by rectifier $R1$. However, if, during acceleration, the loop current rises sufficiently to make potential mn larger than ab, current *can* pass through rectifier $R1$ and the current-limit field in the direction of the solid arrow from m to a to b to n. With the latter field properly polarized with respect to the control field, the output voltage of the amplidyne rises slowly enough to oppose a rapid increase in generator field excitation.

When the generator voltage is quickly reduced by field weakening to lower the motor speed, the cemf of the motor may exceed the armature emf momentarily and reverse the loop current as indicated by the dashed arrow. Since the resistance of the connected armature circuits is very low, the loop current may attempt to exceed the commutation limit and the *reversed* potential nm may rise above that of voltage ca. Current can now flow through rectifier $R2$ and the current-limit field in the direction of the dashed arrow from n to c to a to m with the result that the amplidyne output voltage will drop slowly enough to oppose a rapid decrease in generator field excitation.

Amplidyne current-control circuit. It is sometimes necessary to control very closely the armature current in a shunt motor that is used in a continuous manufacturing operation. A good example of such an application is a reel that winds up (or unwinds) sheet steel as the latter is processed. Here the rapidly moving steel must be kept at a constant tension, and this is generally accomplished by having a regulator maintain the armature current at some fixed value and, at the same time, reduce the motor speed as the reel winds up to keep the *strip speed* constant. The relations indicated can be analyzed as follows. Since tension is equal to the ratio of torque T to the radius of the reel R and, by Eq. (7·2), $T = kI_A\phi$,

$$\text{Tension} = k\frac{\phi}{R}I_A$$

it is seen that the tension on the strip will be constant if the armature current and the flux-to-radius ratio are both constant. This means, of course, that the flux must rise as rapidly as the radius of the reel increases with buildup. The motor speed in revolutions per minute, therefore, diminishes as the field is strengthened to keep the linear speed of the strip, in *feet per minute*, constant for all radial dimensions of the reel. In equation form

$$\text{fpm} = 2\pi RN = 2\pi R\left(k\frac{V - I_A R_A}{\phi}\right) = k'\left(\frac{R}{\phi}\right)$$

where N is the motor speed in revolutions per minute, V is the impressed voltage, and I_A is assumed to be constant. Thus, the strip speed will remain unchanged if the ϕ/R ratio or its inverse does not vary.

A circuit diagram illustrating how an amplidyne can be used for current control is given in Fig. 9·10. Note particularly that it differs from previous arrangements because the motor field is energized here by the *difference* between a reference voltage and the amplidyne output. Also, there are two control fields, one of them a *reference field* whose current direction establishes the proper amplidyne polarity to buck the reference voltage and the other a *current field* which is energized, through a rectifier, by the difference in emfs between that across a dropping resistor R and a potential ab across a control rheostat. The voltage drop across resistor R is proportional to the armature current I_A, and the resultant mmf that excites the amplidyne is the difference between the reference- and current-field mmfs.

When the reel-up operation is started, the control rheostat is set so that little or no current passes through the current field. This is because potential ab is about equal to or greater than the voltage drop mn and the blocking rectifier prevents a current flow in a direction opposite to that shown by the dashed arrow. Thus, with only the reference field excited, maximum amplidyne output opposes the reference voltage, the motor-field excitation is weak, and the motor speed is high; also, the armature current I_A adjusts itself so that sufficient torque, represented by the equation $T = kI_A\phi$, is developed. As the reel builds up and its

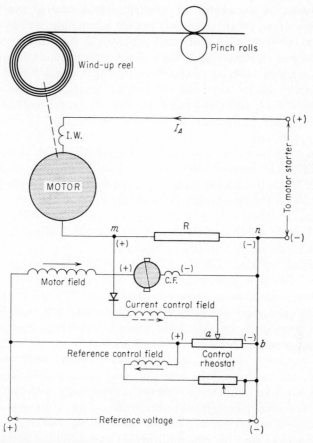

Fig. 9·10 Amplidyne used to control the armature current in a drive for a windup reel.

radius increases, a higher value of torque must be developed. The armature current therefore rises momentarily and causes the potential drop across mn to exceed the voltage ab. The current field is thereby energized in the direction of the dashed arrow, the resultant mmf and the output of the amplidyne decrease, the motor field is strengthened, and the motor speed diminishes. Since the flux and the armature current are both larger than before and the motor torque is higher than it need be, the armature current drops to its original value. Also, the linear speed of the strip is maintained constant because the lower rpm speed of the motor is balanced by the larger reel radius. Should the armature current drop below the required value, the motor speed will increase because its field will be weakened.

Amplidynes in excavator or electric-shovel control systems. The adjustable-voltage (Ward Leonard) system is used almost exclusively to control the speed of the motors that provide the three major motions in an excavator or electric shovel. Each of the motors—for hoisting, crowding (digging), and swinging—is powered by a specially designed generator that has three main fields: the self-excited shunt field, the separately excited shunt field, and the series field. Since the service of these rugged material-handling equipments is extremely severe and the motors stall frequently, the series field of the generator is connected *differentially* with respect to the other two fields and, in conjunction with the amplidyne current-limit field, serves to limit the maximum current flowing from the generator. With the motor stalled, a condition that is equivalent to a short circuit on the generator, the bucking action of the series-field mmf reduces the terminal voltage to a very low value, the self-excited field strength is very small, and the separately excited field can be varied to adjust for a safe maximum (stall) current.

All control functions for the excavator originate at master switches where the direction and magnitude of the currents in the *separately excited shunt fields of the generators* are controlled. This is done by varying the output voltages of the amplidynes, one of which is used for each of the three generators, i.e., the hoist, crowd, and swing generators. The amplidynes generally have four control fields that are specially designed for their purposes, these being the reference control field, the current-limit field, the antihunt field, and the compensating field. The current-limit and antihunt fields are connected differentially with respect to the reference and compensating fields. Since the output voltage of the amplidyne is applied to and directly excites the separate field of the generator, it should be clear that the control fields act to control the voltage applied to the motor. These fields are therefore used to obtain the desired values of main generator no-load voltage, stall current, and peak power.

The compensating field, which carries the current delivered to the generator field, is adjusted to give the proper compounding to the amplidyne; that is, the compensating field regulates the current furnished to the generator field and thus determines the no-load voltage of the generator. Adjustment is made by varying a resistor in parallel with the field and/or using a suitable tap on the winding.

The current-limit field is excited by the voltage drop across a combination of the generator-motor series and commutating fields. This voltage, called the *current-limit voltage,* excites the current-limit field so that its mmf opposes that of the reference field and thereby controls the maximum current of the main generator. If the current-limit field were excited continuously, the generator voltage would drop linearly from its no-load value to zero, the latter being the value when the motor is stalled and is carrying the maximum allowable current. The peak power under this condition of operation would then be very low. To improve performance and obtain much higher values of peak power, a special biasing circuit is employed which does not permit current-limit field current to flow until the generator current exceeds about 40 to 50 per cent of maximum. The point at which the current-limit field starts to carry current, called the "cutoff" point, is indicated by G on the *generator volts versus generator current characteristic* of Fig. 9·11. Beyond this point the voltage decreases

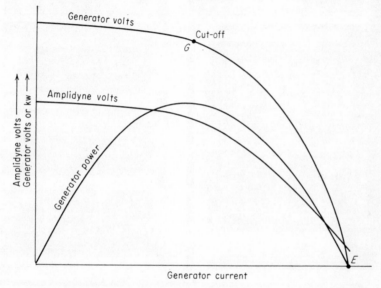

Fig. 9·11 Characteristic curves for amplidyne-controlled hoist motor in an excavator.

rapidly along the curve from G to E until, in the stalled position of the motor, maximum permissible current flows in the loop.

As previously explained (Fig. 9·4a), the antihunt field is excited by the output current of the amplidyne through a capacitor. Current passes through the field only when the voltage of the amplidyne is *changing,* and since it is connected differentially, it serves to limit the armature-voltage surges and decreases oscillations and hunting.

A photograph showing a rear view of the control cabinet of an amplidyne-controlled excavator is illustrated by Fig. 9·12. Note the three amplidyne units for the hoist, crowd, and swing motions at the lower right.

Fig. 9·12   Rear view of control cabinet for an excavator showing the three amplidyne units in the lower right.

A wiring diagram of an amplidyne control circuit for one of the adjustable-voltage systems of an excavator—the hoist—is given in Fig. 9·13. Note that an exciter energizes the motor field and the reference control field of the amplidyne, while the output of the latter excites the separate field of the generator. Voltage and polarity control are obtained by the operator at a master that automatically actuates accelerating and reversing contactors (not shown). Also, the differential current-limit field of the amplidyne is connected to a special resistor-rectifier bridge network which is energized by the voltage drop across the generator-motor series and commutating fields. Since this voltage drop is directly proportional to the loop current I_L, current will pass through the current-limit field only when its emf is greater than that across one of the bridge resistors.

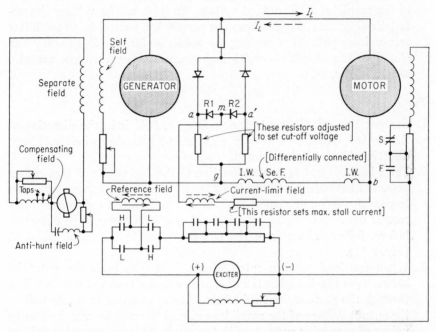

Fig. 9·13 Amplidyne-controlled adjustable-voltage system for the hoist drive of an excavator.

Before the system is put into operation, the bias voltage E_{ag} is permanently set to equal voltage E_{bg} at cutoff, after which the following operating conditions prevail:

For Hoisting

1. Current through the control, or reference, field is in the direction of the *solid arrow*, and voltage adjustment is made at the master through which accelerating contactors are actuated to short-circuit accelerating resistors. The H directional contacts are closed.
2. For a load current I_L in the direction of the solid arrow that makes the potential between a and g higher than from b to g, current attempts to pass through the current-limit field from a to m to b to g to a but cannot do so because of blocking rectifier $R1$. The voltage curve (Fig.

9·11) therefore follows along from the no-load value to the cutoff point G.

3. However, when the load current reaches cutoff value G (Fig. 9·11), the potential between b and g exceeds the emf from a to g. Current can now flow through the current-limit field from b to m to a to g to b. Since this current has a demagnetizing effect upon the resultant field of the amplidyne, the output voltage diminishes. The voltage characteristic (Fig. 9·11) then follows the curved line from G to E, the latter zero-voltage point representing the condition for a stalled motor. The characteristic amplidyne-volts and generator-power curves are also shown in Fig. 9·11.

For Lowering

1. Current through the control, or reference, field is in the direction of the *dashed arrow*, and voltage adjustment is made as before. The L directional contacts are closed.

2. For a load current I_L in the direction of the dashed arrow that makes potential between g and a' higher than from g to b, current attempts to pass through the current-limit field from b to m to a' to g to b but cannot do so because of blocking rectifier $R2$. The voltage curve, as before, follows along from the no-load value to the cutoff point G, (Fig. 9·11).

3. When the load reaches cutoff, point G in Fig. 9·11, the potential between g and b is higher than between g and a'; current can now flow through the current-limit field from a' to m to b to g to a'. As before, the output voltage of the amplidyne diminishes rapidly with increasing values of generator current, with the result that the generator voltage characteristic again follows along the curve from G to E in Fig. 9·11.

During the hoisting operation of the hoist-motor cycle, the drive must develop high torque to pull the loaded dipper through the bank. To perform such heavy-duty service it is generally necessary to *force* the motor field, i.e., overexcite or strengthen the field. This is accomplished by closing contact F across the open portion of the rheostat in the motor-field circuit. Moreover, when the unloaded dipper is being lowered, it is essential that a fast drop be made so that the dipper be in position by the time the boom is swung back to the digging point. This feature of the operating cycle requires that the motor field be weakened by opening the N.C. contact S across the short-circuited portion of the motor-field rheostat. At the end of the lower travel the motor field is strengthened again, after which the drive is plugged in preparation for another digging cycle.

A photograph of a 60-cu-yd excavator stripping overburden from coal in an open-pit mine is shown in Fig. 9·14.

Fig. 9·14　A 60-cu-yd excavator stripping overburden from coal in an open-pit mine.

The Rototrol—principles of operation. Another type of rotating exciter that has amplifying properties is the *Rototrol* (a Westinghouse Electric Corporation development), a trade name that is formed by combining the first and last syllables of the words *ROT*ating and con*TROL*. Fundamentally similar in design and construction to a standard generator (*not* an armature-reaction machine like the amplidyne), it is equipped with a number of field windings that act simultaneously—once properly adjusted—to control and regulate power equipment. Like the amplidyne it has been used to control the voltage and speed of generators and motors, to provide current, power, and current-limit functions, and to regulate such operational quantities as torque, tension, position, power factor, and others.

The principal advantage of this unique method of control is that, acting as a generator, the Rototrol is made to operate on the straight-line portion of the magnetization curve where relatively small variations in excitation produce considerable output voltage changes. This is clearly seen in Fig.

9·15, which shows that the slope of the magnetization curve is greatest and constant below the knee. Since each of the several fields has many

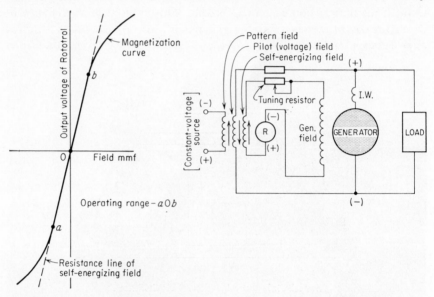

Fig. 9·15 Magnetization curve and circuit diagram for Rototrol used to regulate the voltage of a d-c generator.

turns of fine wire, it occupies a comparatively small space in a composite field coil, Fig. 9·16, and requires rather low values of current for excitation; also, reasonably large mmf changes result from small current variations. In its simplest form the Rototrol has three fields, the functions of which will be understood by referring to an actual elementary application like the wiring diagram of Fig. 9·15, which represents a control scheme that regulates the voltage of a d-c generator. The following three fields and their functions are: (1) the *pattern field*, which is usually energized from a constant-voltage source and serves to set the pattern for the desired results; (2) the *pilot field*, which provides a means for measuring the quantity under regulation—in this case the generator voltage—by comparing it with the set pattern (the value of the regulated quantity is usually adjusted by means of a rheostat in the pattern-field circuit); (3) the *self-energizing field*, which is usually connected in series in the Rototrol armature circuit—for some applications this field can be connected in parallel with the armature and acts as a shunt field, or a combination of series and shunt fields can be used. For most applications the pattern and pilot fields are connected so that their mmfs oppose each other, and any net difference in field ampere-turns produces a highly amplified corrective action that causes the Rototrol to make the necessary readjustment.

If the mmf required to produce the output voltage of the Rototrol resulted from the difference between pattern and pilot fields alone, a small error in regulation would be unavoidable. This is because any change in operating conditions that calls for a change in Rototrol excitation requires *some change* in net ampere-turns between the two fields; as a consequence

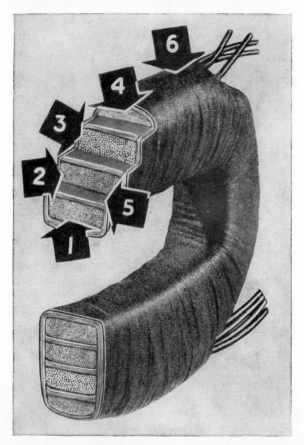

Fig. 9·16 Cutaway view of one of the composite coils of a four-field Rototrol: 1, 2, 3, and 4 are control fields; 5 and 6 are insulations.

the quantity under regulation would have to change. To obtain high steady-state accuracy it is, therefore, necessary to produce the mmf actually required for excitation of the Rototrol by other means. This is the function of the self-energizing field, which is the *heart* of the regulator. In Fig. 9·15, for example, an increased load momentarily reduces the terminal voltage of the generator, with the result that the pilot field (in this case a voltage field) is weakened and the net difference between pattern-

and voltage-field mmfs becomes larger. Since this change is accompanied by a higher Rototrol voltage, the generator field is strengthened and the load voltage returns to its original setting. The increased difference between pattern and pilot fields now disappears, but the higher level of excitation of the self-energizing field maintains the proper generator emf.

For some applications where great sensitivity is required, the effects of the pattern and pilot fields can be combined into one field, which then becomes the control field. As illustrated by Fig. 9·17, this is accomplished

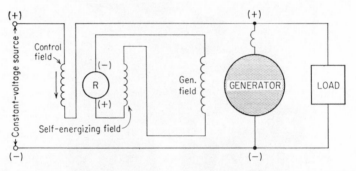

Fig. 9·17   Special high-sensitivity Rototrol circuit to regulate the voltage of a d-c generator.

by joining the negative terminals of the constant-voltage and generator sources and connecting a control winding between the positive lines, which then measures the voltage difference directly.

Figure 9·18 illustrates a three-unit set consisting of a driving motor in the center, a small exciter on the left, and a four-field Rototrol on the right.

Fig. 9·18   Three-unit Rototrol set consisting of an a-c driving motor (center), an exciter (left), and a four-field Rototrol machine (right).

Rototrol speed-control circuit. A control scheme that employs a Rototrol to maintain the speed of a motor constant for all values of output torque is given in Fig. 9·19. In addition to the self-energizing and pattern

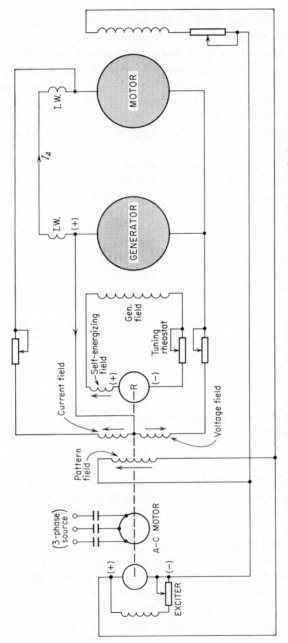

Fig. 9-19   Rototrol-controlled adjustable-voltage system to regulate the speed of a motor.

315

fields there are two pilot fields, in this case called *current* and *voltage fields* because they are energized by current and voltage signals that originate in the power circuit. Since the speed of the motor is given by the equation $N = k(V - I_A R_A)/\phi$ and, for this system, the flux is constant, it follows that $N = k_1 V - k_2 I_A$. As a result, this quantity can be measured by the combined effect of two properly adjusted Rototrol fields connected *differentially*, one of them across the armature terminals to register the voltage factor V and the other across the interpole fields to signal variations in the current factor I_A.

To understand how the system functions, assume that the speed-adjusting rheostat is set to give the desired motor speed for a given torque. Now then, should the load torque increase, the armature current I_A must do likewise, since $T = kI_A$, the flux being constant. This is followed by a normal drop in generator terminal voltage V and a momentary slowdown of the motor. These reactions then proceed to reduce the mmf of the *opposing voltage field*, raise the mmf of the *aiding current field*, and, in combination with the pattern field, increase the resultant flux and terminal voltage of the Rototrol. With the generator field thereby strengthened, the terminal voltage rises to reestablish the motor speed at its original setting. Under the new load-torque conditions, the increased current-field and self-energizing-field excitations are just sufficient to compensate for the higher $I_A R_A$ drop. A momentary drop in motor speed will, of course, be accompanied by an oppositely directed set of reactions and a return to preset operation.

Rototrol arc-furnace regulator circuit. In electric arc-furnace operation, a close relationship exists between the electrode position and the power input. Since furnace conditions change continuously and electrodes burn away with use, it is essential that some means be provided automatically to maintain proper electrode position above the molten steel if power consumption and furnace temperature are to be regulated. The Rototrol control circuit of Fig. 9·20 has been used successfully to perform the service indicated.

Referring to the diagram, note that the Rototrol is equipped with four fields labeled *voltage field, current field, self-energizing field,* and *differential field*. Also, the output of the Rototrol energizes the field of a small generator which, in turn, drives a motor and, through a cable-pulley arrangement, raises or lowers the electrodes. During operation, rectified voltages, proportional to the electrode current and the arc voltage, are impressed on the differentially connected current and voltage fields, respectively. The latter are so designed and adjusted that their net mmfs are zero for the desired value of power in the arc. Then, should there be a change in current or voltage, a resultant mmf will be produced in the Rototrol to cause the electrode motor to turn in the proper direction to position the electrode until balance is restored. The purpose of the differential field, which is energized by a voltage drop across the interpole winding of the generator, is to slow down the motor as it approaches the correct electrode position; this prevents overshooting and hunting.

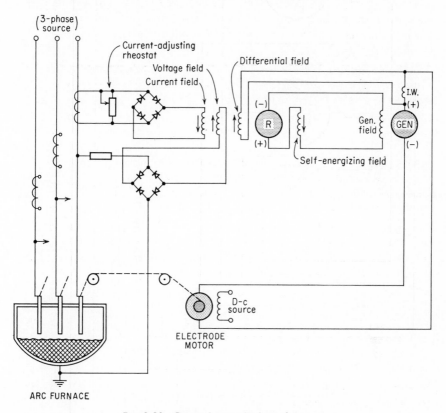

Fig. 9·20   Rototrol-controlled arc furnace.

Rototrols in electric-shovel control systems. A circuit diagram showing how a Rototrol is used to control the operation of one of the three major motions in an electric shovel is illustrated by Fig. 9·21. Performing essentially the same functions as those described for the amplidyne system of control (Fig. 9·13), the Rototrol is equipped with four fields, and the main generator has a differentially connected series field which acts, in part, to limit the stall current. Drawn to represent the hoist section of the equipment, the arrows indicate the current directions for hoisting, assuming that the *H* contacts in the master are closed. Note also that the self-energizing field *shunts* the Rototrol armature rather than being connected

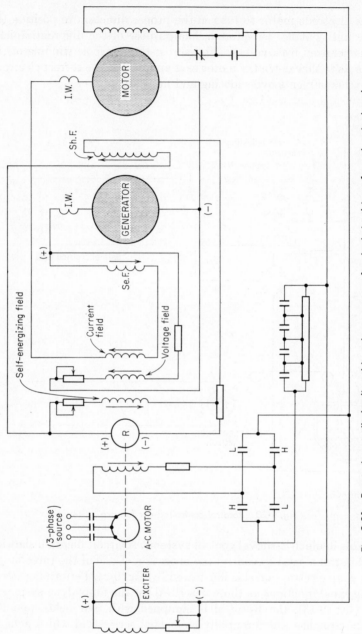

Fig. 9·21 Rototrol-controlled adjustable-voltage system for the hoist drive of an electric shovel.

in series in its output circuit. Moreover, the current in this field is generally adjusted to set the operating point of the system for the condition that the pattern-field mmf is originally made equal to the sum of the voltage- and current-field mmfs. There is also provision for motor-field forcing so that high torque can be developed when the digging is heavy, and for motor-field weakening so that a fast empty-bucket drop can be made.

The action of the voltage field is to maintain a reasonably constant motor speed for any given master position under varying conditions of load. Thus, when the load increases, the generated emf of the main generator tends to drop and reduce the motor speed; it also has the effect of lowering the mmf of the voltage field, with the result that the net mmf and the output voltage of the Rototrol increase. Since the field of the main generator is strengthened by the latter change, the motor speed readjusts itself to its original value.

Increased load, on the other hand, causes the mmf of the differentially connected current field to increase, and this action, which gradually tends to *reduce* the Rototrol output voltage, has the effect of weakening the separately excited shunt field of the main generator. At very heavy load, and particularly in the stalled position of the motor, the Rototrol voltage drops to a value that is low enough to limit the stalled current. Also, the differentially connected series field of the generator tends to reduce further the generated emf under heavy or stalled load conditions.

A system of control that is similar in many respects to the Rototrol is called *Regulex*. Made by the Allis-Chalmers Manufacturing Company, it employs a special quick-response, self-energized exciter whose laminated magnetic structure operates in the unsaturated region of the magnetization curve. Like the rotating machine previously described, it has several control fields that serve to provide many regulating functions.

Rotating regulator for constant-speed motor operation. Another rotating amplifier scheme that is designed for an adjustable-voltage (Ward Leonard) system was developed by Reliance Electric and Engineering Company. Called a VSA drive and illustrated in elementary form by Fig. 9·22, it incorporates two small generators $G1$ and $G2$ which operate in tandem to energize the shunt field of the main generator. Note particularly that the regulating functions are centered in three sensing shunt-field windings, designated as *antihunt*, *signal*, and *reference* fields, which excite generator $G1$. Also observe that the output of the latter machine, the input stage, excites the field of generator $G2$, the output stage. Since very small changes in signal emf at the tachometer generator produce comparatively large voltage changes at the armature terminals of $G2$, amplifying properties are exhibited by the system.

It is important to understand that the algebraic sum of the mmfs of the sensing fields sets the pattern of flux for the first generator $G1$, and this, in turn, determines the output voltage of generator $G2$ and, eventually, the potential at the main generator terminals. This means, therefore, that the speed of the motor can be adjusted at the speed-adjusting rheostat. During normal operation the mmf of the reference field is slightly higher than the oppositely directed mmf of the signal field, and it is this small *mmf difference* that is amplified through the regulator. Thus, should the speed of the motor rise (or fall) momentarily, the signal voltage

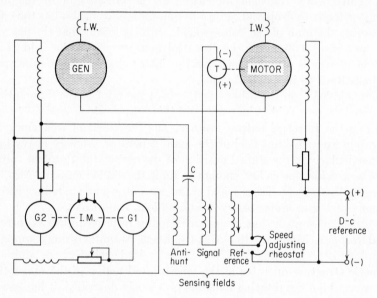

Fig. 9·22 Rotating regulator in an adjustable-voltage (Ward Leonard) system for constant-speed motor operation.

will likewise increase (or decrease). Since the error will, under this condition, be affected in an inverse manner, the main generator voltage will drop (or rise) to prevent the original speed change. Also, with the antihunt field connected directly to the regulator output, all transients in that circuit are delivered to the input where they effectively subdue rapid mmf changes. The blocking capacitor C opposes steady-state feedback.

The magnetic amplifier—principles of operation. When a saturable reactor, not unlike the familiar static transformer in construction, is used in combination with a set of high-grade rectifiers, it exhibits power-amplification properties in the sense that small changes in control power result in considerable changes in power output. Called a *magnetic amplifier*, it is manufactured under several trade names, two well-known designs of which are commonly referred to as the *Magamp* (Westinghouse Electric Corporation) and the *Amplistat* (General Electric Company). Because it is a completely static device and can be made to develop suitable regulating functions with high power gain, it is widely employed in many kinds of industrial control systems.

The "heart" of the magnetic amplifier is the saturable reactor, which generally comprises a closed core of laminated iron and a group of three or more coils of wire that are wrapped around the core limbs. The property of the reactor that makes it behave as a power amplifier is its ability to change the degree of saturation of the core when the control-winding mmf, established by d-c excitation, is changed. The output winding (or windings) which is connected in the a-c load circuit will then have a high impedance if the core is unsaturated and varying values of lower impedance as the core is increasingly saturated. When the core is completely saturated, the output-coil impedance becomes negligibly small. It should be clear, therefore, that small values of direct current in a control winding that has many turns of wire will determine the core saturation and, in turn, control the flow of alternating current through the reactor.

In the simple sketch of Fig. 9·23a, the a-c load in the output circuit is controlled by altering the magnitude of the d-c excitation in the control winding because the latter determines the degree of saturation in the core. Moreover, if the ratio of control-coil turns to output-coil turns is large, an extremely high value of output current can be controlled by a rather small current in the control winding. But if no rectifier is used as in the arrangement shown, the power gain is substantially reduced, since the core is partially desaturated in the half-cycle in which the output-coil mmf *opposes* the control-coil mmf. The difficulty indicated is, however, corrected by employing a rectifier in the output circuit as in Fig. 9·23b. Here, the desaturating half-cycle of output current is blocked by the rectifier, whereas the output- and control-coil mmfs aid each other to amplify saturation in the half cycle in which current passes to the load. It is, in fact, this very difference in operation that distinguishes the *desaturating saturable* reactor of Fig. 9·23a from the *self-saturating magnetic amplifier* of Fig. 9·23b. The inductors shown in the circuits indicate that alternating currents are not induced in the control winding, although these are not needed in properly designed magnetic amplifiers.

Referring to Fig. 9·24a, assume that the core is constructed with high-permeability iron and yields a steep, straight saturation curve with sharp changes from unsaturated to completely saturated states at the knees, points a and b. In the region from a to b, where the core is unsaturated, the inductance is high, since $d\phi/di$ is large in the equation $L = Nd\phi/di$. The inductance is, however, extremely low (or nearly zero) in the saturated regions (to the left of a and the right of b), where $d\phi/di$ is very small (or zero). Thus, if a sinusoidal voltage is impressed across the output terminals of the device and the d-c control winding is unexcited, the wave of load current will be as shown in curve O of Fig. 9·24b. This is because

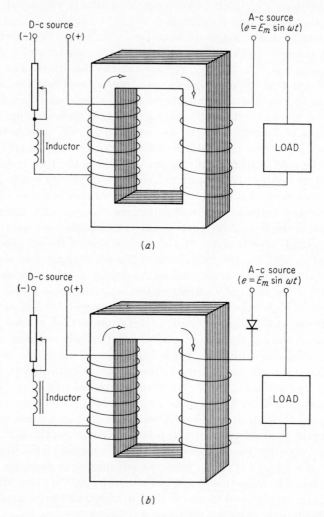

Fig. 9·23 Sketches illustrating difference between a simple saturable reactor (a) and a magnetic amplifier (b).

the impedance is high as the mmf builds up from point O to point b in Fig. 9·24a, after which it drops abruptly to a very low value. Beyond the knee of the saturation curve the load current is proportional to the varying values of the alternating emf and, neglecting the small IR drop in the impedance coil, is equal to e/R_L. During the negative half of the a-c cycle the rectifier blocks passage, except for a return of current to the circuit that is a measure of the energy which was stored in the magnetic field of the reactor at the beginning of the cycle. It is important to understand that the reactor can be in either of two states and that load current *will not* or *will* flow depending upon whether the core is unsaturated or saturated. Moreover, the *average* load current is determined by the *length* of the conducting period under the voltage wave which, in turn, depends upon the so-called *firing point*, i.e., the instant at which the core changes to the saturated state.

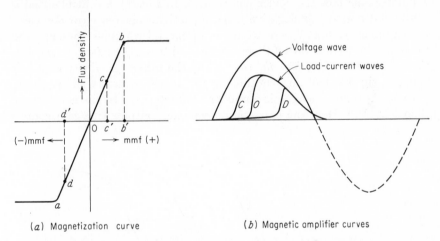

(*a*) Magnetization curve (*b*) Magnetic amplifier curves

Fig. 9·24 Characteristic curves for a simple magnetic amplifier.

If the core is initially premagnetized by exciting the d-c control winding, the firing point can be advanced or retarded to change the conducting period and, therefore, the average load current. For example, an initial magnetization c and an mmf represented by point c' in Fig. 9·24a will permit the core to saturate sooner, lengthen the firing time, and increase the average load current (curve C, Fig. 9·24b). On the other hand, reversing the direction of the control current and premagnetizing the core to point d will make it necessary for the mmf to change by an amount equal to $d'b'$ before saturation occurs; this shortens the firing time and decreases the average load current (curve D).

The foregoing discussion should therefore make it clear that the saturable reactor, through its control winding, can be made to serve as a regulating device because it can be adjusted to control the current in the load circuit. Furthermore, by proper core and winding design and by the use of circuit arrangements that employ rectifiers, extremely low excitation currents can control high values of load current. Considerable amplification is thus possible with these magnetic amplifiers. Another point that should be emphasized is that load current does not flow until the device is triggered ON, and once so fired, load is delivered until the a-c potential drops to zero. The magnetic amplifier acts, in this respect, like the electronic thyratron when the grid loses control as soon as the tube fires and load current flows until the voltage becomes zero.

Control of magnetic amplifiers. Since the output of a magnetic amplifier is a function of the net mmf of control signal, the algebraic sum of the mmfs of several control windings that are energized simultaneously will determine how the device will behave. In general, a control signal is considered positive or negative if it, respectively, increases or decreases the amplifier output. Where a number of control windings are employed to regulate a system, one of them can be used to set the initial magnetization of the core, i.e., *bias the core*, while the others provide positive or negative signals that vary continuously while the system is in operation.

Bias. A typical magnetic-amplifier characteristic is given by curve *A* in Fig. 9·25. Here the *quiescent* output, i.e., output with zero control

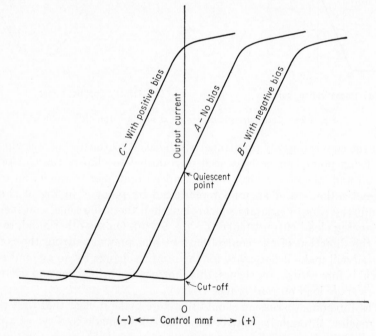

Fig. 9·25 Characteristic magnetic-amplifier curves with different values of bias.

mmf, is rather high, while the cutoff at the lower bend occurs with negative control mmf. If one control winding is used for biasing purposes, it will shift the characteristic to the right (curve B) when negative bias is applied and to the left (curve C) when the bias is positive. Moreover, the quiescent point can be made to coincide with cutoff if the negative bias current is properly adjusted; in such cases positive signals are used to control the output. On the other hand, amplifiers that are positively biased to maximum (saturation) employ negative signals to control the output.

Feedback. In many closed-loop control systems, one or more signals are fed back to separate amplifier control windings to exercise certain regulating functions. In general, negative feedback tends to improve the linearity of the control characteristic, Fig. 9·26, curve B, whereas positive

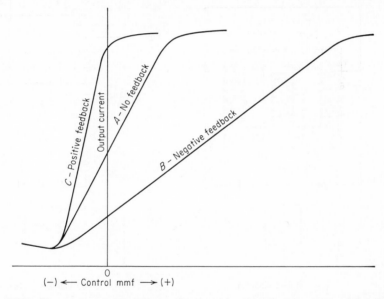

Fig. 9·26 Characteristic magnetic-amplifier curves with different values of feedback.

feedback produces the opposite effect (curve C). Furthermore, a magnetic amplifier can be designed to be bistable for a particular feedback signal, in which case the characteristic curve is almost vertical; the device will then be stable only at the upper and lower extremes of output and will have no intermediate operating point. Such bistable magnetic amplifiers are frequently used as relays, under which conditions they are either *on* or *off* in the sense that the output circuit is closed or open.

Magnetic-amplifier circuits. The cores, windings, and rectifiers of magnetic amplifiers can be arranged in many different ways to operate in single-phase or 3-phase control systems. Moreover, the output can be delivered as direct current or alternating current, and d-c or a-c signals can be applied to the control windings depending on how the winding connections are made. Several of the more common configurations showing core and winding details as well as schematic diagrams are illustrated by the sketches that follow.

Three-legged Core—A-C Single-phase Output. Figure 9·27a represents a basic magnetic amplifier circuit in which a three-legged reactor core has

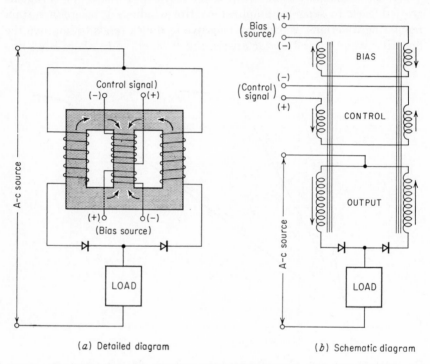

(a) Detailed diagram   (b) Schematic diagram

Fig. 9·27 Magnetic amplifier with three-legged core and a-c output.

the two load (impedance) coils around the outer legs and the control and bias windings over the center leg. For full-wave operation, two rectifiers are inserted in the impedance-winding circuits and are connected so that they function alternately on the two halves of the voltage cycle. Note particularly that the control winding produces a positive signal with respect to either load coil and that, for negative biasing, a negative signal is applied to the bias winding. The small arrows in the cores indicate the flux directions for the respective coils when the latter carry current. Since the load is connected in one of the a-c line wires, outside the influence of the amplifier, it carries alternating current at fundamental frequency. Also, with the control winding excited from a d-c source, the induced emfs in the control circuit will be only even-numbered harmonics of the supply voltage.

A schematic diagram of the same circuit, and one that is simpler to follow when thoroughly understood, is illustrated by Fig. 9·27b. It is suggested that the two sketches be compared and studied carefully because the latter type of drawing, in modified form, generally appears on circuit diagrams of control systems.

Two Rectangular Cores—Bridge-connected Rectifiers—D-C Output. A frequently employed arrangement for single-phase circuits when it is desired to deliver direct current to a load is shown in Fig. 9·28a and its

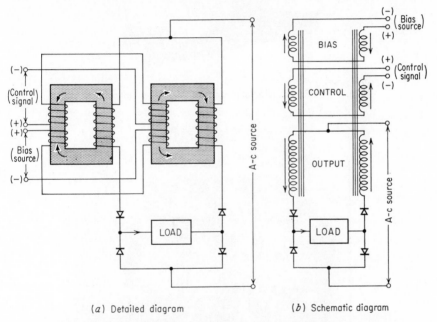

(*a*) Detailed diagram   (*b*) Schematic diagram

Fig. 9·28   Magnetic amplifier with two rectangular cores and d-c output.

schematic counterpart Fig. 9·28b. Here, two rectangular cores are used, with each one having an impedance coil on one leg and the control and bias windings on the other. Although the two cores are effective on alternate halves of the voltage cycle, the control windings are connected in series and are energized by one control signal, while the bias windings are connected similarly and to its d-c source. The familiar full-wave four-rectifier bridge circuit is used to supply direct current to the load.

Figure 9·29 illustrates the constructional details of a magnetic-amplifier reactor unit with a rectangular core, and Fig. 9·30 depicts a cutaway view of four such units in a metal enclosure with terminal leads brought out at the top.

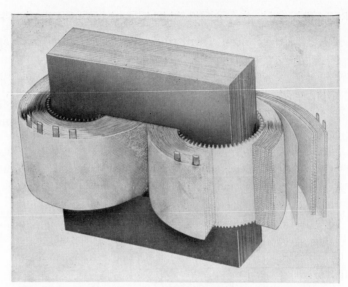

Fig. 9·29　Rectangular core and windings of a saturable reactor.

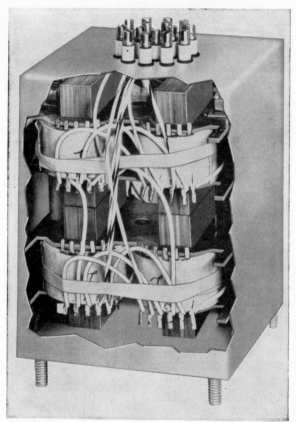

Fig. 9·30　Cutaway view of four magnetic-amplifier reactor units in a metal enclosure.

Three-legged Core—Center-tapped Transformer—D-C Output. A schematic diagram illustrating how a d-c output can be delivered to a load is given in Fig. 9·31. To do this with a three-legged core and two rectifiers

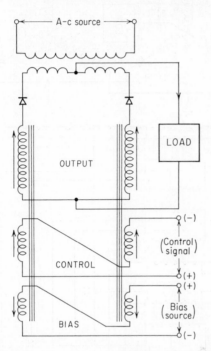

Fig. 9·31 Schematic diagram of a magnetic amplifier with center-tapped transformer and d-c output.

it is necessary to employ a transformer whose secondary has a center tap. As the sketch shows, each of the output coils is connected to a secondary-transformer end through a rectifier and the load is joined to the center tap of the transformer and the junction between the other two terminals of the output coils. The arrangement has the advantage that a single reactor and two rectifiers are needed, in contrast to the two-core four-rectifier configuration of Fig. 9·28; it has the disadvantage that a special center-tapped transformer is required.

Three Three-legged Cores for 3-phase Operation—D-C Output. A scheme that is commonly employed in 3-phase circuits is illustrated by Fig. 9·32. Following essentially the arrangement of Fig. 9·27 for single-phase operation, there are three three-legged cores and three pairs of bridge-connected rectifiers. The d-c load is joined to the two star points $S1$ and $S2$ formed by the three sets of output coils. Assuming a phase sequence of voltages AB-BC-CA, the amplifiers fire in the order of their

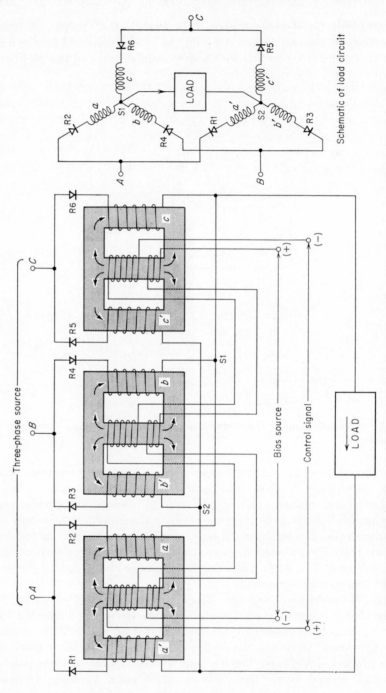

Fig. 9-32 Detailed drawing of a 3-phase magnetic amplifier with bridge-connected rectifiers and d-c output.

Schematic of load circuit

highest positive potentials (as in mercury-arc rectifier systems) as follows: (1) for phase AB, current flows from terminal A through rectifier $R2$ through output coil a to $S1$ through the load to $S2$ through output coil b' through rectifier $R3$ and finally to terminal B; (2) for phase BC, the path is from B through $R4$ and coil b to $S1$ then through the load to $S2$ and finally through coil c' and $R5$ to terminal C; (3) for phase CA, the current passes from terminal C through $R6$ and coil c to $S1$ then through the load to $S2$ and back to terminal A through coil a' and $R1$.

Three Three-legged Cores for 3-phase Operation—A-C Output. The schematic diagram of Fig. 9·33 shows how a magnetic amplifier like that

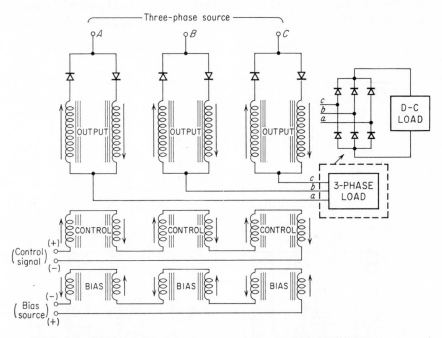

Fig. 9·33 Schematic diagram of a 3-phase magnetic amplifier with bridge-connected rectifiers and a-c output. For d-c load operation, upper right arrangement can be substituted for 3-phase load.

of Fig. 9·32 can be connected to deliver a 3-phase load. During operation, controlled a-c power is supplied to a motor or other electrical equipment through line reactors whose impedances are changed by adjusting the direct current in the control windings. Such adjustment alters the degree of saturation of the cores and, in turn, changes the line impedance and line-impedance drop. This scheme, similar in principle to the line-reactance method of starting a polyphase motor (Fig. 2·20), permits the drive to accelerate smoothly as the reactance is diminished gradually. Also, as the sketch in the upper right of the diagram shows, a d-c load, represented by a motor or other electrical equipment, can be effectively controlled. By this method, varying values of voltage can be applied to the load to fulfill desired power demands.

Positioning control for motor-operated rheostat. When large rheostats must be adjusted repeatedly in control systems, they are frequently operated by small reversing motors. The latter are usually series machines of the universal type (Fig. 8·2) with two field windings. A two-unit magnetic amplifier in combination with a self-balancing bridge circuit, illustrated by Fig. 9·34, provides a satisfactory means of control for such applications. The balancing rheostat and control rheostat (not shown) are

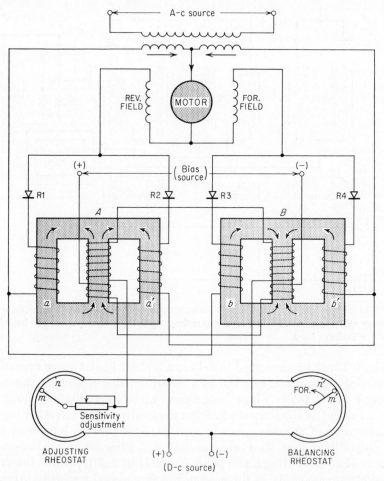

Fig. 9·34 Magnetic-amplifier control in self-balancing bridge circuit for motor-operated rheostat.

geared to the motor shaft and rotate only to restore bridge balance after the adjusting rheostat is moved to a position that disturbs the balance.

When the potential difference between the moving arms of the rheostats, points m and m', is zero (for a balanced bridge) and no current flows in the control windings of the amplifiers, the reactances of the output coils are large and equal. Under this condition the motor does not rotate, because small but equal currents pass into both fields from the armatures. On one-half of the cycle, the REV. FIELD is energized through rectifier $R1$ and reactance coil a, with the FOR. FIELD excited through $R3$ and b; on the other half of the cycle, the REV. FIELD is energized through $R2$ and a' while the FOR. FIELD receives its excitation through $R4$ and b'.

To make the motor drive the arm of the balancing rheostat (and the control rheostat to which it is coupled) to position n', the adjusting rheostat is moved to position n. At the instant this is done, the potential of point n will be more positive than point m' and current will flow in the two control coils of the amplifiers to produce the flux directions indicated. For amplifier A, the mmfs of coil a and the control winding on one half of the cycle, and the mmfs of coil a' and the control winding on the other half of the cycle oppose each other. Amplifier A will therefore be unsaturated, and little or no current will pass into the REV. FIELD. For amplifier B, however, the mmfs of b and the control winding on one half of the cycle, and the mmfs of b' and the control winding on the other half of the cycle will be aiding. Amplifier B will thus be saturated, and a comparatively large current will pass into the FOR. FIELD. The motor will then rotate in a *forward* direction, stopping when the arm of the balancing rheostat reaches point n', where a bridge balance is reestablished. If the arm of the adjusting rheostat is moved in a counterclockwise direction, amplifier A will become saturated, cause the REV. FIELD to carry current, and make the motor rotate in *reverse* direction.

A geared 100-step motor-operated rheostat (M.O.R.) is illustrated by Fig. 9·35.

Magnetic amplifier for dimmer control. The circuit diagram of Fig. 9·36 illustrates how the magnetic amplifier is used for theatre dimming. Two amplifiers are employed, in one of which, the so-called *preamplifier*, the control functions originate and a second, the *power amplifier*, whose control windings are energized by the output of the preamplifier and which, through its output windings, delivers current to the lighting load. The arrangement used makes it possible for milliwatts of control power in the preamplifier to handle as much as 25 kw in the dimming circuit. The degree of saturation of the preamplifier is adjusted by an operator at a console, where numerous other lighting circuits are also controlled. A special feature of the system is the use of feedback windings in the preamplifier which are energized by a negative signal from the load. A voltage drop across a resistor in the load circuit furnishes that signal, through a bridge rectifier, to improve the sensitivity and linearity of the dimmer unit (see Fig. 9·26).

Fig. 9·35 Geared 100-step motor-operated rheostat. (*Cutler-Hammer, Inc.*)

Magnetic-amplifier controller for shunt motor. The wiring diagram of Fig. 9·37 illustrates an adjustable-voltage system for a shunt motor. Direct current for the drive is obtained from a 3-phase source through a conversion unit that consists of self-saturating reactors (magnetic amplifiers), a 3-phase full-wave bridge rectifier (see Fig. 9·32), a variable-voltage autotransformer (*Variac*), and two special transformers. In addition to the usual output, control, and bias windings, the amplifier is provided with windings that function to compensate for the normal IR drop in the armature circuit under conditions of varying load. A current-limit relay connected across a dropping resistor in the armature circuit operates to limit the armature current to about 150 per cent of rated value. Shunt-field excitation is obtained through an autotransformer and a bridge rectifier.

With the speed adjustor set for minimum output voltage, the motor is started at reduced speed by pressing the START button. This energizes the armature at low voltage through the unsaturated line reactors and the bridge rectifier; the shunt field was originally excited when the main switch was closed. To make the motor speed up, it is merely necessary to increase the degree of saturation of the core which, in turn, will reduce the line drop in the saturable reactors and raise the armature voltage; this is accomplished by adjusting the Variac to give a higher output voltage. All motor-speed changes in the adjustable-voltage system are thereafter made by altering the position of the speed adjustor.

In order to maintain the reactor output at the level set by the speed adjustor, a negative feedback voltage is introduced into the control circuit. Since this emf, obtained from a potential divider across the armature circuit, opposes the reference voltage, any departure from the standard results in a changed control signal that reestablishes the desired core saturation. In addition, any tendency on the part of the motor to change speed as the load rises or falls is accompanied, respectively, by a higher or lower voltage drop across armature resistor R. This has the effect of

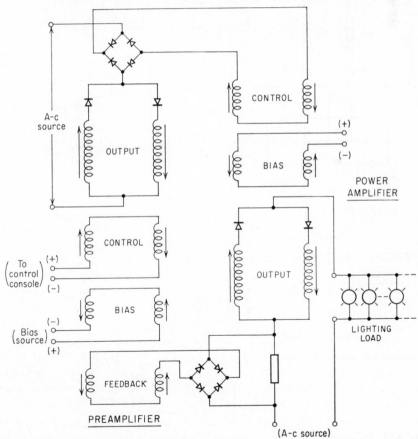

Fig. 9·36   Schematic diagram of a magnetic-amplifier control circuit for a theatre dimmer.

increasing or decreasing the excitation of the IR compensation windings, which, in turn, will similarly affect the core saturation to compensate for line drop. The changed armature voltage then returns the motor speed to its original setting.

The current-limit relay, connected across resistor R, responds to the

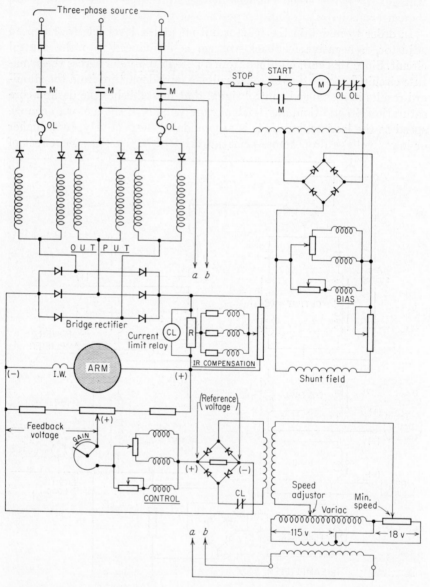

Fig. 9·37   Circuit diagram for operating a shunt motor from a 3-phase source through a magnetic amplifier.

current in the armature circuit. Being a vibrating-type unit, it vibrates alternately to open and close the reference source at the N.C. contact CL until the overload disappears.

A photograph of such a magnetic-amplifier control system for a 20-hp motor mounted in a steel cabinet is shown in Fig. 9·38. The magnetic amplifiers and the bridge rectifiers can be seen at the bottom.

Fig. 9·38 Complete magnetic-amplifier control system for a 20-hp motor mounted in a steel cabinet. (*Cutler-Hammer, Inc.*)

Magnetic-amplifier speed control. Figure 9·39 illustrates how a magnetic amplifier can be used to maintain the speed of a shunt motor constant under varying conditions of load. For the method shown, the motor or generator must be equipped with a special *control field* that is wound directly over the shunt field and is energized from an a-c source through a bridge rectifier and the output windings of the amplifier. Also, a tachometer generator, driven by the motor, feeds a signal to the control winding of the amplifier. In the constant-voltage system of Fig. 9·39*a*, the control field must be connected so that its current direction is the *same* as that of the shunt field, whereas the mmfs of the same two fields on the generator must *oppose* each other in the adjustable-voltage system of Fig. 9·39*b*.

Referring to Fig. 9·39*a*, assume that the field rheostats are adjusted so that the field mmfs produce the required flux for the desired motor speed and that for this operating condition the tachometer generator furnishes a given signal to the control winding of the amplifier. Thereafter, any tendency on the part of the motor to change its speed will be accompanied by a higher or lower tachometer feedback signal that opposes the change. For example, should the speed drop momentarily because of a load increase, the tachometer emf will drop. The following sequence of actions will then take place: (1) The control-winding current will drop; (2) the core saturation will be reduced; (3) the reactance of the output windings of the reactor will rise; (4) the control-field current will be less; (5) the net mmf and flux of the motor will decrease; (6) the motor speed will return to its original setting. A reverse series of effects will occur should the speed of the motor attempt to rise.

In the adjustable-voltage system of Fig. 9·39*b*, a motor-speed change is also accompanied by an equivalent rise or fall of tachometer-generator emf. For this case, however, the effect is to decrease or increase, respectively, the net mmf of the generator and the corresponding voltage in the loop circuit. Thus, if the motor speed should increase momentarily, the feedback signal will rise and the following sequence of actions result: (1) The control-winding current will increase; (2) the core saturation will be higher; (3) the reactance of the output windings of the reactor will drop; (4) the control-field current will be more; (5) the net mmf, flux, and voltage of the generator will decrease; (6) the motor speed will be brought back to its preset value.

Magnetic-amplifier load control. Two circuit diagrams are given in Fig. 9·40 to illustrate how a magnetic amplifier can be used to control the mechanical power output of a shunt motor or the electrical power output delivered by a generator. As in Fig. 9·39, the rotating machines must be equipped with special control fields which are energized, as before, by

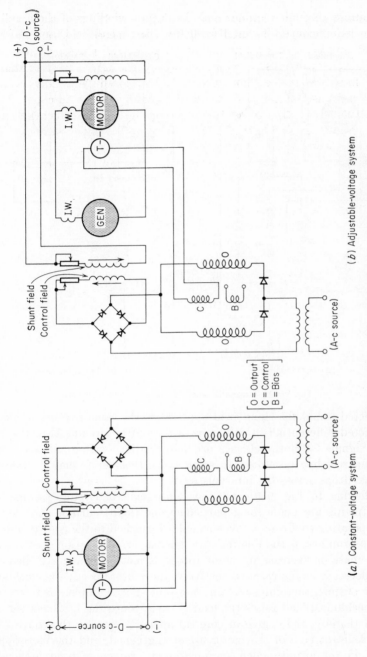

[O = Output
C = Control
B = Bias]

(a) Constant-voltage system

(b) Adjustable-voltage system

Fig. 9-39 Magnetic-amplifier speed-control method for a shunt motor.

a-c sources through rectifiers and the output windings of the amplifiers In the motor control circuit, Fig. 9·40a, the control field mmf must be.

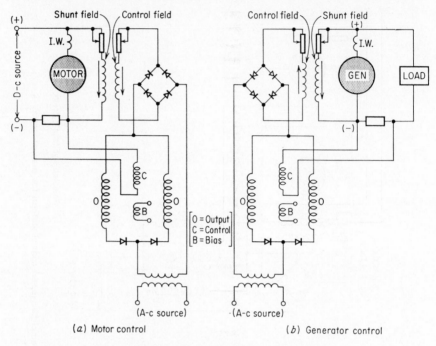

(a) Motor control   (b) Generator control

Fig. 9·40   Magnetic-amplifier load-control methods.

connected to aid the shunt-field mmf, while the same two fields must be connected differentially in the generator control circuit. Also, the feed-back signals for both schemes originate in the load circuits as voltage drops across line resistors. (In some cases, these emfs may be taken as voltage drops across the interpole fields.)

Referring to Fig. 9·40a for the motor control circuit, assume that adjustments are made for a particular line (load) current and that for this operating condition a given signal is furnished to the control winding by the drop across the line resistor. Thereafter, any action that requires an increase or decrease in motor torque is met by changes in field flux that decrease or increase, respectively, the motor speed; the mechanical power output, therefore, remains constant. For example, an increase in the mechanical load raises the load current. Since the feedback signal is raised thereby, the degree of core saturation increases, the control-field and resultant field of the motor are strengthened, and the motor speed drops. Power output, which is a function of torque and speed, is therefore unchanged.

In Fig. 9·40b for the generator control circuit, an electrical load change is also accompanied by an equivalent rise or fall in voltage drop across the line resistor. For this arrangement, however, a load increase or decrease, respectively, reduces or raises the net mmf and terminal voltage of the generator. Thus, if the load should increase, the line-resistor drop and the corresponding feedback signal will be more. The effect of these changes will then be to reduce the net mmf and terminal voltage of the generator. Power output, which is the product of E and I, therefore, remains substantially constant.

QUESTIONS AND PROBLEMS

1. What is meant by amplification when referring to an amplidyne? Why is it comparatively large in this type of machine which takes advantage of armature reaction?

2. What is the overall effect of the simultaneous action of several control fields in an amplidyne?

3. What is meant by *instability* in a regulating system? Under what conditions will a regulating system hunt? When is a regulator said to be unstable? damped? critically damped?

4. In the two antihunt schemes illustrated by Fig. 9·4, why is a capacitor used in sketch a and not in sketch b?

5. List several kinds of regulating functions which may be provided by an amplidyne. Draw simple sketches illustrating how these functions can be accomplished.

6. Describe the operation of the amplidyne in Fig. 9·8 to regulate the voltage of an alternator.

7. Why is *current-limit* control often necessary in a motor while rapid load or speed changes are taking place? Describe how this kind of control is accomplished in Fig. 9·9 with the use of an amplidyne.

8. In the amplidyne control circuit for the windup-reel drive of Fig. 9·10, state in words how the tension of the strip is maintained constant by regulating for constant armature current and constant strip speed. Describe how this is accomplished in the given control circuit.

9. Referring to the excavator control circuit of Fig. 9·13, answer the following questions: (a) What two essential functions are performed by the amplidyne? (b) How must the current in the current-limit field be directed if it is to exercise a moderating influence on the generator output during abnormal operating conditions? (c) What adjustment should be made in the circuit if the cutoff point for current-limit action is to be advanced? (d) What is the purpose of the differentially connected series field? (e) Why does the motor operate under constant-torque conditions? (f) What is meant by *field forcing*, and how is it applied in this circuit? (g) What adjustment should be made to reduce the magnitude of the stalled current? (h) Why does the self-field always build up properly with respect to the separate field for either position of the directional master? (i) Why is the speed of the motor almost directly proportional to the generator voltage? Prove mathematically.

10. Why does the Rototrol generator have amplifying properties? Explain why the self-energizing field in Fig. 9·15 is basic to the proper functioning of the control circuit.

11. Explain the operation of the Rototrol adjustable-voltage system illustrated by Fig. 9·19 to regulate the speed of a motor. Indicate particularly the reason for the two pilot fields, and state why they are connected differentially.

12. Referring to the Rototrol arc-furnace control circuit of Fig. 9·20 answer the following questions: (a) Why are rectifiers essential for the excitation of the two pilot fields? (b) What is the function of the self-energizing field? (c) How must the electrodes be moved as the current attempts to increase? as the voltage attempts to decrease?

13. In the electric-shovel control circuit diagram of Fig. 9·21 explain why (a) the voltage field maintains a constant motor speed under varying load conditions, (b) the current field acts to limit the current output of the generator, (c) the series field of the main generator is connected differentially.

14. Describe the operation of the adjustable-voltage regulator system of Fig. 9·22 that maintains constant motor speed. How could the control circuit be modified to regulate for constant loop current? Why does the system have amplification properties?

15. Distinguish between a saturable reactor and a magnetic amplifier.

16. In a magnetic amplifier, explain why the direction and the magnitude of premagnetization determines the average load current.

17. What is meant by *bias* when referring to a magnetic amplifier? How should bias be applied if a magnetic amplifier is to be controlled by positive signals? negative signals?

18. What is meant by feedback in a magnetic amplifier? What kind of feedback is necessary to improve linearity? to obtain switch action in a control circuit?

19. Describe in detail the action of the magnetic-amplifier control circuit of Fig. 9·34 when the arm of the adjusting rheostat is moved to another position in a counterclockwise direction. Be specific about current and mmf directions and the direction of rotation to establish balance.

20. Why are negative bias and negative feedback used in the preamplifier of the *dimmer* control circuit of Fig. 9·36? What purpose is served by the pre-amplifier? Draw a control circuit for such an application that eliminates the preamplifier.

21. Referring to magnetic-amplifier motor control circuit in Fig. 9·37, answer the following questions: (a) What is the purpose of the *bridge rectifier?* (b) What voltage is available at the control windings when an adjustment is made at this *Variac?* (c) What function is performed by the *output* windings, and how is this accomplished? (d) Why are two rheostats used in the *control* winding and *bias* winding circuits to establish equalization in each of these sections of the magnetic amplifier? (e) How does the negative *feedback voltage* stabilize the motor for a given setting on the speed adjustor? (f) How does the current-limit relay function during an excessive overload? (g) What is the purpose of the *IR compensation* windings, and how is this accomplished automatically? (h) Why is this a constant-torque system? How could it be made to exhibit constant-horsepower characteristics?

22. Modify the magnetic-amplifier speed-control systems of Fig. 9·39a and b so that they eliminate the need for control fields in the motor and generator.

23. Modify the magnetic-amplifier load-control systems of Fig. 9·40a and b so that they eliminate the need for control fields in the motor and generator.

CHAPTER 10

Auxiliary Devices and Special
Control Circuits

Limit switches. A type of pilot device that is extensively employed in control systems to alter the circuit associated with a machine or equipment is the *limit switch*. As usually installed, it is activated mechanically by the driven machine and may operate to stop a motor at its normal limits of travel, may provide overtravel protection in the event a motor attempts to move beyond established limits, may act to interlock and sequence the motions of several parts of an interconnected drive, or may initiate such other functions as reversing, transferring, and cycling. Two general classes of limit switch are available, these being (1) *control-circuit limit switches*, in which the contacts are connected only in the control circuit, and (2) *power-circuit limit switches*, in which the contacts are connected only in the power circuit. Also, when these pilot devices are tripped by levers, plungers, push rods, cams, or other projecting arms attached to the moving machine, they are called *track-type* limit switches. When a shaft on the switch rotates through gears that are coupled to the drive shaft, they are designated *geared-type* limit switches; rotating cams then open and close contacts that correspond to designated points along the line of travel.

Numerous designs and constructions of limit switch are available for many kinds of service. For standard-duty applications such as machine tools and similar equipment, a track-type unit like that of Fig. 10·1 can be used. One or more contacts that are actuated by quick make-and-break mechanisms are mounted in a dust-tight case, and the operating lever, fitted with a rubber-tired roller, can be adjusted to any angular position. For use in the control circuits of a-c or d-c controllers on elevators, hoists, steel-mill auxiliaries, and other mechanisms, the *hatchway-*

type limit switch, Fig. 10·2, is often employed. It usually has two N.C. contacts that open to disconnect both sides of the control circuit and ensure that a motor will stop even though one of the control lines may be grounded. For elevator applications, the switch is positioned with the wheel extending downward so that material, dropping down the hatchway, will not be lodged between the wheel and the wall; it operates when

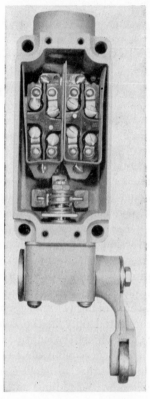

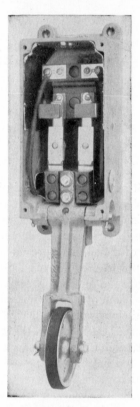

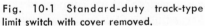

Fig. 10·1 Standard-duty track-type limit switch with cover removed.

Fig. 10·2 Hatchway-type limit switch with cover removed.

the projecting arm is tripped by a cam usually mounted on the elevator car or on the counterweight.

In nonreversible drives that must be started and stopped regularly and frequently, the rotating cam-type limit switch, Fig. 10·3, is generally used. Any number of contacts may be provided to open and close at designated angular positions by properly adjusted cams, where one revolution of 360° represents a complete operating cycle of the machine. For heavy-duty work and particularly where the contacts must carry high load currents, the power-circuit limit switch was developed. As Fig. 10·4

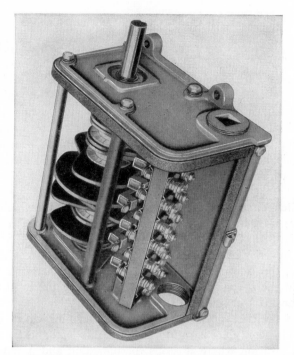

Fig. 10·3   Cam-type limit switch. (*Square D Co.*)

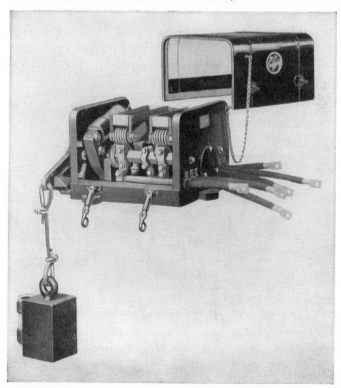

Fig. 10·4   Power-circuit (Youngstown) limit switch. (*Square D Co.*)

illustrates, the unit is very ruggedly constructed and operates to open and/or close large contacts when the operating machine raises the heavy weight. For the unit shown, the limit stop trips when the weight is lifted approximately 4 in. and resets when the weight is lowered to a point about ⅜ in. above the normal position. Used in the manner described to handle main-line currents directly, the power-circuit limit switch is definitely not a pilot device, although it functions to protect electrical and mechanical equipment in much the same way as do the intermediate relay-type switches.

Limit-switch circuits. A simple wiring diagram in which two limit switches are used in a single-phase reversing-motor hoist application is given in Fig. 8·18. Here, the N.C. limit-switch contacts *LSU* and *LSD* are placed in series in the contactor-coil circuits, and for either motion, *up* or *down*, the *U* or the *D* contactors will be deenergized to stop the motor when the proper limit switch is tripped by the hoist mechanism at the upper or lower limits of travel.

A typical cycling application involves a rotating drum unloader in which a loaded container is rotated to dump its load and then returns to an upright position where it stops for reloading and a repetition of a dumping operation. The control circuit for an arrangement such as this requires that the operator press a START button after each load-unload cycle; a motor then starts to rotate the loaded drum and stops automatically when the unloaded drum again reaches its reload (vertical) position. Figure 10·5 illustrates how this is accomplished for a line-start motor with

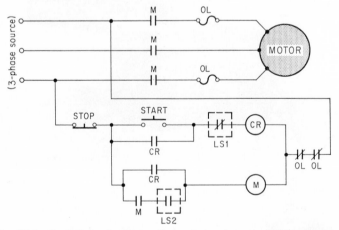

Fig. 10·5   Control circuit with two limit switches for a cycling application.

a control circuit that incorporates two limit switches in addition to a contactor and a relay. With all contacts in their normal positions as shown, the START button is pressed; this energizes the CR relay, which seals itself through a CR interlock and the N.C. $LS1$ limit switch, and causes the M contactor to pick up through another CR contact. After the unloader turns through a small angle, $LS2$ closes its contact and an instant later $LS1$ opens the contact. The control relay now drops out, but the M contactor, energized through the M contact and $LS2$, remains picked up. The motor continues to run until, just before the drum rotates 360°, the $LS1$ contact recloses with no further action. Finally, when the drum makes a complete revolution and reaches its original (vertical) position, $LS2$ opens its contact, the M contactor drops out, and the motor stops. The cycle is repeated after the drum is reloaded.

An application of the power-circuit limit switch (Fig. 10·4) is shown in the reversing dynamic-lowering crane hoist controller diagram of Fig. 10·6. Note that the two N.C. limit-stop contacts are inserted in the

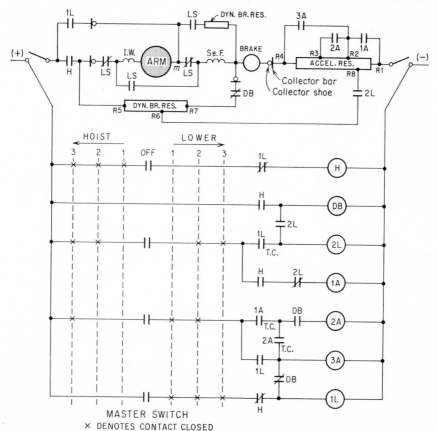

Fig. 10·6 Reversing dynamic-lowering crane hoist with power-circuit limit switch. (See Fig. 10·7 for schematic sketches of various master positions.)

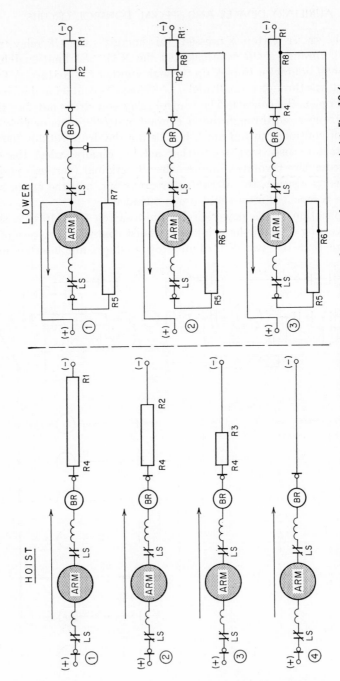

Fig. 10·7 Schematic sketches of various *hoist* and *lower* master positions of master for a crane hoist, Fig. 10·6.

circuit so that they are in series during the hoisting operation (contact $1L$ open) when the machine is connected as a *series motor* and that one contact is in the armature circuit and the other in the series-field circuit during the dynamic-lowering period (contact $1L$ closed) when the machine is connected as a *shunt motor*. Since power current passes directly through the limit-switch contacts, the latter are large and must, moreover, be capable of interrupting high values of current should the crane attempt to enter a forbidden zone while hoisting or lowering. Observe also that four collector shoes riding on collector bars (trolley wires) are provided to permit current to pass into the various electrical components in the various crane structures from the stationary controller cabinet (with its contactors), resistors, and power supply.

The operation of the circuit can be analyzed with the aid of the group of sketches in Fig. 10·7 which show how the motor and associated equipment are connected for each of the three HOIST and LOWER positions of the master switch. As stated above, a series-motor connection is used for hoisting and contactors $1A$, $2A$, and $3A$ close in sequence to give four speeds with a three-point master. Also, the N.C. DB contact is opened by the DB contactor when the H contactor operates; this disconnects the dynamic-braking resistor $R5$-$R7$ at the instant the hoist motion begins.

The $1L$ contactor is energized on all three LOWER points of the master, and this brings current to point m where it divides into two parallel paths. One branch is represented by the armature circuit where reversed current flows with respect to hoist operation, and the other in the series-field circuit where the current direction does not change. Three dynamic-lowering circuit arrangements are obtained as shown.

Pressure and float switches. Pumps and compressors, widely employed in many kinds of industrial, commercial, and domestic installations, are generally driven by electric motors which start and stop automatically as liquid- or air-pressure changes occur at adjusted settings. These systems operate without attention because pressure, vacuum, and float switches act as pilot devices to energize and deenergize the control circuits for the motors. Moreover, unlike the *three-wire circuits* that are controlled with *momentary contact accessories* such as push buttons, the switches indicated above are *maintained contact devices* and function in *two-wire circuits*. Figure 10·8 illustrates how the two types of circuit differ; in the three-wire circuit the motor runs continuously after the START button is pressed momentarily and is stopped by opening the circuit at the STOP button, whereas the motor starts and stops repeatedly in the two-wire circuit to maintain the pressure or liquid level within desired limits.

Figure 10·9 shows a cutaway view of a diaphragm type of pressure

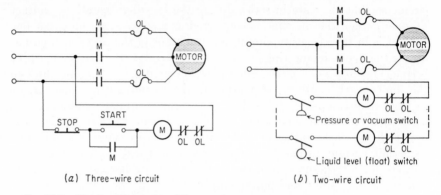

(a) Three-wire circuit   (b) Two-wire circuit

Fig. 10·8   Sketches illustrating difference between three-wire and two-wire circuits.

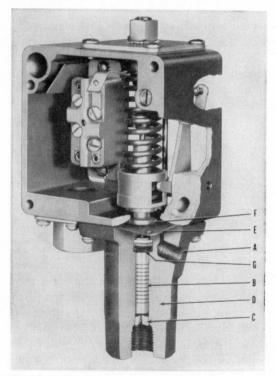

Fig. 10·9   Cutaway view of pressure switch. (F = pressure engagement stud; E = sealing diaphragm; A = oil leakage outlet; G = ring seal; B = steel piston; D = cast iron cylinder; C = pulsation plug.)

switch. Simple in construction with few moving parts, it consists of a diaphragm of oil-resistant (Fairprene) or fire-resistant (Teflon) material that separates the pressure chamber from the pressure system, a lever to magnify the diaphragm action and control the switching mechanism, a compression spring with adjusting screw for regulating the pressure settings, and quick make-and-break switch contacts. During operation, the diaphragm, which is rigidly fastened to the spring, will deflect to open or close the switch contacts should the pressure rise above or fall below the adjusted settings. For standard applications, which are most common, the contacts close to start the motor at low pressure and open to stop the motor at high pressure. Switches that function on reverse operation are also available. When a pressure switch is used for air compressor service where relatively high pressures are involved, it is customary to provide a so-called unloader valve. The latter opens each time the compressor motor stops in order to bleed the discharge line between the compressor and the check valve at the tank so that the motor can restart without load when the pressure switch contacts close again. For reciprocating pumps and compressors, where there are pronounced pressure surges on the system, a pulsation plug serves a useful purpose. As indicated in Fig. 10·9, it is a highly constricted inlet to the diaphragm chamber and effectively dampens the momentary pressure variations so that they will not cause false operation of contacts. Pressure switches are available that will operate contacts at values as low as 1.5 and as high as 5,000 lb per sq in.; particular pressure switches can usually be adjusted to have operating ranges between 3 to 1 and 10 to 1. Also, switches that are designed with large diaphragms and weak springs are often used in vacuum systems where the range of adjustment may be between 3 and 28 in. of mercury.

Gauge-type pressure switches are employed in systems where high pressures, beyond the capacity of the diaphragm design, are encountered. In addition, these units, Fig. 10·10, frequently provide a convenient way to change pressure settings and adjust pressure ranges quickly and include an instrument that indicates the actual pressure in the system at all times. Moreover, such devices make it possible to obtain very accurate settings and wide or narrow differential adjustments. The construction includes a Bourdon tube which

Fig. 10·10   Gauge-type pressure switch.

actuates the gauge pointer and a swinging contact arm, two adjustable stationary contacts, and a relay which handles the pilot circuit of the controller.

Float Switches. These are used with motor-driven pumps to maintain the water level automatically between desired limits. In such devices, pilot switches are actuated by floats whose position indicates the water level. As Fig. 10·11 shows, the switch is operated through a rod (or other

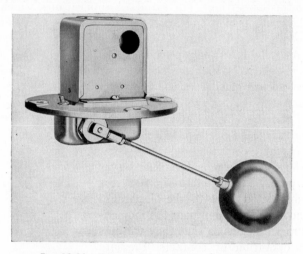

Fig. 10·11   Through-the-tank–type float switch.

mechanism). The switch contacts are closed to start the motor when the water level drops to a low adjustment point and opens to stop the motor when the water level rises to a preset high adjustment point. For small motors, the float switch can usually handle the motor-circuit currents directly, in which case a separate starter is not required. For large motors, however, a separate automatic starter is necessary to control the operation of the motor and the float switch acts as a two-wire pilot device in the control circuit of the starter. The float-switch contacts can be adjusted to open and close between wide or narrow ranges of water level, although a minimum differential is essential if the operating lever is to move sufficiently to actuate the switch. Special float switches have been developed for large industrial pumping stations where several motors, each one driving its own pump, are started and stopped in sequence as the water level changes. They are constructed much like master switches and are usually driven from a float chain through a sprocket; contacts for the individual motor circuits are opened and closed at preset water levels.

Foot switches. Many industrial applications require that the machines or process cycles be started at a time when the operator's hands are both engaged in loading or handling the materials. Foot-operated switches are frequently employed for such purposes, typical examples being resistance welding machines, riveting machines, some types of punch presses, industrial sewing machines, and many kinds of machine tools. These switches are available with such contact combinations as single-pole–double-throw, two-pole–double-throw, and two-stage arrangements, with each stage being single-pole–double-throw. The latter construction is sometimes used for applications requiring the energizing in sequence of two separate but related circuits. For example, in the automatic control of resistance welding operations, closing the first-stage contacts brings the electrodes together, but welding current does not flow through the work until the second-stage contacts are closed. Such an arrangement permits separation of the electrodes before the weld is made if the workpieces do not fit together properly, and a distinct "feel" exists between the first and second stages. Figure 10·12 shows a foot switch of standard design that is available with many contact combinations, and Fig. 10·13 illustrates a construction for a hazardous location.

Fig. 10·12   Standard industrial-type foot switch.

Brakes. Mechanical brakes are used in many kinds of motor application, not only to bring a motor to a quick stop but also to keep a load from moving or drifting under the influence of gravity, wind pressure, or other forces after the drive is brought to rest. Sometimes employed in conjunction with electrical (dynamic) braking, the brake is applied as the drive approaches a stop and acts to hold (or park) the mechanical structure after motion has ceased. Typical examples of installations that require mechanical brakes are hoists, cranes, elevators, and crane bridges (see Figs. 4·7 and 10·6).

Two types of electromagnetically operated brakes are generally used for *industrial* service; these are *band brakes* and *shoe brakes*. A third type, the *disc brake*, has limited application where the load requirements are light and particularly where a compact unit must be mounted directly on the motor shaft. In the latter design, spring-loaded friction plates press against metal discs when the brake coil is deenergized and the plates and discs run free of each other when the brake is energized. In the band brake a strip of friction (asbestos) material is fastened to the inside face of a circular steel band which almost completely surrounds and can be made to touch the outside surface of a brake wheel that is mounted on the motor shaft. With more than 300° contact surface between rubbing surfaces, comparatively large braking torques can be developed, although

Fig. 10·13   Industrial foot switch for hazardous locations.

some difficulty is experienced in maintaining uniform pressures in service and in raising the band completely from the wheel surface when the brake is released.

Shoe brakes. Shoe brakes are widely employed for heavy-duty work because, in addition to being ruggedly constructed, they are reliable, quick-acting, and reasonably trouble-free. As Fig. 10·14 illustrates, they consist primarily of a brake wheel that is mounted on the motor shaft, two brake shoes whose inner surfaces are provided with friction material that covers about one-half of the wheel circumference, and an electromagnet which, when energized, electrically *releases* the normally spring-set brake through a lever mechanism. For a brake that is properly adjusted, the brake shoes move an extremely short distance, about 0.04 in., between the spring-set and electrically released positions, and it is for this reason that operation is fast and shock is reduced to a minimum. When the brake coil is deenergized, a stiff torque spring forces the shoes together so that they press firmly against the wheel and prevent the latter from turning. Adjustments of brake pressure and for brake wear are easily made by turning a loosened lock nut that changes the length of the spring. When the brake coil is energized, the armatures of the electromagnet are attracted to the core with the result that spring force is overcome and the shoes move away from the wheel.

For d-c service, the coils of shoe brakes are designed either to carry motor (armature) power current or to operate directly from the control-circuit source, in which case the action is independent of load conditions. In accordance with NEMA standards, the former, the *series brake*, releases

Fig. 10·14   Shoe brake, showing details of construction.

at 40 per cent or less of rated motor current and, in the released position, will not reset until the current drops to 10 per cent or less of the rated motor current. The latter type, i.e., the *shunt brake*, will release at 80 per cent of rated voltage and will operate satisfactorily without over-heating when the coil voltage runs as much as 10 per cent above rated value. For many kinds of service, and particularly hoists, elevators, and cranes, the series brake has a number of distinct advantages. Wound with comparatively few turns of heavy wire, the coils are extremely rugged and are, therefore, not so subject to breakdown as are constructions that require many turns of fine wire, e.g., shunt coils. Also, because the inductance is low and the flux changes very rapidly when the exciting coil has few turns, the series brake acts quickly to release and set. Moreover, during a power failure or when the load circuit is opened inadvertently by a loose wire or a broken accelerating resistor, the brake will set automatically (no current) to protect an overhauling load application. Shunt brakes, on the other hand, are much slower than their series counterparts but are generally used when the load varies widely and especially under light-load conditions that would not normally release a series brake. They do not, however, provide safety protection against the possibility that a break may occur in the armature circuit, although the difficulty can be avoided practically by connecting a special series relay in the motor armature with its N.O. contact in the brake-coil circuit. With the latter arrangement, the inrush current will energize the series relay and its closed contact will pick up the brake coil, after which the shunt brake will operate; an N.O. interlock on the brake is connected across the relay contact and, being closed after the brake operates, will keep the latter released should the relay drop out on low armature current. Although the arrangement indicated will *not* permit the brake to release with an open armature circuit *while the motor is at rest*, no protection will be afforded should a break occur while the machine is in operation. A shunt brake showing a cutaway view of the exciting coil with its comparatively large number of turns is illustrated in Fig. 10·15.

Fig. 10·15   Shunt brake, showing cutaway view of exciting coil with comparatively large number of turns.

Since the shunt brake is inherently slow acting in contrast with the series type, it is customary to design the shunt coil for half (or lower) operating voltage by using a correspondingly lesser number of turns; the brake will then release more quickly when full line potential is applied to the half-voltage coil. The reason for such improved action is that the time constant L/R of the coil is reduced because the inductance L varies as the *square* of the number of turns whereas the resistance depends *directly* upon the turns of the coil. The quick-acting brake must, however, be protected from overheating after it is released by connecting a resistor in series with the coil. This is accomplished, as in Fig. 10·16, by having a brake relay BR with its N.O. contact across the brake resistor and an N.C. time opening contact M-T.O. in the relay circuit. With the motor at rest and the brake set, the BR relay will be picked up, its contact will be closed, and the brake resistance will be short-circuited. When the motor is started by energizing the M contactor, the brake will release quickly because the resistor is short-circuited and line potential is impressed across the partial-voltage coil. Then, after a slight time delay, the M-T.O. contact will open to drop out the brake relay, which, in turn, will open its contact and insert the resistor in the brake-coil circuit.

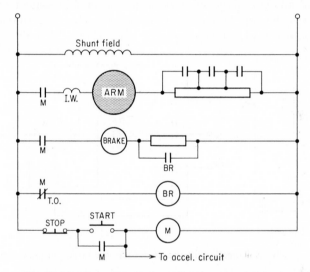

Fig. 10·16   Control circuit for a quick-acting shunt brake.

A brake is frequently selected on the basis of the full-load torque of the motor to which it is coupled, although other conditions of operation may make it advisable to use a unit that is either larger or smaller than that given by the equation for the full-load torque of the motor; i.e.,

$$T_{FL} = \frac{5,250 \text{ hp}}{N} \tag{10·1}$$

On hoist applications, for example, where electrical braking is omitted, a mechanical brake must usually have a torque rating that is somewhat larger than the full-load torque of the motor, i.e., Eq. (10·1), because it must be capable of stopping overhauling loads. The motors for the bridge and trolley motions of a crane, on the other hand, are equipped with rather small brakes, since they function primarily to keep the structure from moving after the latter is brought to a stop by plugging the motors.

The ratings of such brakes, which are used for parking purposes only, are therefore independent of the motor torques, where the latter are determined primarily on the bases of desired rates of acceleration rather than load demands. It should thus be clear that the proper choice of a brake will depend upon the kind of service it must perform as well as the electrical and mechanical equipment to which it is applied.

Brakes in A-C Motor Circuits. Small brakes whose coils are energized directly from an a-c source are operated by solenoids. They are designed to release on 85 per cent of rated voltage and may carry current continuously at an overvoltage of 10 per cent without excessive temperature rise. A sketch showing how the a-c solenoid-type brake is connected in a 3-phase motor circuit is given in Fig. 10·17a. Here, the brake is released

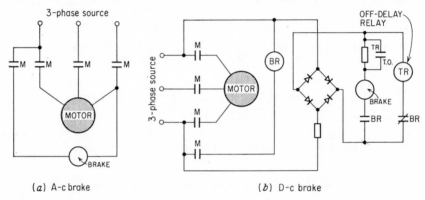

(*a*) A-c brake   (*b*) D-c brake

Fig. 10·17 Sketches illustrating the use of a-c and d-c brakes in a-c motor circuits.

when the brake coil is energized through the auxiliary M contact at the instant the motor is started, and the brake sets very rapidly when the M contactor is dropped out to stop the motor. Direct-current brakes that are energized from an a-c source through selenium rectifiers are frequently used with a-c motors. This arrangement is particularly desirable in the larger brake sizes where the operating mechanism is simpler and more reliable than the a-c constructions. Figure 10·17b illustrates a circuit for an a-c motor and its associated d-c brake. With the motor at rest and the auxiliary M contact open, the brake coil is deenergized and the brake is set; also, under this condition, the TR relay is picked up through the N.C. BR contact, and the TR-T.O. contact is closed to short-circuit the resistor in the brake-coil circuit. At the instant the motor starts and the BR relay is energized, the brake releases quickly because the brake coil is connected directly to the output terminals of the rectifier. Simultaneously, the N.C. BR contact opens to deenergize the OFF-DELAY TR relay which, after a second or so, opens its contact to remove the short circuit across the resistor.

Brake Sizes and Torque Ratings. Mechanical brakes are available in a range of sizes whose torque ratings are generally specified in terms of their ability to prevent loads from moving *after* they are brought to rest. They are, in this respect, *rated* as holding brakes although they normally perform both load-stopping and parking functions. For purposes of retardation, i.e., where a load must be stopped in a specified time, it is, of course, essential that a brake develop sufficient torque to account for the total stored kinetic energy in the moving system.

The holding torque of a brake depends upon the force of the brake shoes against the brake wheel, the diameter of the latter, and the coefficient of friction between the contact surfaces. Since the two brake shoes (Fig. 10·13) cover one-half of the total outside surface the wheel and total force is the product of average pressure and contact area, the braking torque is

$$T = (P \times A) \frac{D}{2 \times 12} c = P \frac{\pi DL}{2} \frac{D}{24} c = \frac{\pi}{48} PD^2Lc$$

where T = torque, *lb-ft*

A = contact area between braking surfaces, *sq in.*

P = average pressure, *lb per sq in.*

D = diameter of brake wheel, *in.*

L = axial length of brake wheel, *in.*

c = coefficient of friction

Assuming an average value of 0.38 for c,

$$T = \frac{\pi \times 0.38}{48} PD^2L = \frac{P}{40} D^2L \tag{10·2}$$

Example 1. The brake wheel of a two-shoe brake has a diameter of 8 in. and an axial length of 3.25 in. If the average pressure between brake shoes and wheel is 20 lb per sq in., calculate (**a**) the holding torque of the brake, (**b**) the horsepower rating of a 1,120-rpm motor to which the brake could be applied.

Solution

a. $T = {}^{20}\!/_{40} \times (8)^2 \times (3.25) = 104$ lb-ft

b. hp $= \dfrac{104 \times 1,120}{5,250} = 22.2$—say, 25 hp

As previously indicated, brakes that are used to stop a moving mechanical system in a specified time must have a torque rating that is capable of converting the kinetic energy of the mass into friction at the brake shoes and brake wheel. Frictional forces may also be present in the moving equipment, but these are generally neglected in making calculations for braking torque unless they are comparatively high. Assuming a rotating load, the average kinetic energy that must be absorbed by the brake in bringing the drive to a stop is

$$\text{KE} = \tfrac{1}{2} T \omega t$$

where KE = total kinetic energy in the moving system, *ft-lb*

T = total retarding torque, *lb-ft*

ω = angular velocity, *rad per sec*

t = time, in *sec*, required to stop the rotating mass *after* the brake sets; it does *not* include the time for the brake shoes to move from released to set positions after the brake coil is deenergized.

But

$$\omega = \frac{2\pi N}{60}$$

where N is the speed of the brake wheel in revolutions per minute. Therefore

$$\text{KE} = \frac{T(2\pi N/60)t}{2} = \frac{\pi N T t}{60}$$

The kinetic energy is also equal to

$$\text{KE} = \tfrac{1}{2} I \omega^2$$

where I = moment of inertia = WR^2/g

W = weight of rotating mass, *lb*

R = radius of gyration of mass, *ft*

g = 32.2, i.e., gravitational constant

Thus

$$\text{KE} = \frac{1}{2} \frac{WR^2}{g} \frac{4\pi^2 N^2}{3,600} = \frac{\pi^2 N^2}{1,800g} WR^2$$

Equating the foregoing two values of kinetic energy,

$$\frac{\pi N T t}{60} = \frac{\pi^2 N^2}{1,800g} WR^2$$

the total retarding torque becomes

$$T = \frac{WR^2 N}{308t} \tag{10·3}$$

Example 2. A 100-hp 650-rpm motor has a WR^2 of 360 lb-ft^2 and drives a direct-connected load whose WR^2 is 600 lb-ft^2. If the installation is equipped with a brake whose torque rating is equal to the full-load torque of the motor, calculate (a) the total time required to stop the motor and its load, assuming that it takes 0.40 sec for the brake to set after the brake coil is deenergized; (b) the number of revolutions the motor makes during the stopping period.

Solution

a. $T_{FL} = \dfrac{100 \times 5{,}250}{650} = 808$ lb-ft

$t_{\text{total}} = \dfrac{(360 + 600)650}{308 \times 808} + 0.40 = 2.91$ sec

b. Revolutions to stop $= $ avg rps $\times t = \dfrac{650}{2 \times 60} \times 2.91 = 15.8$

Table 10·1 lists standard NEMA torque ratings for d-c and a-c brakes. Note that two ratings are given for each size and type of brake, where the

Table 10·1   Standard NEMA torque ratings of brakes

Direct current				Alternating current	
Series brakes		Shunt brakes		Intermittent	Continuous
½ hr	1 hr	1 hr	Continuous		
			3	. . . . .	1.5
		15	10	. . . . .	3
		35	25	15	10
		75	50	35	25
90	60	90	70	75	50
200	135	200	150	160	125
525	350	525	400	400	325
900	600	900	675	800	600
1,800	1,200	1,800	1,350	1,600	1,200
3,600	2,400	3,600	2,700	3,200	2,400

larger values, obtained by using stiffer springs and electromagnets that develop more force, are assigned to the units that are energized less frequently. The different duty cycles are meant to imply that overheating will not occur when the indicated brake services are not exceeded. For the series brakes, the ½- and 1-hr duties generally correspond to those given for series motors and with which they are most frequently used. The shunt brakes and a-c brakes, on the other hand, are usually applied to motors that operate continuously, intermittently, or for 1-hr intervals, and it is for this reason that these duty cycles were adopted for such brakes.

Lifting magnets. In steel mills, railroad yards, foundries, ship operations, scrap yards, and many other allied industries where iron and steel are processed and moved, lifting magnets have become an indispensable part of materials-handling equipment. Made in two general shapes, circular and rectangular, they are used to handle just about every form of steel from scrap, pellets, and pig iron to the various kinds of finished products. A representative group of materials includes castings and molds in foundries; wheels, rails, tie plates, and car springs in railroad yards; ingots, coiled sheet steel, slabs, plates, billets, and pipes in steel mills; and borings, turnings, kegs of nails, and machine parts such as screws, washers, bolts, stampings, and sheets in manufacturing plants.

The circular construction is by far the most popular type of lifting magnet because it permits the operator to handle most classes of material without the necessity of positioning the unit accurately with respect to

Fig. 10·18 Manual-drop magnet controller for a lifting magnet.

the load. Rectangular magnets, on the other hand, are particularly adapted to the handling of regular shapes of iron and steel such as bars, billets, plates, and pipes and are generally used in plants where they can be applied exclusively to this kind of material. Both types are designed so that the magnetic circuit embraces the exciting coil, where the circular construction is a modification of the rectangular configuration in that the two outer poles of the latter are combined into a single continuous pole in the former. The exciting coil is, of course, placed over the center pole so that the mmf produces flux that penetrates into the work to be lifted and returns to the outer pole (or poles) of the magnet. Where the service is very severe and particularly for the movement of regular shapes that are hot, it is customary to use rectangular magnets with *renewable poles*. In addition, special bipolar, duplex, slab- and plate-handling lifting magnets are available for severe, special, and unusual applications.

An important aspect of all lifting-magnet installations is the control circuit that functions to energize and deenergize the highly inductive exciting-coil circuit. For the smaller sizes, *manual drop magnet controllers*, Fig. 10·18, are designed for use with push-button stations, Fig. 10·19, or with three-position masters having OFF, LIFT, and DROP points. Pressing the LIFT button (or moving the master to the LIFT position) closes the *main* contactor in the controller to energize the magnet coil at full line voltage. To drop the load, the DROP button is pressed (or the master is moved to the DROP position); this action deenergizes the *main* contactor and energizes a *drop* contactor to apply reverse potential and discharge the magnet. Release of the DROP push button (or

Fig. 10·19 Push-button station for a manual-drop magnet controller (see Fig. 10·18).

release of the master handle) after the load is dropped removes reverse voltage and allows the magnet to discharge its remaining current through a discharge resistor.

Automatic-drop Magnet Controllers. These are generally employed with the larger magnets and have two-position masters that operate to connect the magnets to the line in the ON position and automatically reverse the current to discharge the magnets quickly in the OFF position; in the completely discharged condition, the reverse current through the exciting coil of a magnet is automatically reversed. A control-circuit diagram for an arrangement such as this is given in Fig. 10·20. When the master switch is moved to the LIFT position and the L contactor picks up,

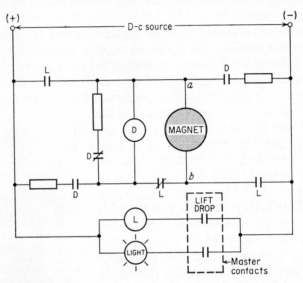

Fig. 10·20 Control circuit for a small lifting magnet with a manually operated master switch.

the two line contacts L close and an N.C. interlock L opens. A lifting operation can then be performed as current passes through the coil from a to b. To drop the load, the master is moved to the DROP position, under which condition the L contactor is deenergized and the three L contacts revert to normal. At the instant power is removed from the magnet and the current starts to decrease, a large voltage is induced in the highly inductive circuit that retards current decay and actuates the D contactor. With the two D line contacts closed, the magnet now proceeds to discharge its energy into the line as the induced emf falls and the current, still flowing from a to b, diminishes. After the current drops to zero, it reverses and, now flowing from b to a, develops an opposing mmf that reduces the residual magnetism. Finally, when the induced emf in the coil falls sufficiently, the D contactor drops out, opens its line contacts, and closes its N.C. D interlock to permit the magnet to discharge its remaining energy into a resistor.

A more elaborate control circuit than that given in Fig. 10·20 is generally used for the larger lifting magnets because, during the lift period, they store considerable energy which must be quickly dissipated without high dielectric stresses when the loads are dropped. Moreover, the extremely high values of exciting current which such magnets take may cause excessive temperature rise unless they are properly applied and protected. The 77-in. unit illustrated by Fig. 10·21, for example, takes currents of about 160 to 120 amp (cold and hot) from a 230-volt d-c source.

Fig. 10·21   77-in. circular lifting magnet operated from an ore bridge.

A control-circuit diagram and the current-time relations for a comparatively large lifting magnet is given in Fig. 10·22. As in the previously described circuit (Fig. 10·20), there are two contactors D and L and a two-position master switch. An important constructional detail of the line contactor L is a copper sleeve over the core which delays the opening of its line contacts when the coil is deenergized. An explanation of the operation of the circuit and the manner in which the current varies with time during the lift and drop periods follows:

1. When the master is moved to the LIFT side, the DM contact opens and the LM contact closes. This energizes the L contactor, which, in turn, closes the main L contacts to excite the magnet coil. Also, two N.O. L contacts close and one N.C. L contact opens, but these actions have no function during the lift operation when the current rises exponentially from O to A in the current-time curves of Fig. 10·22. The line and magnet currents are, of course, the same during this period.

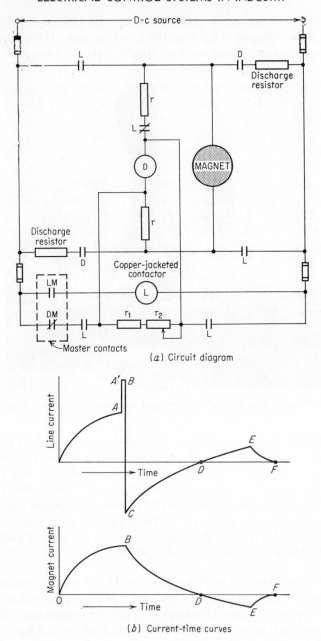

(a) Circuit diagram

(b) Current-time curves

Fig. 10·22 Control circuit and current-time relations for a large lifting magnet with a manually operated master switch.

2. To drop the load, the master is thrown to the OFF-DROP position, under which condition the *DM* contact closes and is followed by the opening of the *LM* contact. Since the *L* contactor drops out with a slight time delay (its core is surrounded by a copper sleeve in which induced eddy currents retard flux decay), the two line contacts *L* and the two N.O. *L* interlocks remain closed momentarily. The *D* contactor, designed for half voltage, therefore, picks up instantly to close its two line *D* contacts, and this action connects the two discharge resistors in parallel and across the line. The line current thus rises from *A* to *A'* in the upper wave sketch, although the magnet-coil current remains unchanged at point *B* in the lower wave sketch.

3. After a slight time delay represented by the period *A'B*, the *L* contactor drops out to open its line and interlock contacts. Simultaneously, the N.C. *L* interlock closes to connect the *D* contactor, in series with two limiting resistors *r*, across the magnet coil. The *D* contactor, therefore, remains picked up because excitation is now provided by a sufficiently high induced emf in the magnet coil. Moreover, since the induced emf is *higher* than, and oppositely directed with respect to, the impressed voltage, the line current reverses (point *C*) whereas the coil current continues to flow in the same direction as it begins its exponential descent. The magnet thus discharges its energy into the line as the negative line current drops along curve *CD* while the coil current diminishes along curve *BD*.

4. As the induced emf falls, it eventually reaches a point when it becomes equal to the impressed voltage. The net circuit voltage is then reduced to zero, under which condition the current drops to zero (points *D* on both curves).

5. The induced emf continues to fall and now drops below the impressed voltage. This causes the line current to resume its original direction, whereas the current in the magnet coil reverses (curve *DE*). A demagnetizing mmf is thereby developed which acts to reduce the residual flux to zero.

6. The induced emf finally reaches a value that is insufficient to keep the *D* contactor picked up; it therefore drops out (point *E*) and permits the magnet to discharge its remaining energy through the *rDr* circuit.

7. The dropout point of the *D* contactor can be adjusted by changing the resistance of the shunt path $r_1 r_2$ at rheostat r_2. If the resistance $r_1 r_2$ is increased, the contactor will drop out later because less current is diverted into this circuit and more current will pass through the *D* contactor to keep it picked up longer. The reverse is true if resistance $r_1 r_2$ is diminished. The proper adjustment is obtained when the residual magnetism is reduced to an extremely low value.

A photograph of a large magnet used in a steel mill to move large rolls of sheet steel is shown in Fig. 10·23.

Fig. 10·23   Circular lifting magnet in a steel mill being used to move rolls of sheet steel

Resistance welders. Two general welding methods are widely used to bond metals together strongly. In one of these, *arc welding*, rodded metal is melted in a high-temperature arc which is then *added* to the heated pieces where they are joined. In a second, *resistance welding*, this function is accomplished *without added metal* by firmly pressing the pieces together where they are to be bonded and having a very high electric current pass through them. By the latter scheme, the secondary of a step-down transformer develops a low voltage (1 to 24 volts) that sends 2,000 to more than 100,000 amp through the workpieces for timed intervals ($\frac{1}{60}$ sec to several seconds) where heat brings the contact surfaces to a plastic state and pressure fuses the metal together. Pressure is usually applied by foot-pedal action in *foot-operated machines;* by toggle, cam, or eccentric action in *motor-operated machines;* or by piston motion in *air-* or *hydraulic-operated machines.*

Depending upon the kind of resistance welding that is to be performed, several types of welders are available, although all operate on the principle that (1) the pieces must be *squeezed*, (2) heat must be localized at definite contact surfaces by the flow of high currents, and (3) pressure must be maintained for a short time after current interruption to permit the plastic metal to solidify. In *spot welders*, copper-alloy electrodes of small diameter ($\frac{3}{8}$ to $1\frac{1}{2}$ in.) with rounded tips clamp the workpieces together and conduct the welding current. The size and location of the welded spot (or spots) are determined by the *projections* embossed on the metal, where several spots can be welded simultaneously using projection welding, or workpieces can be assembled to closer dimensional tolerances. In *seam welders*, copper-alloy welding wheels (width $\frac{3}{8}$ to $\frac{3}{4}$ in.) are rotated under pressure on both sides of the workpieces which move "wringer fashion" as they are bonded together. As current passes between the wheels, a series of overlapping welds or a continuous weld can be made, with a seam width that is determined by the width of the wheel. *Upset* or *butt welders* operate mechanically to butt the workpieces against each other under pressure, and, with current passing between all contact surfaces, welds are made over the entire area. In *flash welders*, which are somewhat similar to upset welders except that the work surfaces touch lightly at first, arcing occurs between the pieces as the metal burns at the surfaces, i.e., *flashes*. The generated heat thus leaves the surfaces in a molten condition, which are then forced together suddenly under high pressure with accompanying increase in current; after a brief "upset time" the current is interrupted.

The system of control for spot and projection welders involves the following three periods: (1) the *squeeze time*, when the electrodes clamp the work in preparation for the weld; (2) the *weld time*, when current flows either for a single weld-time impulse or during heat periods that are separated by cool times within a *pulsation-weld interval;* (3) the *hold time*, when pressure is maintained after current is stopped so that the molten metal can solidify. In repeat welding, electrodes are separated during the *off time* so that the operator can move the work between steps. The foregoing operations are implemented by (1) a *welder contactor*, which must be capable of making and breaking a heavy current (at line voltage) in the primary circuit of the welding transformer, (2) a *timer*, which governs the duration of current flow on foot- or motor-operated machines or electrode motion *and* weld duration on air- or hydraulically operated machines, (3) power equipment that maintains good voltage regulation, circuit breakers, and other safety devices, and (4) accessories that include push buttons, limit switches, foot, air, or hydraulic switches, solenoid valves, and others.

Seam welders have motor-driven wheels that require conventional motor controllers with START-STOP push-button stations. Pressure from an air or hydraulic cylinder is controlled by a solenoid valve that clamps the work between wheels, and a relay under the control of a foot switch energizes the solenoid valve, while a pressure switch initiates the seam-welding cycle as pressure builds up on the work. (Sometimes a two-stage foot switch is used, and welding begins only after the pressure switch and the second stage of the foot switch are both closed.) The seam-welding cycle usually involves alternate heat (current flow) and cool times, repeated rapidly as the wheels make overlapping welds; for this purpose it is generally necessary to employ a synchronous-precision seam timer, an electronic contactor, and an electronic heat-control device. Sometimes it is possible to allow continuous current to flow without building up heat too fast or burning the work; in such cases no timer is required and either electronic or magnetic contactors can be used.

Flash or upset (butt) welders are usually specialized machines that are engineered to fulfill particular specifications. In such installations the control equipment includes cam-operated limit switches whose sequencing and timing actions simulate the mechanical motions.

Welder Circuit Diagram. Figure 10·24 shows the wiring connections of a typical spot welder in which the major components are a welding transformer, two ignitron tubes, two thyratrons, a pulse transformer, and a timer. Other parts include a control relay, resistors, capacitors, rectifiers, and fuses. In addition, a synchronous-precision controller is often used when welding stainless steel, aluminum, nonferrous metals, or small parts and thin sections; operating to prevent variations in weld heat due to saturation of the welding transformer, the current flow for a given weld is always during the a-c half-cycle that is opposite the half-cycle of the previous weld. This is in contrast to nonsynchronous control, which gives random starting welds with accompanying transient conditions that greatly affect the heating action during the succeeding cycles of current flow.

Referring to Fig. 10·24, assume that the CR relay is energized by the timer supply and that the CR contacts are closed in the plate circuits of the thyratrons. Also, with the thyratron transformer excited, one circuit is established by SEC. 1 through $R1$ and 1REC, and another through $R2$ and 2REC. With capacitors $C1$ and $C2$ polarized as indicated, negative potentials of about 55 volts are therefore placed upon the grids of thyratrons $B1$ and $B2$ through resistors $R3$ and $R4$, respectively. Such negative potentials must, of course, be overridden by higher positive potential, superimposed on the grids by the pulse transformer, if the thyratrons are to fire and the welding transformer is to function.

Assume next that the *weld firing means* in the timer makes terminal $H1$ of the *left-side* pulse transformer positive and that a 350-volt peak pulse occurs in the circuit represented by $H1$, $R1$, $R3$, grid-to-cathode of $B1$, and $H2$. This causes thyratron $B1$ to conduct. Current can now pass into the ignitor of the $E1$ ignitron from $L1$ through the CR contact, plate to cathode of thyratron $B1$, $R5$, welding transformer primary, and $L2$. The $E1$ ignitron thus breaks down and conducts from anode to cathode, and a heavy surge of current passes through the welding transformer secondary and the workpieces.

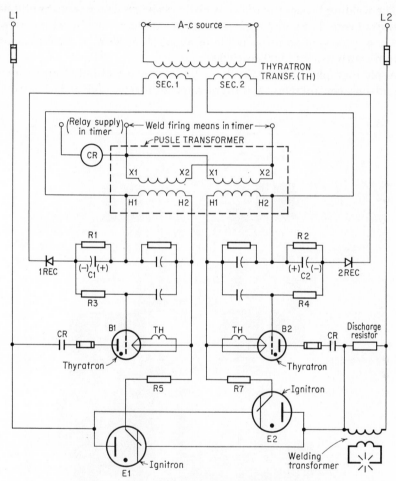

Fig. 10·24 Circuit diagram for resistance-type spot welder.

When $H2$ of the *right-side* pulse transformer is positive, a 350-volt positive pulse is established in the circuit represented by $H2$, $R2$, $R4$, grid to cathode of $B2$, and $H1$. Thyratron $B2$ will now conduct, and current will pass into the ignitor of $E2$ from $L2$ through welding transformer primary, the CR contact, plate to cathode of thyratron $B2$, $R7$, and $L1$. The $E2$ ignitron thus breaks down and conducts, again to permit a heavy surge of current to pass through the welding transformer secondary and the workpieces. The cycle of spot welds is then repeated for as long as the timer is actuated. The discharge resistor, about 400 ohms, permits the welding transformer to discharge its stored energy after each firing operation, since the ignitron cuts off at a voltage equal to the IR arc drop, not zero.

The welding transformer is a specially designed water-cooled unit with a *hypersil* core. For many applications it would have a rating of more than 100 kva, a ratio of transformation of about 20 or 28 to 1, and secondary current outputs of 14,000 to 19,000 amp.

A photograph showing the tube firing panel assembly consisting of thyratrons, control relay, and thyratron transformer is given in Fig. 10·25.

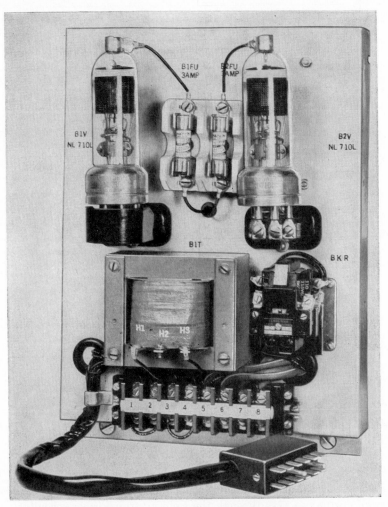

Fig. 10·25 Tube firing panel for spot welder. (*Square D Co.*)

Electric-truck control system. Electric trucks with self-contained storage batteries that operate series motors are widely used to handle and transport material in industrial and processing plants, warehouses, and shipping depots. With its control system designed to function very much like that on an automobile, it contains a foot-operated accelerating master, a hand-operated directional master, a special type of plugging relay, accelerating resistors, and reversing and accelerating contactors. However, unlike most other applications of the series motor, the direction of rotation of the latter is changed by reversing the *field* rather than the armature.

The wiring connections for an electric-truck control system are given in Fig. 10·26. As shown, it consists of a four-position master and a directional master, which permit the operator to control the speed and direction

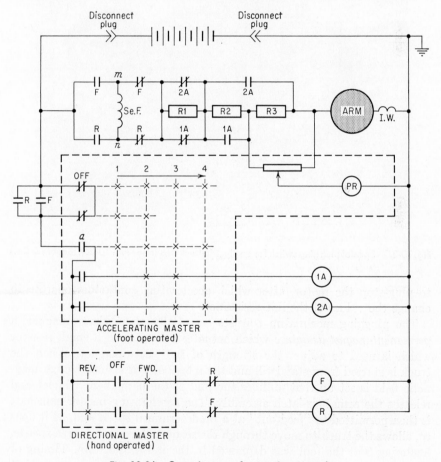

Fig. 10·26   Control system for an electric truck.

of the truck, and a plugging relay that functions either to bring the truck to a quick stop or to decelerate the vehicle rapidly in preparation for a quick reversal. For example, if the truck is moving at full speed in a forward direction, under which condition the foot pedal is pressed down to position *four* and the directional master is turned to FORWARD, the operator can negotiate a reversal quickly and smoothly by (1) moving the directional master to REVERSE, (2) releasing the foot pedal and allowing it to return to the OFF position, and (3) again depressing the foot pedal to position *four*. The actions described cause a unique type of plugging relay, illustrated by Fig. 10·27 and developed by the Square D Company,

Fig. 10·27   Special plugging switch for an electric-truck circuit (see Fig. 10·26). (*Square D Co.*)

to plug-stop the motor, after which accelerating contactors operate to change the values of the line resistance.

The plugging mechanism consists essentially of an electromagnet, a *permanent-magnet armature* which actuates a latch, and a small resistor which is used to adjust the strength of the electromagnet. When the truck is started from standstill and the relay coil *PR* is energized, a magnetic field is set up that nullifies the pull of the permanent magnet and releases the armature-latch assembly. The accelerating master camshaft is then permitted to "feed out" at a predetermined speed and, as it does so, allows the truck to move through the second, third, and fourth speeds, assuming that the foot was depressed to the full ON position. Timing of the relay can be varied between steps by turning adjusting screws that regulate the escape of air in the dashpots; the latter are air escapement devices which control the motion of pistons as the air flows through orifices. During a plug-stop or a reversing procedure, i.e., while the truck is still moving in a given direction, the operator throws the directional master to the other side, removes his foot from the accelerator, and then depresses the pedal again. These actions allow the plugging latch mechanism to pull in a second time and prevent the camshaft from moving out of the first speed until the truck has decelerated to the point where it is safe to apply reversed torque. While the motor is passing through the plugging stage, the cemf in the rotating armature is impressed across the relay coil and, being in a direction to establish an mmf that *strengthens* the permanent magnet field, strongly seals the latching mechanism. As plugging subsides and the motor speed and cemf approach zero, the relay voltage becomes the armature IR drop, which is extremely low. The current through the relay coil then reverses and again weakens the permanent magnet field. The latch now releases, and depending upon the position of the accelerating pedal, the motor speeds up in the opposite direction.

Referring to Fig. 10·26, assume that the directional master is moved to FORWARD and the foot pedal is depressed to the *first* position. The F contactor is thus energized through the N.C. OFF contacts and the closed a contact. With the closing of the N.O. F and the opening of the N.C. F main contacts, current passes through the motor circuit from m to n in the series field, resistors $R2$ and $R3$, and the armature. In position 2, the $1A$ contactor picks up to open the N.C. $1A$ and close the N.O. $1A$ main contacts; this cuts out the $R2$ resistor and leaves the $R3$ resistor in the motor circuit. In position 3, the lower N.C. OFF contact opens and the $2A$ contactor is energized with the latter being picked up through the F interlock which was closed when the F contactor was originally actuated; this opens the N.C. $2A$ and closes the N.O. $2A$ main contacts, short-circuits resistors $R2$ and $R3$, and inserts resistor $R1$ in the motor circuit. Finally, in position 4, the $1A$ contactor drops out to short-circuit the $R1$ resistor, and the motor continues to run on full battery voltage.

To plug-stop and reverse the direction of rotation of the motor, the directional master must *first* be moved to REVERSE. In passing through the OFF position, the F contactor must, of course, drop out and return its main and interlock contacts to normal. Note particularly that this action results because the accelerating master is still on the fourth point and, with the lower N.C. OFF contact open, no circuit is established to the R contactor even though the a contact is closed. However, when the foot pedal is released and the accelerating master returns to the OFF position, the above-mentioned OFF contact recloses and permits the R contactor to pick up as the pedal is again pressed down. Current now passes through the series field from n to m, and as previously explained, the motor plugs with resistors $R2$ and $R3$ in the circuit. Then, at zero speed, the plugging relay permits contactors $1A$ and $2A$ to pick up in timed sequence as the motor accelerates normally in the reverse direction.

Eddy-current brakes and couplings. A useful device that is often used as a brake or to couple a motor to a load magnetically is an *eddy-current clutch* or *coupling*. Its braking or coupling action is based upon the principle that eddy currents, induced in a metal drum that surrounds a set of rotating magnets, create linking magnetic fields whose poles are strongly attracted to the rotating structure. When employed as a brake, it can function to retard an overhauling load, act to maintain constant tension on unwind stands in processing paper, film, textiles, and metal, and provide artificial loading to prevent backlash and chattering on machine tools. As a coupling, it can serve as a flexible link between a constant-speed induction motor and a varying-speed load whose range is wide and stepless.

Figure 10·28 shows a cutaway view of an eddy-current clutch magnetically coupled to a squirrel-cage induction motor for speed-control purposes. As the unit is assembled, a *coupling drum* in the form of a steel ring, smooth on its inner surface and ribbed on the outside to dissipate heat effectively, is fixed to the squirrel-cage rotor. Fastened to an independent shaft that lines up with the motor shaft and drives the load is the field member of the coupling. This consists of an annular coil of wire (doughnut-shaped) enclosed on three sides by magnetic steel, with projecting teeth that are concentric with the inside surface of the coupling drum and separated from it by a small air gap. Note particularly in the illustration that the squirrel-cage shaft—the driver—and the output shaft—the driven member—are structurally independent, with the inside surface of the coupling drum magnetically linked to the radial teeth of the spider. When the coil is energized with direct current, a strong magnetic field is produced which, encircling the coil, passes through the steel casing and crosses the air gap into the drum.

During operation the coupling-drum assembly is rotated by the squirrel-cage of the induction motor at practically constant speed, but the inner field structure remains stationary until voltage is applied to the coupling coil. Then, when the latter is excited, eddy currents are generated in the drum which, in turn, form a pattern of magnetic poles on the inside of the moving member. Forces of attraction are thereby created between the poles of the field and those of the drum, with the result that torque is developed to drive the output shaft. The field therefore tends to follow the drum in rotation.

Since relative motion between the two members of the magnetic coupling is essential if eddy currents are to be generated in the drum, it should be clear that the speed of the output shaft can never exactly match that of the squirrel-cage rotor. This means that an eddy-current coupling is basically a *slip* device, i.e., it can transmit torque only when the load shaft is rotating at a slower speed than the motor shaft. Moreover, since

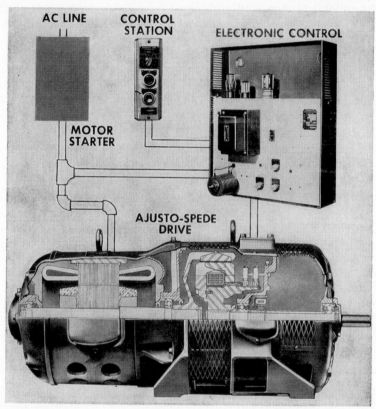

Fig. 10-28 Cutaway view of an induction-motor with eddy-current clutch assembly and an electronic control unit.

the magnitude of the developed torque depends upon the strengths of the two sets of magnets—those created by the excited coil and the poles generated in the drum by eddy currents—speed control is possible by simply adjusting the direct current in the field that excites the coupling coil. The same principle applies, of course, to the braking action of the device when it is used as a brake to retard an overhauling load or to maintain constant tension in a processing line.

Another point that should be emphasized is that slip in this type of magnetic coupling is always accompanied by eddy-current heat losses in the drum. Such losses are in this respect equivalent to those produced in the external resistors of a wound-rotor type of motor when the speed of the latter is reduced. It is for this reason that the coupling drum is usually constructed with an integral fan that provides the necessary flow of air to carry off the slip-power heat loss.

The controlled direct current required for the excitation of the eddy-current clutch coil is generally obtained from an electronic control unit, Fig. 10·29, which is designed to modulate the rectified power and follow

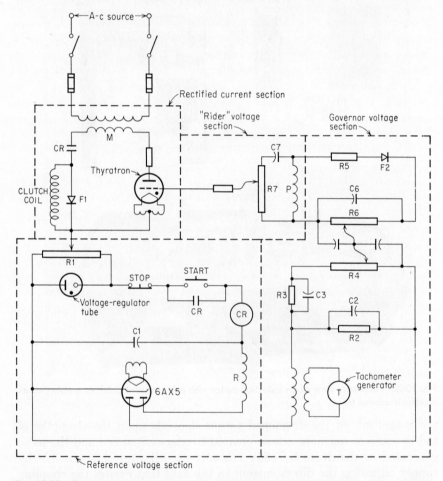

Fig. 10·29 Excitation circuit to control the speed of a squirrel-cage motor with an eddy-current clutch.

automatically the varying load demands; the latter, through a unique feedback system, maintains a constant-load speed for given rheostat adjustments. A circuit diagram showing the complete wiring connections for this method of speed control is given in Fig. 10·29. It consists essentially of four sections, all of which are interconnected to energize the clutch coil for definite and recurring periods of time and at such values of current that a desired speed is maintained. After the main transformer is connected to a single-phase a-c source, voltages appear in the three secondaries labeled M, R, and P. These then energize (1) the rectified current section, (2) the reference voltage section, and (3) the rider voltage (phase-shift) section. A fourth section—the governor voltage section—is energized by the secondary of a small independent transformer whose primary is connected to a tachometer generator mounted on and driven by the output drive shaft. When the motor is in operation with all speed adjustments properly set, the four sections act simultaneously upon a thyratron that permits the proper value of pulsating current to excite the clutch coil. As is well known, the grid of the thyratron initiates the "firing" or conducting period of the current cycle and loses complete control after conduction begins and until the current wave again passes through zero.

Referring first to the rectified current section, observe that the clutch coil is energized during the half-cycle that the thyratron conducts current, i.e., when current passes from anode to cathode; under this condition rectifier $F1$, connected as shown across the coil, behaves like an open circuit. During the second half of the cycle, however, the thyratron does *not* conduct, but the stored energy in the highly inductive clutch coil is released to establish a current through the rectifier, now properly connected to permit the current to pass. Thus, a smooth flow of current through the coupling coil is maintained for the nonconducting half-cycle of the tube.

The major speed-control adjustment of the motor is the function of the reference voltage section, where potential divider $R1$ is manipulated to apply a desired positive potential to the grid of the thyratron with respect to its cathode; as the positive potential is increased by moving the slider to the right, the motor speed is raised. The a-c power for this circuit originates at the transformer secondary R and is rectified by one-half of the twin-diode vacuum tube 6AX5; capacitor $C1$ acts as a reservoir, taking current on the positive half of the cycle and releasing it to the circuit during the "off" half of the cycle. Operation for this section, as well as that of the rectified current section, is started by pressing the START button; this energizes the control relay CR, which picks up to close one contact for the clutch-coil circuit and another to seal in the relay coil. Also, to establish a stable voltage source, uninfluenced by possible line-voltage fluctuations, a gas-filled voltage-regulator tube is included in the circuit as shown. This tube, with relay coil CR as a ballast resistor, is capable of regulating at approximately 105 volts. Thus, voltage across the speed control potentiometer $R1$ is maintained at a constant value.

In series with the reference voltage section, which always supplies a definite *positive* potential on the grid of the thyratron to set the speed at which the drive is to operate, is one section that exercises a speed-governing function and another section that controls the sensitivity or response to load changes. The governing action originates at a small tachometer generator T, mounted on the load shaft, whose alternating voltage varies directly with the speed. After being stepped up by an independent transformer, it is rectified by the other half of the 6AX5 vacuum tube, the latter serving the dual function of rectifying the alternating current in two circuits. The direct voltage which then appears across $R2$ is filtered by capacitor $C2$ so that smooth direct current passes through resistors $R3$ and $R4$. Also part of the governing section is a minor rectified current circuit that originates at transformer secondary P and passes through resistor $R5$ and rectifier $F2$ and then potential divider $R6$. Note particularly that $R4$ and $R6$ are so connected that they tend to *subtract* from the positive potential that is impressed on the grid by $R1$; i.e., their potentials with respect to $R1$ are negative.

After the controls $R1$, $R4$, and $R6$ are set to impress a definite *positive* potential on the grid, the thyratron fires for measured periods of time to energize the clutch coil at the desired effective current; the exciting current is then responsible for the torque that must be developed to drive the load at the selected speed. Moreover, any attempt on the part of the load shaft to slow down (or speed up) is immediately accompanied by a tachometer-voltage change whose reduced (or increased) negative potential raises (or lowers) the positive potential on the grid correspondingly. As a result, the original speed setting is maintained, assuming, of course, that the torque remains unchanged.

Since the thyratron must pass rapidly through successive periods of conduction and nonconduction, the grid voltage must be increased to its critical positive value (at which point the tube fires), once for each positive half-cycle of the voltage wave. This is accomplished by superimposing an a-c "rider" voltage on the grid so that it is 90° out of phase with respect to the anode voltage. Also, to permit this wave the necessary degree of control, it is desirable to use a simple phase-shifting circuit similar to that illustrated by the rider voltage section. Energized by the transformer secondary P, potential divider $R7$ can be manipulated to *raise* or *lower* the rider voltage and make the thyratron fire at any point on the voltage wave. Thus, with a given emf on the grid, the phase position of the rider can be adjusted to neutralize the critical negative potential of about 3 volts at any desired point on the anode voltage wave. When this occurs near the start of the half-cycle, the tube fires sooner and more effective current is delivered to the clutch coil; for shifts toward the finish of the half-cycle, the exciting current is reduced correspondingly.

Although the eddy-current clutch method of speed control is extremely wasteful of energy, particularly at low speeds, it does possess the following advantages: (1) The speed is progressively and steplessly adjustable; (2) the controls are simple and involve rheostats that are small; (3) very little control power is used; (4) the feedback feature of control makes it possible to maintain a constant speed for a given speed-torque setting; (5) starts and stops are soft, with the result that acceleration and deceleration are smooth; (6) speed changes are made without the tendency to overshoot a desired setting.

Alternating-current adjustable-speed drive with rectifiers and coupled d-c motor. Semiconductor rectifiers are being applied increasingly in industrial control systems to improve and extend the operating performance of power equipment. One such application utilizes a bridge-connected set of six silicon rectifiers in the secondary circuit of a wound-rotor motor to supply slip power to a coupled d-c motor for speed-adjustment purposes. Developed by Westinghouse Electric Corporation under the trade name *Rectiflow drive,* it is constructed with the rotors of the a-c and d-c machines on a common shaft and all necessary control components including rectifiers. Depending upon the selection of the motors in the drive, it can provide continuously variable speed control with speed ranges of $1\frac{1}{2}$ to 1, 2 to 1, or 3 to 1, with a maximum speed of about 1,690 rpm.

A diagram showing the circuit connections for the *Rectiflow* control system is given in Fig. 10·30. Note particularly that the electrical slip-frequency power in the wound rotor, normally dissipated in secondary

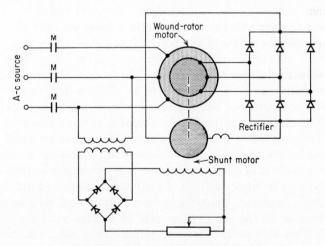

Fig. 10·30   Rectiflow adjustable-speed drive.

resistors when the motor speed is to be reduced, is changed to direct current by the rectifiers and is delivered to the armature of the d-c motor; the field of the latter machine is energized from the a-c source through a transformer and another bridge-connected rectifier. During operation, the cemf of the shunt motor opposes the rectified voltage of the a-c machine, and their difference is just sufficient to circulate the required currents in the wound rotor and the armature for the particular load and speed; under this condition, the mechanical power output of the d-c motor is added to that of the a-c motor with the result that the slip power is made available to the drive instead of being wasted in resistors, as indicated above. Should the field of the d-c motor be strengthened, the cemf will increase, the resultant voltage will decrease, and the armature and wound-rotor currents will diminish. Since these changes cause both motors to develop *less* torque, the drive decelerates to a lower speed; then, as the speed drops, both the rotor-slip voltage and the cemf increase. Deceleration eventually ceases when the two voltages (rotor and armature) differ by an amount that is just necessary to circulate the currents required by the load torque at the new speed.

Conversely, if the d-c field is weakened, the cemf will drop, the resultant voltage will increase, and the armature and wound-rotor currents will rise. The set then accelerates to a higher speed and stabilizes at a value that again permits sufficient current to flow for the torque demanded by the load.

The set can be started either at reduced voltage or full voltage. With the closing of the line contacts M, the shunt field and the a-c stator winding are energized simultaneously, and because the flux in the former builds up with a time lag, a gradual increase in torque permits the machines to accelerate smoothly. Constant-horsepower characteristics are developed by the basic drive, although constant-torque operation is possible if a suitable d-c motor of the proper frame size is selected.

The brushless alternator. When a special source of alternating current must be provided for such applications as offshore drilling operations, field oil-drilling equipment or, in general, where explosive or extremely wet ambient conditions prevail, it is desirable or even essential that a brushless type of alternator be employed. This eliminates all sliding contacts and avoids the destructive effects that result from arcing.

The unique feature of brushless alternator is its use of a 3-phase bridge-connected silicon rectifier (see Fig. 10·30) which, mounted on the shaft of the rotating machine, connects the a-c output of a *rotating* 3-phase exciter alternator to the d-c *rotating* field of the energy converter. The entire excitation circuit, i.e., exciter, rectifier, and d-c field, therefore, rotates as a unit and eliminates the need for slip rings and brushes. A diagram showing how the various parts of such a system are interconnected is given in Fig. 10·31.

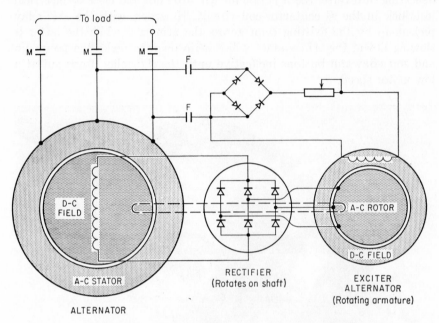

Fig. 10·31   Connections for the brushless alternator.

The primary source of mechanical power for the alternator is generally a diesel engine which drives the field-rectifier-exciter assembly, while the stationary d-c field of the exciter, energized from the output terminals of the alternator, builds up through a bridge-connected rectifier. A rheostat in the latter circuit is used to adjust the voltage output of the exciter which, in turn, regulates the d-c excitation and the a-c output voltage of the alternator.

The rotating section, a brushless alternator, clearly showing the large d-c field and the small a-c armature, is illustrated by Fig. 10·32.

Electronic control circuit for d-c motor. For the purpose of starting and controlling the speed of a d-c shunt motor when the latter is operated from an available 3-phase source, an electronic drive is sometimes convenient. A circuit diagram showing how this can be accomplished using three thyratrons and three diodes, together with such other auxiliary equipment as transformers and a bridge-connected rectifier (Figs. 9·37, 10·30, and 10·31), is given in Fig. 10·33. Here, provision is also made for reversing, dynamic braking, and jogging in a *forward* direction only. In addition, an antiplug relay AP is included to prevent plugging from a *reverse* running direction. If the STOP button is pressed while the motor is operating in reverse, the R contactor will drop out and close its electrical interlock in the F contactor-coil circuit. However, with the AP relay picked up by the existing cemf across the armature while the latter is slowing down, the AP contact will remain open to make the FORWARD and JOG FORWARD buttons ineffective until the AP relay drops out at a low motor speed.

Fig. 10·32   Rotating section of brushless alternator. (*Kato Engineering Co.*)

The shunt field is excited from the 3-phase supply through three transformers connected delta-wye and a bridge-connected rectifier. A field-economy resistor is used to limit the field current and heating while the motor is at rest. After the F or R contactor is energized, an F or R interlock closes to short-circuit the field-economy resistor and that portion of the field rheostat that is cut out.

During operation, the three grid-controlled thyratrons "fire" in succession at 120° intervals and, with corresponding diodes, permit current to pass through the armature in one direction only—from *left* to *right* for *forward* running or from *right* to *left* for *reverse* running. Thus, when phase $L1$-$L2$ is more positive than $L2$-$L3$ and $L3$-$L1$, current will flow from $L1$ through thyratron $T1$ (from anode to cathode), through the armature, and back to $L2$ after passing through diode $D2$ (from anode to cathode). Thyratron $T2$ will "fire" next when phase $L2$-$L3$ becomes most positive,

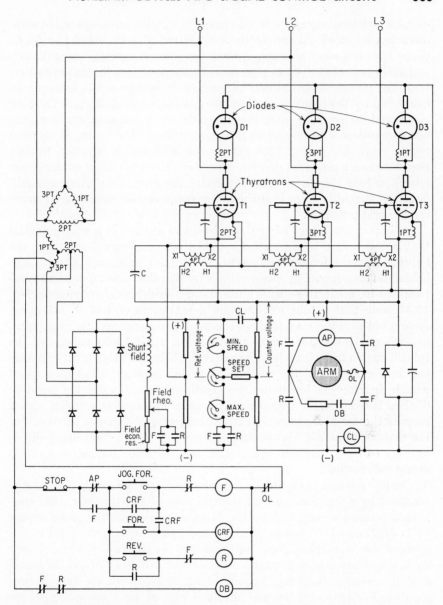

Fig. 10·33 Circuit diagram for operating a shunt motor from a 3-phase source through electronic (thyratron and diode) devices.

under which condition current will pass through the armature in the same direction, as tubes $T2$ and $D3$ become active. Finally, when the $L3$-$L1$ phase becomes most positive, $T3$ will fire and, in conjunction with $D1$, will again permit current to pass through the armature in the same direction. The period of current flow through each thyratron will, however, be controlled by the point on the positive portion of the half-cycle the tube is allowed to conduct, and this depends upon the magnitude of the *reference voltage* with respect to the *countervoltage*. The former is supplied by the same d-c output that excites the field and is adjusted by two rheostats and a potential divider. Moreover, since the firing points determine how long the current passes through the armature during each 360° period, it should be clear that the motor speed can be controlled by varying the grid potential.

An important aspect of the speed-control system is a grid potential that is represented by a constant a-c rider potential superimposed on the adjustable d-c reference emf. With the former so chosen that it lags behind its corresponding anode voltage by 90°, the firing point of a thyratron can be made to depend upon the d-c grid bias. Referring to Fig. 10·34, note that a tube can conduct a maximum of 120° in a 3-phase system but that the magnitude of the d-c bias determines the firing point, i.e., where the a-c rider wave intersects the critical grid voltage of the tube. Thus, in (a), with a large negative d-c bias, the firing point m occurs late in the cycle and the conducting period is small; the motor speed would, therefore, be low under this condition. In (b), where the negative d-c bias is reduced, the thyratron fires earlier in the cycle, at point n, and permits the armature to carry current longer; this increases the speed of the motor. If the d-c bias is made positive as in (c), the firing point is advanced to point p; with nearly 120° of conduction the motor speed is raised still further.

The 90° relation between the a-c rider emf and the anode potential is obtained by using three special transformers $4PT$, with their delta-connected primaries energized by the filament secondaries of transformers $2PT$, $3PT$, and $1PT$ and their star-connected secondaries applied to the grids of the thyratrons. Note particularly that the grid of $T1$ is excited by a secondary whose primary is connected across $1PT$ and $2PT$ and that the latter voltage lags behind $L1$-$L2$ by 90° while the tube is conducting. Furthermore, the grid of $T2$ is excited by a secondary whose primary is connected across $2PT$ and $3PT$, and that voltage lags behind $L2$-$L3$ by 90° during the firing period of the tube. The grid of $T3$ is excited by a secondary whose primary is connected across $3PT$ and $1PT$, and that voltage lags behind $L3$-$L1$ by 90° when the tube fires.

A useful component of the circuit is a capacitor C that provides a cushion for the application of the reference voltage; its action is to smooth acceleration and limit inrush currents to safe values for the thyratrons and armature until the current-limit circuit assumes control. The grids are negatively biased when the motor is started and gradually build up a positive bias through capacitor C from the d-c source. Current-limit operation is provided by introducing a negative bias to limit the armature current to a predetermined value during acceleration or when the drive is overloaded. The operating coil of the current-limit relay CL is connected across a resistor in the armature circuit, and the voltage drop across that unit is directly proportional to the armature current. The relay is generally set to pick up on an armature current of about 150 per cent of full-load value; when it does so, a negative bias is placed on the grids by the closing of the CL contact.

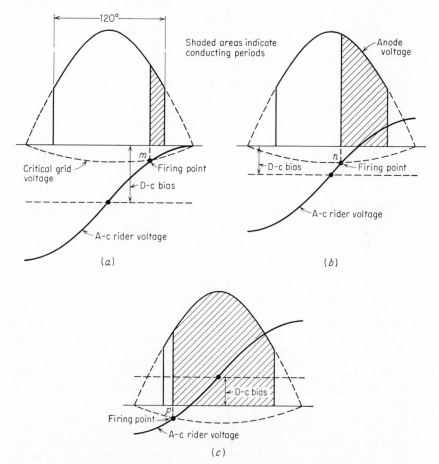

Fig. 10·34 Curves illustrating how d-c bias determines the firing point and conduction period of a thyratron.

QUESTIONS AND PROBLEMS

1. Describe the operation of the control circuit of Fig. 10·5 for the drum unloader application. Assuming the two limit switches to be in their normal positions at the instant the START button is pressed, approximately how many degrees should the drum rotate before (a) $LS2$ closes? (b) $LS1$ opens? (c) $LS1$ recloses? (d) $LS2$ reopens? What types of limit switch are used?

2. Referring to Fig. 10·6 for the dynamic-lowering crane hoist circuit, why is it necessary to have *two* power-circuit limit switches?

3. Distinguish between two- and three-wire circuits, indicating when each type is used. Give several practical examples of two-wire circuits.

4. List three general types of mechanical brakes, and indicate their fields of application.

5. State the standards under which *series-* and *shunt-type* shoe brakes must operate.

6. Indicate which type of brake, *series* or *shunt*, is more desirable with regard to the following: ruggedness; speed of operation to release or set; application to widely fluctuating loads; control-circuit simplicity.

7. Prove that the time constant of a shunt brake, i.e., the time to release, is shortened by designing the exciting coil for reduced-voltage and applying full-voltage at the instant the motor is started (see Fig. 10·16).

8. It is desired to use a series brake that has a 1-hr torque rating of 350 lb-ft with motors whose full-load operating speeds are about 400, 600, 900, and 1,200 rpm. What should be the horsepower ratings of the motors at the given speeds?

9. A two-shoe brake (Fig. 10·14) has a brake wheel whose diameter and axial length are, respectively, 12 and 3.5 in. If the developed torque is 300 lb-ft and the coefficient of friction is 0.35, calculate the average applied pressure.

10. A series brake has a torque rating that is 10 per cent more than the full-load torque of the 75-hp 575-rpm motor to which it is applied. If the WR^2 of the motor is 180 lb-ft² and the direct connected load has a WR^2 of 320 lb-ft², calculate (a) the total time required to stop the rotating system, assuming a brake setting time of 0.35 sec.; (b) the number of revolutions the motor makes during the stopping period.

11. If the load in Prob. 10 is geared to the motor and operates at one-third the speed of the latter, calculate the stopping time.

12. Why is it necessary to discharge the stored energy in a lifting magnet when the load is dropped?

13. Referring to the control circuit of Fig. 10·20 for a small lifting magnet, answer the following questions: (a) What causes the D contactor to pick up when the master is moved to DROP? (b) What is the direction of the current through the magnet and to the line while the magnet potential is greater than the source voltage? when the magnet potential is less than the source voltage? (c) How does the magnet discharge its remaining energy after the D contactor drops out?

14. Referring to the control circuit of Fig. 10·22 for a large lifting magnet, answer the following questions: (a) Why is it desirable to have contactor L copperjacketed so that it drops out with a time delay? (b) What is the purpose of the two discharge resistors? (c) Why does the line current reverse at the instant the D contactor picks up? At what instant does the magnet current become zero? When does the magnet begin to discharge its energy? (d) What happens to the remaining energy in the magnet after the D contactor drops out?

15. Distinguish between arc and resistance welding; spot, seam, flash, and butt welders.

16. Referring to the control circuit of Fig. 10·24 for a resistance welder, answer the following questions: (**a**) Under what conditions will thyratron $B1$ or $B2$ fire? (**b**) After a thyratron fires, say $B1$, what actions take place and what circuit is established to perform a momentary welding operation? (**c**) What is the purpose of capacitors $C1$ and $C2$, and how must they be polarized?

17. In the control-circuit diagram of Fig. 10·26 for the electric truck, list the main contact openings and closings as the master is moved from the OFF position to position 4. For each position state which resistors are cut out of the armature circuit.

18. In Fig. 10·26 with the truck moving in a *forward* direction and the master in position 4, explain in the proper sequence what happens: (**a**) when the directional master is moved to REV; (**b**) when the foot pedal is released and the accelerating master returns to the OFF position; (**c**) when the foot pedal is again depressed to position 4.

19. Carefully describe the operation of an eddy-current clutch, stressing particularly (**a**) the relative speeds of the induction motor and coupling drum, (**b**) speed-control means, (**c**) the way in which torque is developed, (**d**) power losses.

20. Referring to control-circuit diagram Fig. 10·29 for the eddy-current clutch answer the following questions: (**a**) Why does the clutch coil carry pulsating direct current during both halves of the a-c cycle? (**b**) How does the position of the slider of potential divider $R1$ determine, in the main, the positive potential applied to the grid of the thyratron? Trace the circuit for this portion of the control. (**c**) What functions do the tachometer generator and its associated components have in this system of control? Exactly how is this accomplished? (**d**) Why does the firing period of the thyratron determine the operating speed of the motor? (**e**) How does the "rider" voltage section function to cause the thyratron to fire at a particular point on the conducting half-cycle?

21. In the adjustable-speed drive (Rectiflow) of Fig. 10·30, how is the slip-frequency power of the main motor converted to useful power? What is the effect of increasing the excitation in the shunt-field circuit? Is this similar to the action that occurs when the field of a shunt motor is strengthened? What should be the rating of the shunt motor compared with that of the wound-rotor motor if the speed of the latter is to be adjustable down to about 50 per cent of synchronous speed?

22. Complete the control circuit for the brushless alternator application of Fig. 10·31, and indicate the sequence of operations to develop voltage at the output terminals.

23. Referring to the electronic control-circuit diagram of Fig. 10·33 for a d-c shunt motor, answer the following questions: (**a**) How does the antiplug relay AP prevent plugging from a reverse running direction? Trace the circuits carefully. (**b**) How is the field energized? (**c**) Which tubes are active when phase $L1$-$L2$ is most positive? when phase $L2$-$L3$ is most positive? when phase $L3$-$L1$ is most positive? (**d**) How are the potentials on the grids of the thyratrons adjusted to control the firing time of these tubes and the speed of the motor? Refer to Fig. 10·34 in discussing this important control function. (**e**) What purpose is served by the capacitor C? (**f**) How does the CL relay exercise its current-limiting function? Trace the circuit to the grids of the thyratrons for a closed CL contact.

CHAPTER 11

Regulators and Regulating Systems

General aspects of regulators. Industrial control systems for processing installations that handle, manufacture, and move materials generally involve several (or many) independent motorized sections whose operations are properly coordinated. The driving machines must, in addition, be very closely regulated, so that electrical quantities such as current, voltage, and power, and mechanical functions like speed, torque, tension, and position are held within narrow limits. Moreover, since the motors are usually separated widely in a long line, as, for example, in the annealing, tinning, tempering, and slitting of sheet steel or in packaging and conveyor installations, the regulating equipment must make interdependent sections operate as a unit if material is to move smoothly. To fulfill the conditions indicated, several basic types of closed-cycle regulator are employed which, as explained in Chap. 9 in connection with rotating and magnetic amplifiers, are energized in part by some function of the output and in such a manner as to minimize deviations from desired values.

A regulator consists essentially of five fundamental parts all of which are interrelated and perform simultaneously to maintain required output. These are (1) one or more sources of power that serve both to drive the various components and to excite the control circuit, (2) a signal-sensing device such as a tachometer generator, a photoelectric tube, or a properly selected voltage drop which varies directly with the measured quantity and provides an electrical feedback signal, (3) a reference source such as a constant-potential transformer or a closely regulated voltage supply against which the feedback signal is compared, (4) an error-sensing device which accepts the error, i.e., the difference between the feedback and reference signals, and (5) an amplifier which, after magnifying a compara-

tively weak error signal, returns sufficient power to the system to regulate
the latter properly. It should be pointed out in connection with error-
sensing devices that the amplidyne and Rototrol circuits illustrated and
described in Chap. 9 are *electrically* operated, since *voltages* are compared,
while the magnetic amplifier circuits in the same chapter compare
magnetic quantities to perform similar regulating functions. Both arrangements have definite advantages as well as limitations and are widely used;
in some rather extensive systems both are often employed, each type
being applied where it serves most effectively.

Current and voltage regulators. Two magnetic-amplifier types of
regulator that are widely used in industrial processing systems are *voltage
regulators* and *current regulators*. These are generally provided, often with
others that may also regulate such mechanical functions as speed and
position, to keep voltages and currents at reasonably constant levels in
the various portions of a complex electrical installation. As previously
indicated, a reference winding in such a regulator receives a signal from
a constant-potential source somewhere in the system, and this quantity
is compared with a normally varying feedback emf or current in another

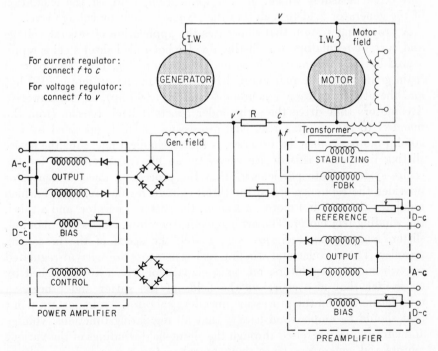

Fig. 11·1 Wiring connections for a current or voltage regulator in a Ward Leonard
system.

winding to yield a corrective signal that is enlarged by the magnetic amplifier. The output of the latter then energizes the field of a generator or motor whose emf or speed is to be regulated. Moreover, in systems that involve considerable values of power, i.e., large generators and motors, it is customary to employ preamplifier–power-amplifier combinations (see Fig. 9·36) to obtain comparatively high levels of output power as well as good response.

Figure 11·1 shows the connections of a two-stage magnetic amplifier, i.e., preamplifier and power-amplifier sections, that is designed to regulate either current or voltage in the loop circuit of a Ward Leonard control system. For current regulation the feedback signal is derived from a voltage drop across a resistor R as indicated, whereas voltage regulation results when the signal originates at the armature terminals vv' of the generator. In either case, the output of the preamplifier and, in turn, the power delivered to the generator field by the power amplifier are determined by the relation between the feedback and reference mmfs. After all currents have been carefully adjusted in the various windings, any actions which attempt to change the current (in the current regulator) or the voltage (in the voltage regulator) are accompanied by appropriate corrective measures which, after amplification, readjust the excitation of the generator field to maintain the proper current or voltage level.

A processing system that illustrates the application of several voltage and current regulators in a slitting line for hot-rolled sheet steel is represented by the elementary (single-line) diagram of Fig. 11·2. Two pairs of main generators, one pair being driven by an induction motor (250 hp) and the other pair by a synchronous motor (1,000 hp), supply power to five motors that drive the strip under constant back tension from the mandrel to the winding reel. The processing uncoiler, operated by two 200-hp motors connected in series, is the mechanical pacer for the entire slitting line, since this device serves as an electrical reference for the entire system. Note particularly that the processing uncoiler generator provides the reference signal for three voltage regulators, one of which regulates the mandrel motor, a second, the slitter generator, and a third, the winding-reel motor. The arrangement, therefore, causes the mandrel, slitter, and winding-reel speeds to match the speed of the processing uncoiler. The uncoiler and winding-reel generators are current-regulated by two current regulators, reference signals for which are furnished by an exciter that also energizes the fields of the slitter and processing uncoiler motors on the processing uncoiler generator. An important point that should be emphasized here is that all regulating functions, voltage and current, are exercised through the shunt-field windings of the various motors and generators, since they control motor speeds and, by voltage adjustments, generator output currents.

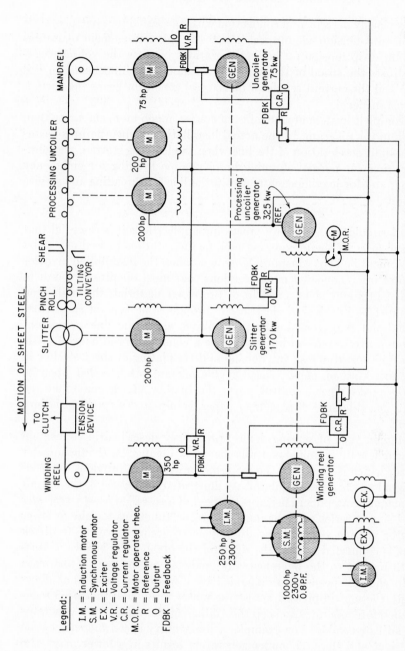

Fig. 11-2 Elementary diagram of a slitting line for processing hot-rolled sheet steel. Regulation is provided by three voltage regulators and two current regulators.

Legend:

I.M. = Induction motor
S.M. = Synchronous motor
EX. = Exciter
V.R. = Voltage regulator
C.R. = Current regulator
M.O.R. = Motor operated rheo.
R = Reference
O = Output
FDBK = Feedback

As previously described in connection with an amplidyne control system for a constant-tension windup reel (Fig. 9·10), this arrangement maintains a constant strip tension for coiling operations because the voltage regulator adjusts the speed of the winding-reel motor to compensate for coil buildup and the current regulator on the reel generator keeps the current at a constant level.

The complete system of control is, of course, much more elaborate than is shown in the diagram. In the actual installation, it incorporates means for time-limit acceleration of the line when the latter is started, regenerative braking for normal stopping and dynamic braking for emergency stopping, and for jogging each motor in the line for threading and service operations.

Speed regulators. In many processing applications it is essential that the linear speed of the moving material or the rotational speed of revolving equipment be kept constant. Moreover, when the installation involves a group of interconnected motor sections that must operate as a unit to move a continuous sheet or strip of processed material, the individual drives must be interlocked electrically through some type of *speed regulator* that not only coordinates the motions along the line to prevent pile-up and excess tension but maintains a constant speed as well. Also, in such mill operations as galvanizing and tin plating of sheet steel or in the manufacture of paper, speed regulators are often called upon to maintain *precise* speed control, since the quality of the manufactured product, i.e., thickness of plating or paper, is impaired if speed variations do occur.

A common type of sensing device, previously described in connection with the magnetic-amplifier control schemes of Fig. 9·39, is the tachometer generator which is driven by the main-motor drive and provides the feedback signal to the speed regulator. As illustrated by Fig. 11·3, the output of the magnetic amplifier energizes the field of a so-called *booster generator* which, being in series with the main generator, functions to raise or lower the motor voltage as speed variations tend to occur. Since the feedback winding is connected so that the signal of the tachometer is always *negative* with respect to the reference, any tendency on the part of the main drive to change speed is accompanied by a feedback signal that alters inversely the emf impressed across the motor terminals; the speed regulator, therefore, operates to keep the voltage of the tachometer generator at the adjusted value. For example, a momentary speed increase of the main drive (and the tachometer generator) results in a lower magnetic-amplifier output, a reduced booster-generator voltage, a drop in the potential across the main motor, and a return to the desired speed.

As noted, the speed regulator behaves very much like a voltage regulator for which a tachometer generator originates a signal that acts to maintain proper system operation. It is for this reason that the rotating sensing device is designed to develop a comparatively high value of volts-per-revolution, since the latter quantity affects the response and degree of sensitivity of the regulator.

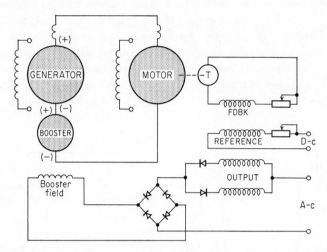

Fig. 11·3   Wiring diagram for a speed regulator employing negative tachometer-generator feedback and a booster generator.

Position regulators. Another device that is frequently employed in a continuous-processing system in which material must be made to "track" accurately as it travels over rolls and moves along a line and/or through a machine is a *position regulator*. Its function in such installations is to shift the rolls laterally as the work attempts to drift from a true centered position. The requirement indicated is particularly important in printing applications where the moving sheet paper or webbing must be made to "line up" exactly with the machine through which it passes. Also, when the printing process involves several colors, each one applied in an independent section, it is essential that the position or register control be very precise. The feedback signals that actuate the position regulators in these presses originate at assemblies of photoelectric cells and lights that are focused on marks or dots along the edge of the moving strip.

In still other continuous-processing operations it is often necessary to maintain a slack loop at one or more places in the line. Here, the strip hangs down loosely, festoonlike, between pairs of rolls, and a position regulator keeps the vertical length of the loop constant by maintaining the same entrance and exit speeds. Then, during a shutdown on the in-going side to change reels and splice a new end to one that is finished, the processing continues uninterrupted as it uses up a large part of the material in the so-called *accumulator*. When the splice is completed, the regulator proceeds to increase the entrance speed until the loop again reaches its full length, at which point normal operation continues. A similar procedure would take place at the windup end of the line where the loop in the accumulator is permitted to lengthen as the exit end is stopped to replace a full reel with a new (empty) one; under this condition the position regulator would increase the exit speed at the loop after reels are changed and proceed to readjust the vertical distance to its normal position. An elementary wiring diagram showing how a magnetic-amplifier type of position regulator is connected into the entrance end of a continuous processing line is given in Fig. 11·4. Note that the magnitude of

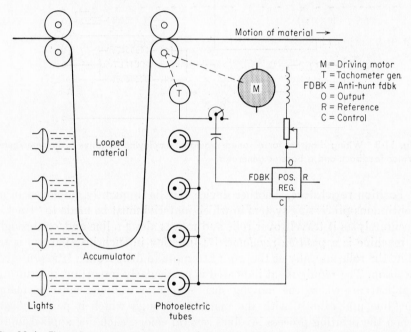

Fig. 11·4 Elementary wiring connections of a position regulator to control the loop length in an accumulator for a continuous process line.

the control signal is determined by the number of light beams that are able to pass across the loop to the corresponding photoelectric tubes and that the magnetic-amplifier output regulates the speed of the motor through shunt-field control. Also, a tachometer generator, coupled to the drive rolls, furnishes a feedback signal that effectively prevents hunting.

Photographs illustrating a four-unit light source and its corresponding photoelectric cell assembly for a loop-control system are depicted in

Figs. 11·5 and 11·6; note the small electronic amplifier at the bottom of the phototube cabinet with the cover removed.

Fig. 11·5   Four-unit light source for a loop-control system. (*Cutler-Hammer, Inc.*)

Fig. 11·6   Four-unit photoelectric assembly with electronic amplifiers for a loop-control system. (*Cutler-Hammer, Inc.*)

Counter-voltage regulator. In a constant-tension processing system, previously described in Fig. 9·10 in connection with an amplidyne control scheme, two operating conditions must be kept constant; these are (1) the driving motor current and (2) the strip speed in feet per minute. This is because tension is equal to T/R and is also proportional to $I(\phi/R)$, since $T = kI\phi$, where T and R are, respectively, torque and reel radius. Thus, if the motor current is not permitted to change and the ratio of ϕ/R is kept constant, the strip tension will not vary as the reel unwinds at the beginning of the line. For the latter condition, a decrease in radius is always accompanied by a corresponding increase in motor speed (in revolutions per minute) that results from a reduced value of field flux.

A current regulator is used to perform the first of the two functions; it is connected as shown in the elementary diagram of Fig. 11·7 to act on

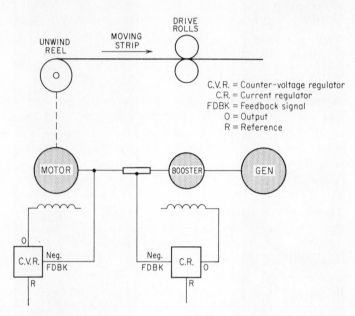

Fig. 11·7 Elementary diagram showing a counter-voltage regulator and a constant-current regulator in a continuous process line for sheet steel that maintains constant tension.

the field of a booster generator (in series with the main generator), receiving a negative feedback signal from a dropping resistor located in the circuit between the booster and motor. As the current attempts to drop with the unwinding reel (torque demands are reduced), the feedback signal to the magnetic-amplifier regulator increases the booster field excitation which, in turn, raises the voltage to the motor and readjusts the current to its original setting.

The *countervoltage regulator* as shown acts on the field of the motor and in such a manner as to weaken the excitation, to increase the speed, as the reel unwinds. Thus, as ϕ and R diminish simultaneously and by equal amounts, the ratio of ϕ/R remains constant; the strip speed, which is equal to fpm $= 2\pi R \times$ rpm, therefore, remains unchanged, since rpm rises as R decreases. This type of regulator is, in reality, a special adaptation of the voltage regulator (Fig. 11·1) because its output is an emf that varies with the potential applied to the driving motor. The reason for the countervoltage designation is that the current in this control arrangement is kept constant by its associated current regulator. Since $V = E_c + I_A R_A$ and $I_A R_A$ is constant, it follows that $E_c = V - k$, where $k = I_A R_A$; thus, any voltage change V is effectively a countervoltage change E_c.

During operation here, when the booster emf increases the motor voltage as the reel unwinds, a negative feedback signal to the counter-voltage regulator causes the output of the latter to decrease. This reduces the field excitation to the motor and, in turn, raises the rpm speed to maintain a constant strip speed.

Processing line involving five basic types of regulator. The foregoing articles described five basic types of regulator that are frequently used in continuous-processing systems. Depending upon the complexity of the installation and the extent to which the power equipment must be regulated, one or more units may be required. Moreover, when the system is highly automated and involves a multiplicity of motors that are electrically interlocked, several of each type of regulator may be necessary if precise regulating functions are to be provided. This is particularly true in manufacturing plants where sheet steel is annealed, tempered, tinned, and slit in long, continuous lines. In such cases it is generally necessary to maintain constant tension at the unwind and windup reels as the material moves smoothly from one coil to the other through various stages, and this requires not only that the proper regulators be carefully selected, matched, and located but that they be adjusted when the equipment is installed.

The elementary diagram of Fig. 11·8 illustrates a portion of a complex processing line for steel strip in which voltage, current, speed, position, and countervoltage regulators are employed. Note particularly that each unit (a magnetic amplifier in this control system) acts on the *shunt field* of an appropriate rotating machine and that a constant-potential source provides reference signals to the first three of them. As previously explained, the current and countervoltage regulators function to maintain constant tension in the steel as it unwinds, while the speed regulator, through its negative feedback signal from the tachometer generator, causes the drive motor to run at constant speed and thus provide constant strip speed. Since one main generator furnishes the power to the three operating motors (in each case through a booster generator) and its output voltage must not be permitted to vary, a voltage regulator is used as shown. Observe also that coupled motor-operated rheostats (M.O.R.) are included in the reference-signal circuits of the voltage and speed regulators because these units are, in effect, interrelated "pace setters" for the line and must, therefore, work together. The speed regulator functions to keep a given length of free-hanging sheet steel in the accumulator so that material will be available and fed out to the outgoing section when the ingoing section is shut down for a reel change.

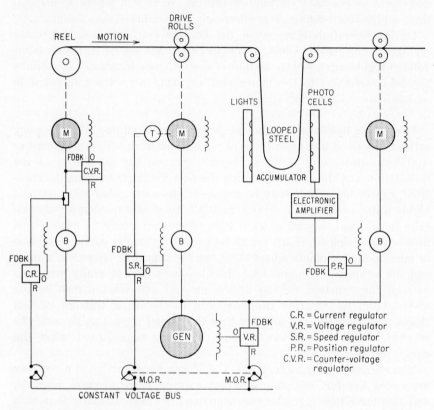

Fig. 11-8 Elementary diagram showing five basic types of regulator in a continuous-process line for sheet steel.

Continuous-processing system for linoleum printing. It should be clear from what has already been said about continuous-processing systems that each of the motors in a multimotor drive, although operating its own section of the line, must be electrically interlocked with all the other machines. Such interdependent mode of operation is essential if a smooth flow of material at carefully regulated tensions is to be ensured. As explained, a practice that frequently accomplishes this objective is one that employs several types of regulator (Fig. 11·8) whose outputs, determined by the interaction of reference and feedback signals, energize motor- and generator-field windings.

Another system of control that illustrates how a group of motors can be interconnected electrically in a continuous-manufacturing operation involves a so-called rotary printer for linoleum. Since this scheme of regulation is unique in that it incorporates several familiar components that were previously considered as well as others of more recent development, it will be desirable to discuss it in some detail. It does indeed illustrate the many typical problems that are involved in the design of a complete control system. The very elaborate equipment for this installation includes five major motors and coupled drive rolls that produce a steady flow of material as the latter passes from unwind to rewind stands through in-feed and out-feed accumulators, a web-cleaning unit, and a printing press. During normal operation, the press section of the processing line prints colored patterns on asphalt-impregnated felt base, called *web*, and printing is not interrupted when rolls are changed at the unwind and rewind stands. Figure 11·9 shows the general arrangement of the power and mechanical equipment and indicates, without the control units, the flow sequence from left to right.

Unwind Stand and In-feed Rolls and Accumulator. The unwind stand consists of two mandrels to permit the operator to make a quick change from an unwind reel to one that is full as uninterrupted processing continues. A special type of drag brake torque regulator is applied to each mandrel and, receiving a low-voltage (50-mv) signal from an armature-circuit shunt, maintains a constant tension of 2 lb per inch of strip width at the incoming portion of the line. Power to drive the in-feed pull rolls is provided by a 5-hp 1,750-rpm shunt motor, and a disc brake is coupled to the machine to execute a quick stop when necessary. A packaged magnetic-amplifier control unit, similar to that of Fig. 9·38 and represented by the wiring diagram of Fig. 9·37, is used for the motor, and provision is made so that the operator can select manual, splice, or automatic position. The speed of the drive is preset when the line is run in the automatic position, and splice and manual speed adjustments are made possible by speed-setter rheostats in the splice or manual positions. During automatic operation, the pull roll speed is determined by the position of a so-called *dancer-roll rheostat* which measures the height of the material in the in-feed accumulator and continuously adjusts the motor speed to keep the same length of material in the storage compartment at all times.

An elementary diagram showing the arrangement of mechanical and electrical components in the incoming section of the line is given in Fig. 11·10. Note that the magnetic amplifier has four inputs, three of which set the mode of operation [manual, automatic, jog (splice)] at a selector switch (not shown) and a fourth, a voltage feedback signal, which regulates the voltage output of the amplifier. An interesting addition to the control circuit is the above-mentioned dancer-roll rheostat which acts to change the motor speed as the vertical length of material in the

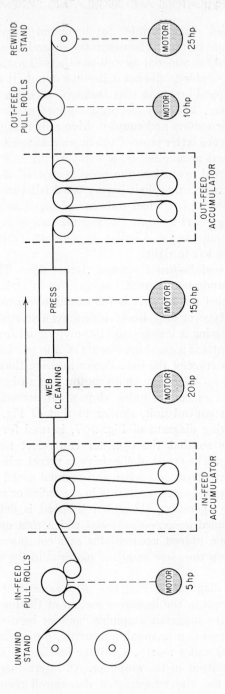

Fig. 11·9 Arrangement of power and mechanical equipment for a linoleum-printing processing line.

402

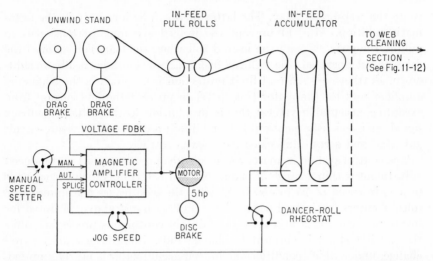

Fig. 11·10 Elementary diagram of the incoming section for a linoleum-printing processing line.

accumulator attempts to lengthen or shorten. During normal automatic operation it is desirable to keep a given amount of strip stored in the in-feed accumulator, to do which the motor speed must speed up or slow down as changes occur. Thus, for example, should the loop lengthen, the rheostat arm will move down, diminish the shunt-field resistance, and slow down the motor; for a shortened loop, the rheostat arm will move up to increase the shunt-field resistance and speed up the motor. Figure 11·11 illustrates a rear view of a dancer-roll rheostat and indicates how a chain-and-gear mechanism turns the movable arm.

For the purposes of safety, a web-break detector relay is incorporated in the drive to shut down the in-feed pull rolls automatically should the web break before it comes to the end of the roll or if the tail end of the web unwinds from the mandrel and is unnoticed by the operator; under these conditions, tension loss will ac-

Fig. 11·11 Rear view of a dancer-roll rheostat.

tuate the web-break relay. The latter device functions only in the automatic position. Also, to prevent the in-feed accumulator tension from pulling the strip through the in-feed roll when the main contactor of the drive is not operated, the pull roll disk brake sets automatically on shutdown. Furthermore, current-limit protection is provided in the magnetic-amplifier unit for the motor (Fig. 9·37) to prevent the line current from exceeding acceptable values; this is in addition to the voltage feedback signal, as previously mentioned, to maintain a constant armature-circuit potential and a constant speed for a given setting.

The control system is, moreover, flexible enough to permit motor-speed adjustment within a range of about 292 to 1,850 rpm, which corresponds to a strip speed of 75 to 450 fpm. After the web leaves the in-feed pull rolls, it enters the in-feed accumulator and, as noted above, is stored for future use when it is necessary to maintain continuous processing while the unwind stand is stopped to change rolls. Immediately after a roll change, under which condition the in-feed accumulator is nearly emptied, the motor accelerates to maximum speed, 1,750 rpm, and the in-feed speed reaches 450 fpm. Since the latter speed is higher than press speed, the accumulator begins to fill, and when it gets within 30 in. of its full position, the dancer-roll rheostat slows down the in-feed speed section to match that of the press section.

Web Cleaning Brush and Press Sections. The web cleaning brush section is powered by a 20-hp 1,750-rpm d-c shunt motor and, as in the in-feed pull roll section, is controlled by a static (magnetic-amplifier) unit. The latter is fed by a preamplifier whose reference is the output of a tachometer generator coupled to the motor that operates the succeeding press section. A similar arrangement of preamplifier and magnetic-amplifier control unit is provided for the 150-hp d-c press-drive shunt motor. Since the tachometer generator also furnishes a closed-loop feedback signal for the press section, the two motors tend to operate at approximately the same speed. In addition, a voltage feedback from the web cleaning motor (20 hp) is magnetically compared with the tachometer reference signal to give a regulated output with respect to input. Refer to Fig. 11·12 for the circuit diagram.

As in the uncoiler section, the speed range of the web cleaning motor is about 6 to 1, i.e., from 292 to 1,750 rpm, and acceleration is controlled by a current-limit relay in the magnetic-amplifier circuit (Fig. 9·37). However, in practice the speed of this motor is usually preset and is permitted to remain unchanged to match the press speed approximately; under this condition the web cleaning brush will not rotate too rapidly at low press speed or too slowly at high press speed. An added feature is a TEST-RUN selector switch which permits the operator to test the web

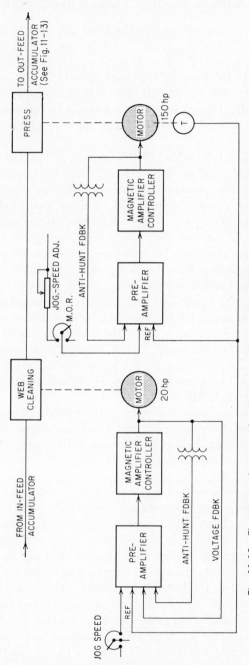

Fig. 11-12 Elementary diagram of web-cleaning and press sections for a linoleum-printing processing line.

cleaning brush; in the TEST position the press will be shut down while the brush is running.

An important aspect of the press-motor controller is a *motor-operated rheostat* M.O.R. in the preamplifier input circuit. The latter unit receives a timed signal through the M.O.R. that sets the acceleration rate of the motor at approximately 20 sec from low speed to full speed. Moreover, with the tachometer feedback as indicated, speed regulation is kept within ± 1 per cent.

Ample protection is provided in this portion of the line against excess current and possible malfunction. As in all magnetic-amplifier controllers, a current-limit relay prevents overcurrents during motor-acceleration periods and when overloads occur. In addition, two kinds of stop are provided, namely, (1) a *normal* stop which permits the motor to coast to a standstill and (2) an *emergency* stop which incorporates 150 per cent dynamic braking. The press is, moreover, interlocked electrically with the web cleaning unit so that the press cannot run when the former is in the test position. Furthermore, malfunction is avoided by interlocking the press section with the in-feed and out-feed accumulators. Thus, if the in-feed drive is at rest and the accumulator is emptied, a limit switch will initiate a normal press stop, and if the limit switch fails, an emergency dynamic braking stop will occur. Again, should the out-feed accumulator be filled beyond its normal range, the press will be brought to rest on a normal stop or will be given an emergency dynamic-braking stop in the event of a limit-switch failure.

The press is normally controlled from the operator's panel but can be jogged or stopped from several other locations spread along the line. Tensions in the press are held at approximately 10 lb per in. by mechanical means, and maximum speed is set for 300 fpm.

Out-feed accumulator and pull rolls. After leaving the press section, the web progresses into the out-feed accumulator, where the material from the line is again stored to permit continuous processing in the web and press sections when it is necessary to make a change from a full rewind reel to an empty rewind reel. The out-feed pull rolls which continue to move the web to the rewind stand are operated by a 10-hp 1,750-rpm d-c shunt motor, the latter being fitted with a disc brake as in the case of the in-feed rolls motor. However, unlike the first pull-rolls section which is equipped with a magnetic-amplifier controller (Fig. 11·10), the second one incorporates an M-G (Ward Leonard) set as shown in Fig. 11·13 because the motor is frequently required to regenerate. The tension of the web as it enters the pull rolls is 2 lb per inch of strip width but is raised to 5 lb per in. between pull and windup rolls to ensure tightness in the latter.

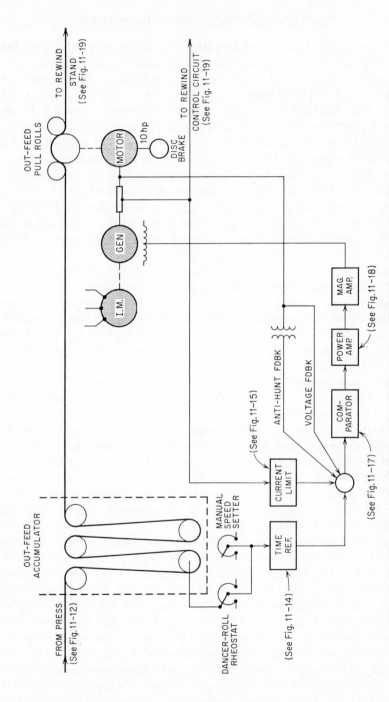

Fig. 11-13 Elementary diagram of out-feed accumulator and out-feed pull rolls for a linoleum processing line.

The speed signal for the out-feed drive can originate at either a dancer-roll regulator that is actuated by the out-feed accumulator rolls or a manually operated rheostat. This signal is then fed into a so-called *comparator* (Fig. 11·17) through a *time-reference* module, Fig. 11·14, the latter being designed to accelerate and decelerate the motor at definite preset times. Time-limit provision is necessary for this drive because it is essential that the reel accelerate faster than the out-feed pull roll. (Unless this condition prevails, i.e., if the pull rolls accelerate more rapidly than the rewind reel, a slack loop will be thrown in the web between these two stands.) After the signal passes through the comparator where it undergoes both voltage and current gains, it is further amplified by a succeeding power amplifier (Fig. 11·18); this component then drives the magnetic amplifier whose output energizes the field of the generator that supplies power for the out-feed pull-rolls motor.

Another feature of this section is the use of a *current-limit module* (Fig. 11·15) which measures the current in the armature loop of the M-G set. A modified signal then passes in turn through the comparator, the power amplifier, and the magnetic amplifier to limit the armature current to adjusted values of 100 to 200 per cent of the full-load rating. That is, the current-limit module is adjusted so that, during motor acceleration or when other transients occur, the armature voltage is not permitted to reach a value that is too high for the existing motor speed.

The Time-reference Module. The purpose of this component is to convert a *step* input voltage to an output voltage that changes *linearly* with time and, in addition, provide a means whereby the rate of voltage rise (or fall) can be adjusted. It can, in this respect, replace a motor-operated rheostat (Figs. 9·35 and 11·8) and regulate the rate of acceleration (or deceleration) of a drive motor. The time-reference module performs this function in Fig. 11·13 by impressing its linear changing output voltage, after amplification, across the field of the generator that controls the speed of the out-feed pull-rolls motor.

The device operates on the principle that the voltage across a capacitor will change linearly if it is charged or discharged at a constant rate. As illustrated by Fig. 11·14, it utilizes one transistor $T1$ to control the charging current and another transistor $T2$ to control the discharging current. During the charging period a definite potential difference is applied between collector and base of $T1$ (base positive), and this results in a collector current that is proportional (or equal) to the emitter current. During this same period a potential difference is applied to $T2$ between collector and base *in the reverse direction*, under which condition it behaves like a diode that conducts in a forward direction. To discharge the capacitor, a fixed emf is connected to collector and base of $T2$ (base positive) while a reverse voltage is applied to the same elements of $T1$; for this mode of operation the capacitor is made to discharge at a constant rate, since $T1$ now behaves like a diode that conducts in a forward direction. The emitter currents of the transistors, and therefore the capacitor charging and discharging currents, are established by setting potentiometers $P1$ and $P2$ in the auxiliary rectifier power supply section.

Referring to Fig. 11·14, assume that a 50-μf capacitor is initially uncharged and that $P1$ is adjusted to establish an emitter-base current of

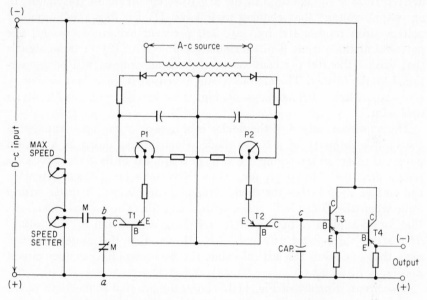

Fig. 11·14   Wiring connections for a time-reference module.

0.75 ma in transistor $T1$. If, next, an emf of, say, 30 volts is applied to terminals ab (by pressing a RUN button to open an N.C. M contact and close an N.O. M contact), the capacitor will charge at a constant current of 0.75 ma and the voltage between a and c will rise linearly to 30 volts in 2 sec. This is because $t = RC = (E/I)C$, from which

$$t = \frac{30 \times 50 \times 10^{-6}}{0.75 \times 10^{-3}} = 2 \text{ sec}$$

When the potential across the capacitor reaches 30 volts (equal to the input voltage), charging current will cease abruptly. During this interval, transistor $T2$ will act as a short circuit.

To alter the charge or discharge rate it is merely necessary to adjust potentiometers $P1$ and $P2$. As the foregoing equation and example indicate, a decrease in charging current will increase the charging time, and vice versa.

Transistors $T3$ and $T4$ are emitter-follower current amplifiers which deliver current outputs that are many times their inputs at slightly lower voltage levels.

The Current-limit Module. This device is used in conjunction with the magnetic-amplifier control circuit of Fig. 11·13 to convert an input signal, derived from a voltage drop in the armature circuit of the d-c motor, to an output voltage that changes as follows: (1) Up to a predetermined voltage drop represented by, say, 100 per cent armature current, the potential at the output (load) terminals will be zero; (2) for input signals that exceed the 100 per cent value, the output voltage will be proportional to the former. The characteristic indicated is somewhat similar to the current-limit arrangement of Fig. 9·13 for the hoist drive of an excavator.

The nonlinear output of the device is obtained by taking advantage of the unique property of a *Zener* diode which breaks down at a critical potential when an increasing d-c voltage is applied to its terminals *in the reverse direction.* Moreover, for values above the breakdown potential, the voltage drop across the diode remains nearly constant *at the critical value* when the Zener is in series with a resistor. This means, therefore, that for all applied voltages E that exceed the critical emf E_z, the voltage drop across the resistor will be $E - E_z$. However, in the range of applied emfs that are below the critical value, the Zener acts like an open circuit and $E_z = E$.

The circuit diagram of Fig. 11·15 shows the internal connections of the current-limit module, where its input is a voltage across a dropping resistor in the loop of the M-G set. As designed for the system of Fig. 11·13, no voltage will appear across the load resistor (a magnetic-amplifier winding) until the impressed emf exceeds 6.5 volts. Above the 6.5-volt input value, the output potential will rise toward a maximum of about 1 volt, under which condition current-limit operation begins. Thus, when adjustments are properly made for a minimum-current-limit setting of $I_{A_{FL}}$, the resistance of the dropping resistor will be $R_s = 6.5/I_{A_{FL}}$.

Since these devices are designed for magnetic-amplifier applications where the originating signal may be an alternating emf, it is necessary to employ a filter (an LC section) to minimize the ripple. If an unfiltered signal were applied to the Zener diode circuit, the peaks of the varying wave that are higher than 6.5 volts would appear at the output terminals and give a false current-limit signal. For d-c systems such as this linoleum drive, however, the filter can be readily removed by using jumpers and disconnect links.

A photograph illustrating a current-limit module is given in Fig. 11·16.

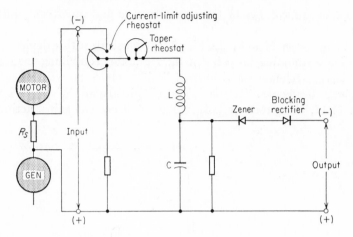

Fig. 11·15 Wiring connections for a current-limit module. (Polarities shown for deceleration.)

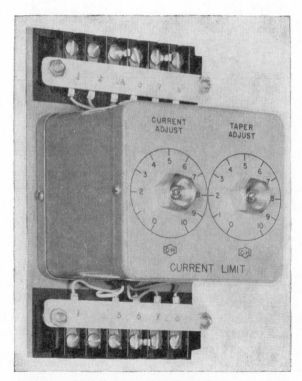

Fig. 11-16 Current-limit module. (*Cutler-Hammer, Inc.*)

The physical appearance of the time-reference and other components is similar.

The Comparator Module. As Fig. 11·13 illustrates, this device is used at the signal summing up point of the regulator, combining the outputs from the time-reference and current-limit modules as well as antihunt and voltage feedbacks. The module, a circuit diagram of which is given in Fig. 11·17, also produces both voltage and current gains, acting as a

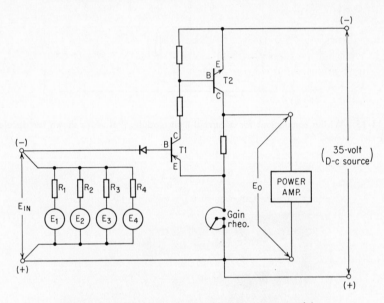

Fig. 11·17   Wiring connections for comparator module.

two-stage transistor amplifier with adjustable feedback around the first stage and around both stages. Under normal operating conditions, a 0.35-volt input signal produces base and emitter currents, respectively, of 0.11 and 5 ma when the GAIN rheostat is set for, say, 25 ohms and carries 10 ma. Thus, with a 0.1-volt drop between emitter and base of $T1$ and a drop of $0.01 \times 25 = 0.25$ volt across the GAIN rheostat,

$$E_{in} = 0.1 + 0.25 = 0.35 \text{ volt}$$

The output voltage and current, with properly selected fixed and load (magnetic-amplifier winding) resistances, will then be about 30 volts and 10 ma. For the values given, therefore, the voltage gain will be

$$E_{out}/E_{in} = 30/0.35 = 86$$

and the current gain will be $I_{out} = I_{in} = 10/0.11 = 91$. $T2$ is a power transistor which has a maximum current rating of 10 ma. All signals sources (E_1, E_2, E_3, and E_4) and their respective series resistors (R_1, R_2, R_3, and R_4, each usually made equal to 500 ohms per input volt) are connected in parallel and to the input terminals as indicated.

The Power-amplifier Module. This device is a linear power amplifier that is used to couple the low-power-output signals from the comparator (Fig. 11·17) and a magnetic amplifier (Fig. 11·13) which, in turn, energizes the shunt field of the main generator.

It consists of fixed resistors, a Zener diode, a blocking rectifier, and a germanium-type power transistor, Fig. 11·18. The latter is connected as a current regulator to minimize output variations that result from line-voltage and temperature changes.

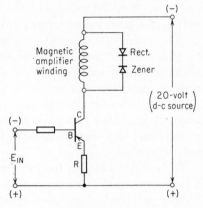

Fig. 11·18 Wiring connections for power-amplifier module.

The collector current that feeds the load (the magnetic-amplifier winding) is directly proportional to and many times greater (about ninety times for this application) than the base current. The emitter and collector currents are, in fact, nearly equal, since the base current is extremely small. Also, neglecting the emitter-base voltage drop, the required input voltage, for a required output current, will be the product of the desired current and the feedback resistance R.

Note particularly that the magnetic-amplifier winding is connected in series with the collector of the transistor and is shunted by two diodes. One of these is a Zener diode, which, as previously explained for Fig. 11·15, has a critical breakdown potential, say, 15 volts, and the other is a simple blocking rectifier. The Zener has two functions: (1) It protects the transistor from high induced emfs in the magnetic-amplifier winding, since the voltage in the latter is limited to the peak ("spillover" potential of, say, 15 volts) of the Zener diode; (2) speed of response of the magnetic amplifier is improved by the forcing action of the Zener, since the latter maintains a constant 15-volt potential across the magnetic-amplifier terminals during the "turn-off" transient condition. The rectifier is a conventional diode which functions to block current through the Zener in a forward direction.

Rewind Stand. The final stage of this complex control system that processes linoleum continuously is the rewind stand. This section is operated by two 25-hp d-c shunt motors, which, through a transfer switch, permits the operator to change from a completely wound reel to an empty reel without shutting down the rest of the line. Referring to

Fig. 11·19, note that the rewind-stand tension reference in the running position is set by the operator at a potential divider and that the signal is fed into a comparator similar to that of Fig. 11·17. The output signal, amplified by the module, is then sent into a *crossover* network (Fig. 11·20), which, in turn, controls two power amplifiers, one for the armature circuit and another for the field current of each of two rewind-reel motors.

The output of the comparator is adjustable over a considerable range, i.e., about 0 to 30 volts. For the first half of the voltage variation the output energizes the *armature* power amplifier (after which the input emf to the latter remains substantially constant) and the upper-half voltage range is fed into the *field* power amplifier. Note particularly that: (1) the armature power amplifier drives a preamplifier which, in turn, excites a magnetic-amplifier controller to provide an adjustable voltage source (0 to 500 volts) for the reel motor; (2) the field power amplifier drives a magnetic amplifier which, energizing the field of the motor, is biased ON to the "full-field" condition until the field weakening range of the crossover is reached.

To keep the running tension properly adjusted it is necessary to have a feedback signal from the out-feed pull-rolls motor drive in addition to the one previously mentioned. This is obtained from the dropping resistor in the loop circuit of the M-G set of Fig. 11·13 where the signal results when the current is due to an overhauling load condition. A current-limit module (Fig. 11·15) is also provided in the circuit to prevent the reel motor from drawing excessive current during the accelerating period or under transient conditions; note that the feedback signal is taken from a dropping resistor in the armature circuit of the motor.

To permit the operator to change reels, a *selecter transfer switch*, as shown, is provided. With the switch in the *down* position, the *left* motor is energized, while the *up* switch position energizes the *right* motor.

In order to maintain constant web speed in feet per minute, the speed of the motor in revolutions per minute must diminish as the reel builds up. The speed range of the motor is, in fact, about 9 to 1 for a normal change from empty to full reel. Since the motor speed can also be varied manually over a 6-to-1 range, apart from the automatic speed change indicated, the total effective range of the reel-motor drive is 54 to 1.

It is important to understand that the reel and out-feed pull rolls do *not* operate independently except while the line is jogged or when the strip is moved for threading purposes. Under normal conditions the two sections work together to maintain the proper tension selected by the operator and set by the out-feed pull rolls.

The Crossover Module. The reference input to this module is split into two parts so that (1) with full field, the voltage across the armature of a shunt motor can be adjusted to give a below-rated speed range and

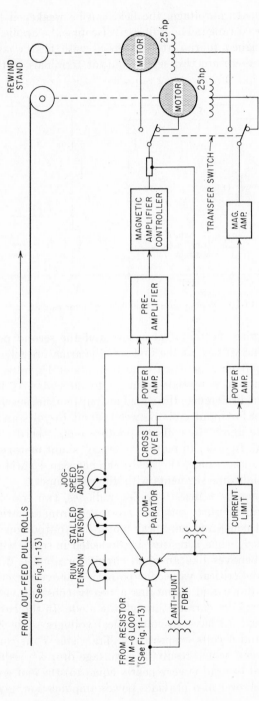

Fig. 11·19 Elementary diagram of rewind stand for linoleum processing line.

(2) with rated armature voltage, the field can be weakened to produce an above-rated speed range. The unit can, of course, be applied only to a motor that is designed to run at high speed with field weakening. As Fig. 11·20 shows, there are two sets of output terminals, with one pair

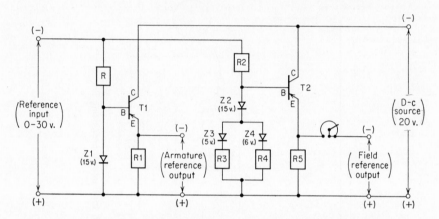

Fig. 11·20   Wiring connections for crossover module.

connected to energize the motor armature and the second pair feeding the motor field. The voltage at the first set, the armature reference, rises linearly as the input emf increases from 0 to about 15 volts, thereafter remaining constant for potentials from 15 to 30 volts. At the second terminal set, the field reference, the signal output remains nearly constant at a very low level for input emfs between 0 and 15 volts and increases at a nonlinear rate above 15 volts to produce equal speed changes per unit of input signal. In order to match "average" shunt motors that operate over a 4-to-1 speed range, the field-reference signal variation passes through three straight-line segments with different slopes.

The crossover module utilizes four Zener diodes, two transistors, and several resistor components to establish crossover (from armature-voltage to field-flux adjustment) and nonlinear motor-field output operation. As previously explained, when a resistor is connected in series with a Zener diode, the latter behaves like an open circuit for applied d-c reversed voltages below some critical value; for potentials above the critical emf, the Zener breaks down and maintains a nearly constant voltage drop regardless of the value of current through the diode. In the circuit shown Zener diodes $Z1$ and $Z2$ have 15-volt critical voltages while $Z3$ and $Z4$ break down at 5 and 6 volts, respectively. Transistor $T1$ is connected as an "emitter-follower," which results in a voltage drop across resistor $R1$ that is proportional to (and is very nearly equal to) the emf across diode $Z1$. The emitter-follower also provides power amplification, giving maximum output for the armature reference of about 15 ma with only 5 ma through $Z1$.

The input reference is also applied across the series-parallel circuit consisting of the $R2$-$Z2$ branch in series with two parallel branches $R3$-$Z3$ and $R4$-$Z4$. With an input reference voltage below 15 volts, Zener $Z2$ acts like an open circuit and little or no voltage appears across transistor $T2$. However, at potentials that exceed 15 volts, $Z2$ breaks down and the excess appears at the base of $T2$. Moreover, at an input voltage of 20 volts $Z3$ breaks down and resistors $R2$ and $R3$ act as a voltage divider to change the slope of the potential characteristic applied to the base of $T2$. Then, at 24 volts, $Z4$ breaks down to connect resistors $R3$ and $R4$ in parallel. This last circuit change which connects $R2$ in series with the parallel combination of $R3$ and $R4$ further reduces the slope of the varying potential that is applied to the base of $T2$. Since $T2$ is connected in an emitter-follower circuit, the voltage drop across $R5$ is proportional to (and very nearly equal to) the base potential; this means, of course, that the motor field-reference output signal is a nonlinear function of the input signal voltage for inputs *above* 15 volts.

The Complete Control System. The student will now find it helpful to fit together the succession of circuit diagrams (Figs. 11·10, 11·12, 11·13, and 11·19) to form a complete system. It will be desirable to do this in connection with the original sketch of Fig. 11·9, which is shown without the electrical control circuits.

QUESTIONS AND PROBLEMS

1. List the various fundamental sections of a regulating system, and indicate how they are interrelated to control such output quantities as speed, position, tension, torque, current, voltage, and others.
2. Explain how the regulator in Fig. 11·1 functions to maintain current or voltage at some desired setting.
3. What is a tachometer generator? How is it ordinarily connected into a system to keep the speed of a motor constant? Explain its action in the simple arrangement of Fig. 11·3.
4. What is a *position regulator*? Why is it important in such applications as color-printing, steel-slitting, and coil-winding operations, and others?
5. Explain the operation of the position regulator illustrated by Fig. 11·4. How does the tachometer generator function to prevent hunting?
6. Explain the action of the *countervoltage regulator* in connection with the unwind reel application of Fig. 11·7. How does the current regulator function in this circuit?
7. Redraw the single-line diagram of Fig. 11·7 to include the complete wiring connections of voltage and current regulators, all rheostats, and necessary rectifiers. (Use Fig. 11·3 as a guide in making the diagram.)
8. What is a dancer-roll rheostat? Explain its operation in the circuit of Fig. 11·10.

9. Describe how a processing line like that of Fig. 11·9, in which material is unwound from a reel at one end and rewound on another reel at the other end, can be kept in continuous operation, without shutdown, while either end is stopped when it is necessary to replace an empty reel (at the entrance end) or a full reel (at the finished end). Indicate clearly the function of the in-feed and out-feed accumulators and the manner in which motors are speeded up or slowed down during change-over periods.

CHAPTER **12**

Special Control Circuits and Problems

Industrial control engineers are frequently called upon to solve unusual problems, design special circuits that must fulfill definite specifications, modify existing circuits to improve operating conditions, or develop circuit arrangements that employ new types of equipment. Considerable experience as well as specialized training are generally necessary if the assignments indicated are to be carried out so that the circuits not only are safe and reliable but do not malfunction under the most adverse conditions or even when the operator fails to follow instructions properly. Also, to reduce cost and minimize servicing requirements, the designer must use as few components as possible.

This chapter will concern itself with a great many problems and circuits of varying degrees of complexity and will attempt to show, by illustrative examples, how circuits and numerical solutions satisfy given specifications. Although the operation of the various circuit diagrams will not be explained, it is recommended that the student make every effort to analyze them or, perhaps, suggest alternate solutions of his own. Such exercises will certainly serve to develop confidence and skill in the handling of industrial control systems.

Problem 1. Three line-start motors are to be started and stopped from a single push-button station. Each is to have its individual across-the-line starter, and there is to be a small time delay between starts as the machines accelerate in sequence. Also, the tripping of the overload relay on one starter must stop the particular motor and not affect the other two.

Design a control circuit to fulfill the given specifications.

Solution (See Fig. 12·1)

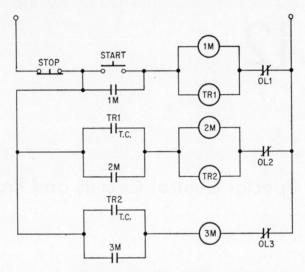

Fig. 12·1   Solution to Prob. 1.

Problem 2. It is desired to start two line-start motors from a single START-STOP station simultaneously under the following conditions:

    a. Depressing the STOP button or the tripping of the overload relay of motor No. 1 shall stop that motor only, and 4 sec later, motor No. 2 is to stop automatically.

    b. The tripping of the overload relay on starter No. 2 shall stop both motors immediately.

Design a control circuit to fulfill the given specifications.

Solution (See Fig. 12·2)

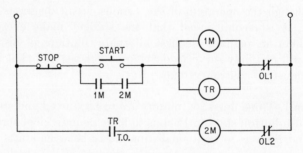

Fig. 12·2   Solution to Prob. 2.

Problem 3. It is desired to control two motors in the following manner:

a. Depressing the START button at the START-STOP station will start motor No. 1. This motor will continue to run until it is stopped either by the pressing of the STOP button or by the tripping of its overload relay.

b. When motor No. 1 stops, motor No. 2 is to start and, after running for 1 min, will stop.

Draw a line diagram for the control circuit only.

Solution (See Fig. 12·3)

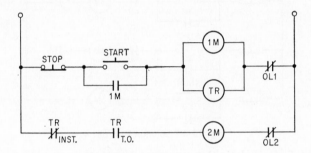

Fig. 12·3   Solution to Prob. 3.

Problem 4. Design a control circuit that will permit five motors to operate in the following manner:

a. When motor No. 1 is started, motor No. 2 will also start after a short time delay and will run only while No. 1 is in operation.

b. It shall not be possible to start motors Nos. 3, 4, or 5 unless motors Nos. 1 and 2 are running.

c. It shall not be possible to start motor No. 3 when No. 4 is running or No. 4 when No. 3 is running.

d. The tripping of the overload on motor No. 1 or 2 shall stop all motors.

e. The tripping of the overload in motor No. 5 shall stop the latter *and* motor No. 3 or 4, whichever is running.

Solution (See Fig. 12·4)

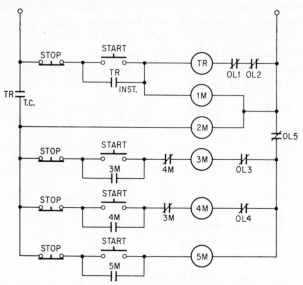

Fig. 12·4 Solution to Prob. 4.

Problem 5. It is desired to start and stop four motors from a single START-STOP station. There are to be four independent across-the-line starters, and the circuit is to function as follows:

a. Each overload relay is to carry the current of one contactor coil.

b. The START or STOP button is to carry no more than the current of one contactor coil.

c. The tripping of any overload relay shall stop all operating motors.

d. The auxiliary contact of any starter shall not carry more than the current of one contactor coil.

Design a control circuit that will fulfill the given specifications.

Solution (See Fig. 12·5)

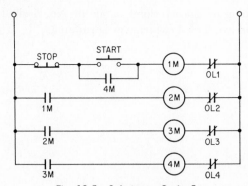

Fig. 12·5 Solution to Prob. 5.

Problem 6. Design a control circuit for a five-motor conveyor system that will operate in accordance with the following specifications:

a. Each motor is to have its own across-the-line starter and its own START-STOP station.

b. It shall be necessary to start motors Nos. 1 to 5 in numerical sequence. This shall be done by depressing the START buttons in succession, and only one motor at a time shall start when a START button is depressed.

c. If the pressing of the STOP button or the tripping of the overload relay of motor No. 5 stops the latter, it shall be the only one to stop.

d. The stopping of motor No. 1 shall stop *all* motors.

e. The stopping of any motor by the pressing of a STOP button or the tripping of its overload relay shall stop the particular motor and, in addition, all higher numbered motors.

This control scheme is frequently used in conveyor systems to prevent a pile-up of material should one of the "take-away" sections stop.

Solution (See Fig. 12·6)

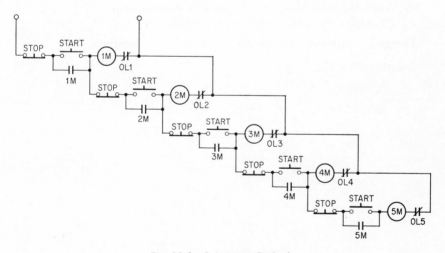

Fig. 12·6   Solution to Prob. 6.

Problem 7. The control circuit for a reversing compound motor (Fig. 3·15) is to be modified so that the machine can be made to slow down while running normally in *forward* only. One method of accomplishing this is to provide an armature shunt as in Fig. III-*C* of Table 7·1, page 207. Show schematically how this can be done by incorporating a special contact and a resistor in such a circuit.

Solution (See Fig. 12·7)

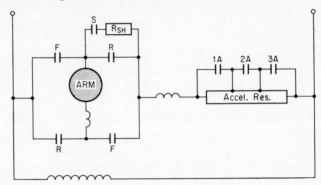

Fig. 12·7 Solution to Prob. 7.

Problem 8. A small synchronous motor rotates a cam which repeatedly closes and opens a contact every 15 sec. It is desired to make this contact act on two independent circuits A and B in the following manner:

a. For the first contact closing, circuit A is energized.

b. When the contact opens, circuit A is deenergized.

c. For the second contact closing, circuit B is energized.

d. When the contact opens, circuit B is deenergized.

e. The above cycle of operations repeats itself.

Design a control circuit for the listed cycle of operations.

Solution (See Fig. 12·8)

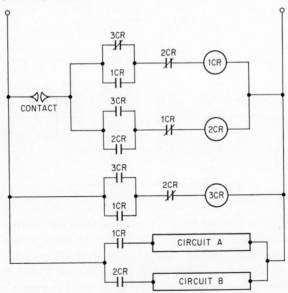

Fig. 12·8 Solution to Prob. 8.

Problem 9. It is desired to operate two motors from a single START-STOP station under the following conditions:

a. The pressing of the START button shall start motor No. 1, and 30 sec later motor No. 2 shall start.

b. After motor No. 2 has been running for 1 min, it shall stop; motor No. 1 shall continue to run.

c. The pressing of the STOP button shall stop both motors.

d. If, during operation, the overload on motor No. 2 trips, both motors must stop.

e. The operating cycle of motor No. 2 must not be affected by the tripping of the overload on motor No. 1.

Design a control circuit to fulfill the given specifications.

Solution (See Fig. 12·9)

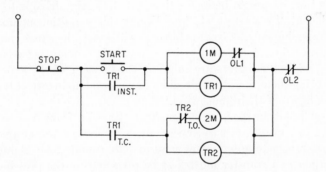

Fig. 12·9 Solution to Prob. 9.

Problem 10. Six grid resistors are available. Four of them are tapped at center points, and two of them have taps one-third from the ends.

a. If the total resistance of each of the first four sections is 6 ohms (3 ohms between taps) and the last two sections has a total of 9 ohms (3 ohms between taps), how should the resistors be connected to yield a total resistance of 2 ohms?

b. 12.75 ohms?

Solution (See Fig. 12·10)

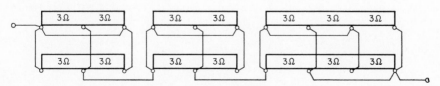

Fig. 12·10 Solution to Prob. 10.

Problem 11. A small material-processing drive is to be equipped with three line-start motors whose operations are to be controlled at a panel containing (1) a two-position selector switch marked AUTOMATIC-MANUAL, (2) a MASTER START-STOP station, and (3) an individual START-STOP station for each of the three motors. The following operating conditions are to prevail:

a. When the selector switch is set on MANUAL, each motor may be started and stopped in the conventional manner by pressing the appropriate button. Since dynamic braking will be provided, a running motor will come to rest quickly when its STOP button is pressed.

b. In the MANUAL position of the selector, only one motor may be operated at a time, and neither of the other two machines may be started until an operating machine is shut down.

c. When the selector switch is set on AUTOMATIC, the individual START-STOP buttons are to be ineffective and the entire drive is to be controlled at the MASTER station. Under this condition of operation, the pressing of the START button will start the three motors in sequence, i.e., 1, 2 and 3, with a small time delay between starts. Also, while the motors are running, the pressing of the STOP button will bring them to a stop. Should an overload trip in any of the motor circuits, all of them must stop.

d. It shall not be possible to make a transfer from AUTOMATIC to MANUAL control, or vice versa, unless all motors are stopped. If a transfer is attempted while motors are running, no change shall take place; a transfer must first be preceded by the pressing of the appropriate button.

Design a control circuit to fulfill the specifications listed.

Solution (See Fig. 12·11)

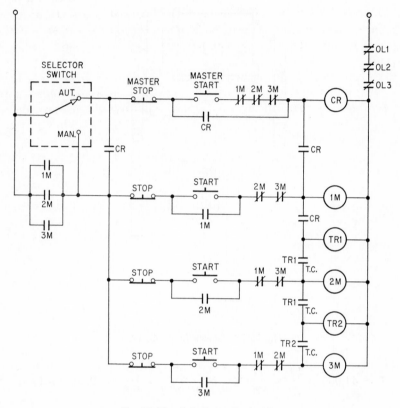

Fig. 12·11   Solution to Prob. 11.

Problem 12. The following information is given in connection with a hoist application in which a 380-rpm d-c series motor drives a drum, two gear trains, and a rope suspension:

Total weight to be lifted (load plus hook and cable) = 22 tons
Gear ratios = 51 to 90 teeth and 17 to 120 teeth
Rope reduction = 2 to 1
Diameter of drum = 27 in.
Estimated overall efficiency of hoist mechanism = 85 per cent

It is desired to specify the horsepower rating of a suitable motor for the application and to submit a sketch illustrating the arrangement of electrical and mechanical equipment.

Solution (See Fig. 12·12)

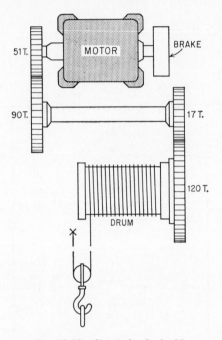

Fig. 12·12   Sketch for Prob. 12.

$$T = 44,000 \times \frac{51}{90} \times \frac{17}{120} \times \frac{1}{2} \times \frac{27}{12 \times 2} \times \frac{1}{0.85} = 2,340 \text{ lb-ft}$$

$$\text{hp} = \frac{2,340 \times 380}{5,250} = 169.5 \qquad \text{Use 175-hp motor}$$

Problem 13. In a certain shearing-machine application, the procedure is to move a steel billet into place gradually and in steps by starting and plugging a 5-hp 220-volt line-start 3-phase squirrel-cage induction motor repeatedly about five or six times. During each plugging operation it is found that the inrush current is approximately 10 per cent higher than the normal starting current of about 90 amp. Although the motor is permitted to rest about 5 to 8 min after each shearing operation, its temperature exceeds permissible values.

It is, therefore, suggested that the plugging current be reduced to 85 amp by connecting permanent resistors in the line wires. This would, of course, reduce the starting torque, but it is felt that the latter would be more than ample, the 220-volt starting torque being $1.6T_{FL}$.

Determine the value of each of the three line resistors and the starting torque of the motor under the new operating conditions. Assume a plugging-current power factor of 0.5 lagging.

Solution (See Fig. 12·13)

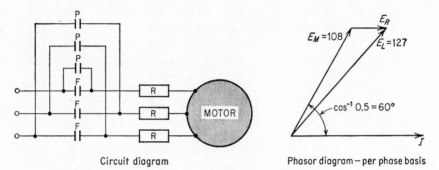

Circuit diagram   Phasor diagram – per phase basis

Fig. 12·13   Sketches for Prob. 13.

I_{plugging} (at 220 volts) $= 1.1 \times 90 \cong 100$ amp

E_{motor} (for a plugging current of 85 amp) $= \dfrac{85}{100} \times 220 = 187$ volts

E_{line} (per phase) $= \dfrac{220}{\sqrt{3}} = 127$ volts

E_{motor} (per phase) $= \dfrac{187}{\sqrt{3}} = 108$ volts

Referring to the phasor diagram, Fig. 12·13,

$(127)^2 = [(0.5 \times 108) + E_R]^2 + (0.866 \times 108)^2$
$E_R = 32$ volts

Therefore

$R = {}^{32}\!/_{85} \cong 0.38$ ohm
Per cent T_{ST} (at 187 volts) $\cong 160 \times ({}^{187}\!/_{220})^2 = 115.5$

Problem 14. A dual-voltage 220/440-volt 3-phase squirrel-cage induction motor is available with nine terminal leads $T1$ to $T9$ brought out and labeled in the standard manner (see Fig. 2·7). It is proposed to operate the machine on 220 volts but to connect the windings in series for starting and in parallel for running.

Design an automatic switching circuit for the motor, and determine the starting current and starting torque in terms of the normal values at 220 volts.

Solution (See Fig. 12·14)

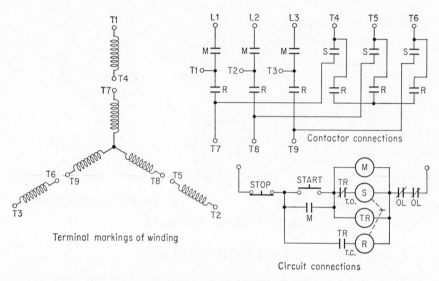

Fig. 12·14   Diagrams for Prob. 14.

$$I_{ST} \text{ (series-star)} = \tfrac{1}{4} I_{ST} \text{ (two-parallel-star)}$$
$$T_{ST} \text{ (series-star)} = \tfrac{1}{4} T_{ST} \text{ (two-parallel-star)}$$

Problem 15. A control circuit is to be designed for a water-pumping installation in which a pressure switch actuates an auxiliary contact that starts and stops a motor at adjustable values of low and high water pressures. Since surges are likely to occur during starting and stopping and "backspin" is possible when a pump is shut off, timing relays should be included in the circuit for protective purposes. It should be pointed out in this connection that the force of a sudden stop of a long column of water produces surges which, operating the pressure switch, tend to subject the motor starter to chattering. A similar situation prevails on starting when the pressure switch tends to drop out as low pressures occur during surges. Such improper operation can be avoided by using off-delay and on-delay timers whose contacts bypass the pressure switch momentarily during starts and stops. Backspin tends to occur when a head of water runs back through a centrifugal pump to turn the latter in a reverse direction just after a turnoff. Should a pump be started during a backspin, the pump or motor might be damaged. The use of the relays indicated above automatically provides backspin protection.

Solution (See Fig. 12·15)

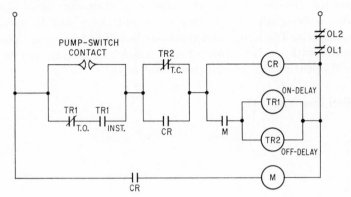

Fig. 12·15 Solution to Prob. 15.

Problem 16. In a two-wire control circuit for a motor which is turned ON and OFF at the contacts of a thermostat, an intermediate step must be used between starter and thermostat; this is because the contacts on the thermostat are generally incapable of handling the contactor-coil current. Recognizing this limitation, design a thermostatically controlled circuit for a small line-start motor that will make the thermostat contacts carry an extremely low relay-coil current and where the operation of the relay will start and stop the motor.

Two solutions are given for this problem.

Solutions (See Fig. 12·16)

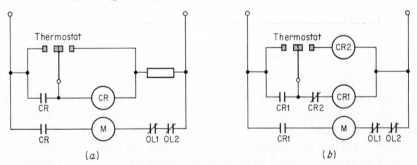

Fig. 12·16 Solutions to Prob. 16.

Problem 17. It is desired to design a control circuit for a two-speed single-winding 3-phase squirrel-cage motor to operate in accordance with the following specifications:

a. It shall be possible to start the motor in *low* speed only.

b. The speed of the motor can be advanced from *low* to *high* immediately thereafter by pressing the HIGH button.

c. With the machine running in *high,* a shift may be made to *low* by pressing the LOW button, but the LOW contactor must pick up after a designated time delay to permit the motor to decelerate sufficiently. The latter condition must be fulfilled by using a plugging switch with an N.C. contact and which can be adjusted to close (and open) at any desired speed (not zero).

Solution (See Fig. 2·17)

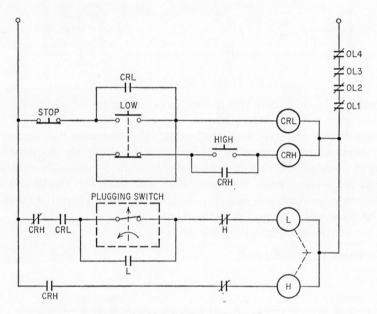

Fig. 12·17   Solution to Prob. 17.

Problem 18. A sequencing control system is to be designed for a machine tool that is equipped with three operating motor drives and a coolant pump motor. A master START-STOP station is to be provided with, and the starters of each of two operating motors $1M$ and $2M$ are to be actuated by, three-wire pilot devices. The starting and stopping of the third operating motor $3M$ are to be controlled by a two-wire pilot device, in this case a pressure switch. Moreover, when any one of the motors $1M$, $2M$, or $3M$ is started, it must automatically start coolant pump motor $4M$ and the tripping of the overload in the latter circuit must shut down the system. The operating motors shall be protected by independent overload relays, the tripping of any one of which shall stop the corresponding machine only.

Solution (See Fig. 12·18)

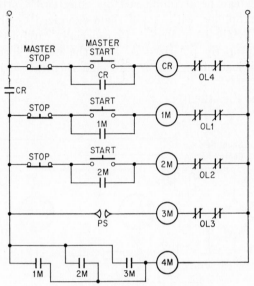

Fig. 12·18   Solution to Prob. 18.

Problem 19. Specify the proper horsepower rating of a 3-phase 1,140-rpm squirrel-cage induction motor that performs a machining operation on the following repetitive duty cycle:

a. $T_1 = 33$ lb-ft for $4\frac{1}{2}$ min
b. $T_2 = 30$ lb-ft for 50 sec
c. $T_3 = 48$ lb-ft for 3 min
d. Rest period for 4 min

NOTE: The selection of a motor for an application whose duty cycle repeats itself at regular intervals must be made on the basis of the *rms* torque or horsepower. This value is then a measure of the thermal capacity of the motor, i.e., its ability to maintain a reasonable temperature rise. Since there is a considerable reduction in cooling while the motor is at rest, it is customary to use one-third of the rest-period time in making calculations.

$$T_{\text{rms}} = \sqrt{\frac{[(33)^2 \times 270] + [(30)^2 \times 50] + [(48)^2 \times 180]}{(270 + 50 + 180 + {}^{240}\!/\!_3)}} = 36$$

$$\text{hp} = \frac{36 \times 1,140}{5,250} = 7.8 \qquad \text{Use 7.5-hp motor}$$

Problem 20. Design a magnetic-amplifier control circuit similar to that of Fig. 9·39b for controlling the speed of a shunt motor in an adjustable-voltage system (Ward Leonard) by eliminating the tachometer generator. This can be accomplished by using two control windings on the magnetic amplifier, with one of them energized by the loop voltage and the other by the loop current. Be especially careful that the control windings are properly polarized so that the motor speed will remain constant for all load changes at a particular rheostat setting.

Solution (See Fig. 12·20)

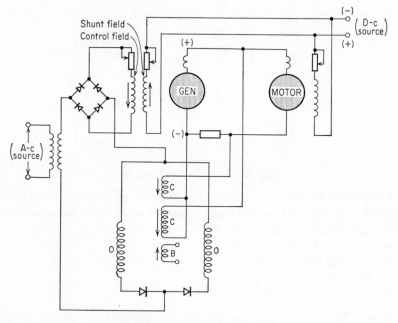

Fig. 12·20   Solution to Prob. 20.

Problem 21. A control circuit is to be designed for a two-winding three-speed constant-horsepower 3-phase nonreversible squirrel-cage motor to fulfill the following operating conditions:

    a. With the motor at rest, it shall be possible to start at any of the three speeds (low L, medium M, or high H) by pressing the appropriate button.

    b. It shall be possible to advance to a higher speed directly, i.e., from L to M, L to H, or M to H.

    c. To change to a lower speed, i.e., from H to M, H to L, or M to L, it shall be necessary first to press the STOP button. This procedure permits the motor to slow down slightly before an electrical change to lower speed connection takes place.

The winding connections, with the proper terminal markings, should be included with the control-circuit diagram.

Solution (See Fig. 12·21)

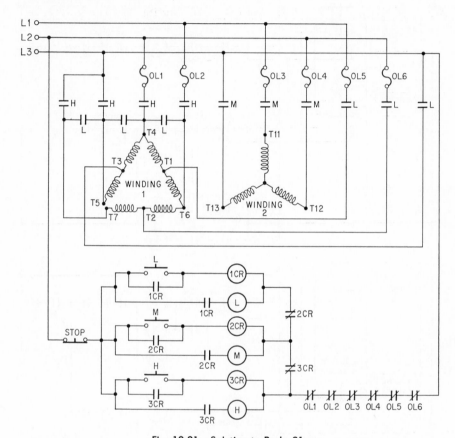

Fig. 12·21   Solution to Prob. 21.

Problem 22. A complete control circuit is to be designed for a two-winding four-speed constant-torque 3-phase nonreversible squirred-cage motor to comply with the given specifications as listed:

a. With the motor at rest, it shall be possible to start the motor at any of the four speeds by pressing the appropriate button. However, the following limitations are imposed: (1) on *high*, 1,200 rpm, immediately upon closing the main control-circuit switch; (2) on *3rd*, 900 rpm, 10 sec after the switch is closed; (3) on *2nd*, 600 rpm, 20 sec after the switch is closed; (4) on *low*, 450 rpm, 30 sec after the switch is closed.

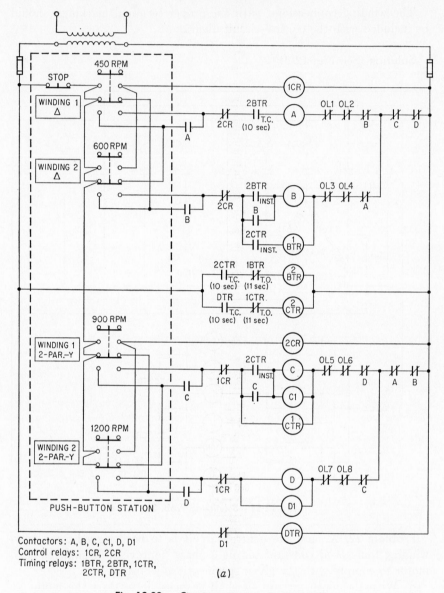

Contactors: A, B, C, C1, D, D1
Control relays: 1CR, 2CR
Timing relays: 1BTR, 2BTR, 1CTR,
2CTR, DTR

(a)

Fig. 12·22a   Circuit connections for Prob. 22.

b. With the motor operating at any speed (in accordance with the conditions specified in **a**) it shall be possible to *advance* to a higher speed; e.g., it shall be possible to advance directly from *low* to *2nd*, *low* to *3rd*, or *low* to *high*, from *2nd* to *3rd* or *2nd* to *high*, and from *3rd* to *high*.

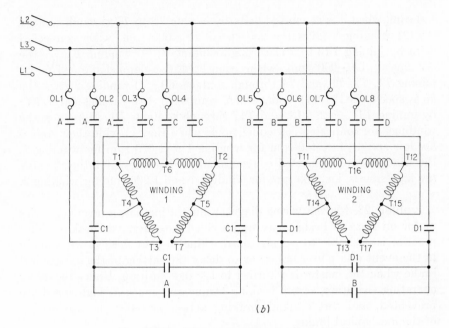

Fig. 12·22b Wiring connections for Prob. 22.

c. With the motor running in *high*, i.e., 1,200 rpm, it shall be possible to shift (1) to *3rd* after a 10-sec delay, (2) to *2nd* after a 20-sec delay, or (3) to *low* after a 30-sec delay. This may be done in either of two ways: (1) by keeping the appropriate button depressed for the time periods indicated or (2) by first pressing the STOP button and then, after a suitable time delay, pressing the appropriate button.

d. With the motor running in *3rd*, i.e., 900 rpm, it shall be possible to shift (1) to *2nd* after a 10-sec delay or (2) to *low* after a 20-sec delay. This may be done in a manner similar to that indicated in **c**.

e. With the motor running in *2nd*, i.e., 600 rpm, it shall be possible to shift to *low* after a 10-sec delay in either of the two ways indicated in **c**.

Solution (See Fig. 12·22a and b)

NOTE: The winding-connections diagram of Fig. 12·22b illustrates how the two windings are normally open-circuited at the *T3-T7* and the *T13-T17* terminals. Thus, when winding 1 is energized, inductively coupled winding 2 can carry no current, and no induced current can pass through winding 1 when winding 2 is energized. Also observe that for this *constant-torque motor:* (1) winding 1 gives the *low* speed, i.e.,

450 rpm, when it is connected delta by joining $T3$ to $T7$ through contact A; (2) winding 2 gives the *2nd* speed, i.e., 600 rpm, when connected delta by joining $T13$ to $T17$ through contact B; (3) winding 1 gives the *3rd* speed, i.e., 900 rpm, when connected two-parallel-star (2**PY**) by joining $T1$, $T2$, $T3$, and $T7$ through contacts $C1$; (4) winding 2 gives the *high* speed, i.e., 1,200 rpm, when connected two-parallel-star (2**PY**) by joining $T11$, $T12$, $T13$, and $T17$ through contacts $D1$. Moreover, the windings are connected consequent-pole when they provide their *respective* low speeds, i.e., 450 rpm for winding 1 and 600 rpm for winding 2; the windings are connected conventionally when they give their *respective* high speeds, i.e., 900 rpm for winding 1 and 1,200 rpm for winding 2.

Problem 23. Make a wiring diagram illustrating how a reversing series motor on a crane bridge may be provided with dynamic braking. This practice is desirable to permit the machine to shut itself down quickly in the event of a power failure or to delay the setting of the mechanical brake when the master is returned to the OFF position. Under the latter operating condition, severe braking torque and shock at high speed are prevented, and soft dynamic-braking action precedes the application of the mechanical brake.

Unlike the dynamic-braking procedure for shunt motors (Fig. 3·21), where the shunt-field excitation is maintained by the source and the armature is connected to a *DB* resistor, a slightly different method must be employed for series motors. Here, the armature, series field, and resistor are connected into a series loop, and, during the dynamic-braking period, the series-field excitation is maintained *in the same direction* by the generated voltage in the armature, i.e., the cemf. Although the application of this method is comparatively simple for nonreversing series-motor drives, it is somewhat more involved to obtain dynamic braking for machines that operate in both directions. A simple sketch illustrating how a nonreversing series motor is connected for emergency dynamic braking is given in Fig. 12·23a. Note the use of N.C. *DB* contacts which are spring closed (see Fig. 3·22).

Controller wiring connections that provide emergency dynamic braking for a *reversing* series motor are given in Fig. 12·23b. Note the *four* N.C. *DB* contacts, spring closed, and the *DB* resistor. To understand how the system operates it will be necessary to recognize that: (1) the *DB* contactor coils are energized at the instant the master is moved out of the OFF position and, with the motor running in *either* direction, *all DB contacts are open*; (2) in the event of a power failure or if the master is returned to the OFF position, only two contactors are permitted to close and the other two are locked out *mechanically*; (3) with the motor running in a *forward* direction, a camshaft mechanism locks out the 3*DB* and 4*DB* contactors but permits the $1DB$ and $2DB$ contactors to close freely; thus, when there is a power failure, only the $1DB$ and $2DB$ contacts close while the armature is rotating, and dynamic braking occurs as current passes through the loop made up of the armature, the series field, the DB resistor, and the $1DB$ and $2DB$ contacts; (4) with the motor running in a *reverse* direction, the camshaft mechanism locks out the $1DB$ and $2DB$ contactors but permits the $3DB$ and $4DB$ contactors to close freely; thus, when there is a power failure, only the $3DB$ and $4DB$ contacts close while the armature is rotating, and the dynamic-braking loop is made up of the armature, the series field, the DB resistor, and the $3DB$ and $4DB$ contacts; (5) for dynamic braking from forward or reverse rotation, *the direction of the current in the series field remains the same.*

Solution (See Fig. 12·23a and b)

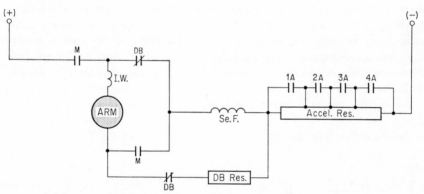

(a) Emergency dynamic-braking connections for a non-reversing series motor

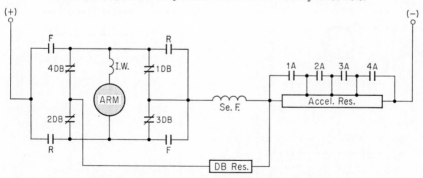

(b) Emergency dynamic-braking connections for a reversing series motor

Fig. 12·33   Circuit diagrams for Prob. 23.

Problem 24. Make a wiring diagram illustrating how a duplex, series-motor drive for a crane bridge may be provided with dynamic braking by the *Wilson-Ritchie* method. The scheme, named after its inventors, employs four standard spring-closed *DB* contactors for the braking circuit (as in Fig. 12·23) and the usual combination of resistors and contactors. However, unlike the duplex control system of Fig. 4·12, where reversing the two series motors is accomplished by changing the current directions in *both* armatures, the Wilson-Ritchie method reverses the motors by interchanging the armature terminals in one machine and the series-field terminals in the other. The unique feature of this arrangement is that special mechanisms are unnecessary (as in Fig. 12·23*b*) and conventional devices are used. The four N.C. *DB* contacts of Fig. 12·24 open at the instant the motors are started and close during the braking period.

When the bridge crane is moving *forward* or *reverse,* the two motor circuits are completely independent, with the armature and field of motor *A* in series in one path and the armature and field of motor *B* in series in the other. However, while the motors are undergoing a dynamic-braking stop and all directional contactors are open, one loop circuit consists of armature *A*, series-field *B*, and a *DB* resistor while a second loop circuit comprises armature *B*, series-field *A*, and a *DB* resistor. Moreover, an especially important aspect of the system is that the series-field current directions in both motors do *not* reverse, although the armature currents do, as the machines brake to a stop.

Solution (See Fig. 12·24)

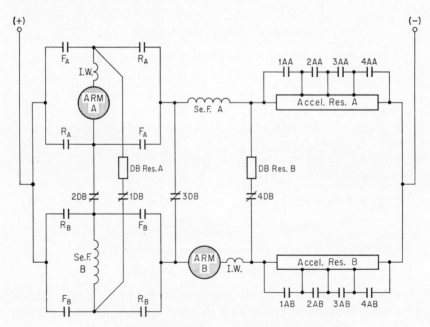

Fig. 12·24   Wilson-Ritchie circuit diagram for Prob. 24.

Problem 25. As explained in Chap. 6 (see Figs. 6·25 to 6·27), the speed of a phase-wound rotor motor can be adjusted by inserting varying values of resistance in the secondary. For installations requiring considerable speed ranges *with a minimum number of contactors*, it is sometimes found desirable to employ sets of resistors having a succession of values that vary according to a geometric ratio. The master controller can then be wired to insert resistances equal to one base unit at each step of the secondary controller, varying the magnitude of the phase resistance from a maximum value (represented by the sum of all sections) to zero. This scheme of speed adjustment has been applied satisfactorily to large printing presses where the mode of operation is to set the master at the desired speed point before energizing the primary; a push-button starter is then actuated to start the motor and bring it up to a preset speed. An added feature for such controllers is to include a so-called high-torque contactor in the secondary so that the motor will always develop sufficient starting torque regardless of the controller setting. The latter action is implemented by making the high-torque contactor pick up at the same time the line contactor closes and have it energized as long as the operator keeps the START button depressed; the release of the START button after the machine has accelerated causes the high-torque contactor to drop out and the insertion of the required secondary resistance.

Make a wiring diagram showing the connections for a wound-rotor motor with four sections of resistance in each of the three phases graduated according to a geometric ratio. Show in tabular form how 16 different speeds can be obtained with this arrangement.

Solution (See Fig. 12·25)

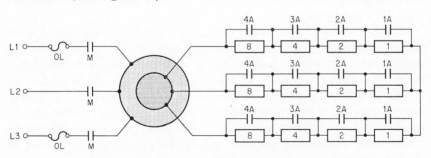

CONTACTORS	MASTER POSITION															
	1	2	3	4	5	6	7	8	9	10	11	12	13	14	15	16
1A		⊠		⊠		⊠		⊠		⊠		⊠		⊠		⊠
2A			⊠	⊠			⊠	⊠			⊠	⊠			⊠	⊠
3A					⊠	⊠	⊠	⊠					⊠	⊠	⊠	⊠
4A									⊠	⊠	⊠	⊠	⊠	⊠	⊠	⊠

Fig. 12·25 Circuit connections and table for Prob. 25.

CHAPTER 13

Static-switching Circuits and Control

Logic functions in static-switching systems. Conventional electrical systems are controlled, for the most part, by two general types of components, namely (1) *pilot devices*, which act as sources of information, and (2) *relays*, which perform *logic*, i.e., decision-making functions that are based on the given sources of information. Resulting signals are then transmitted to *contactors* which provide energy at power levels to electrical operating equipment such as motors, generators, solenoids and solenoid valves, lifting magnets, and many other kinds of electromagnetic actuators. Pilot devices, previously considered, include push buttons and limit switches and temperature (thermostatic) and pressure (air and water) sensing units.

In contrast to relays, which act to establish or disconnect electric circuits by physically closing or opening contacts, static switching devices carry out similar logic functions *without moving parts*. The latter types of switch are usually specially designed magnetic amplifiers or transistorized components which respond to *digital* signals, i.e., sharply defined, discrete pulses of definite duration. Such signals differ greatly from *analog* quantities—like those originating at a tachometer generator—that follow and simulate continuously varying changes in an electrical circuit.

Five basic logic functions will fulfill practically all static switching requirements in industrial control systems. These are (1) the AND, (2) the OR, (3) the NOT, (4) the MEMORY, and (5) the DELAY. The first three are the simplest to recognize and understand, since they express familiar concepts.

The AND function, symbolized by Fig. 13·1a, provides an output signal only when *all* input signals are applied. Thus, if the component is provided with, say, three input signals A, B, and C (it can be designed to handle from two to four inputs), no output will be delivered unless A AND B AND C are present; it is, in this respect, equivalent to a simple relay circuit consisting of three *open contacts in series* and in which the operating unit will be energized when all contacts are closed by corresponding relay actions. Moreover, with the circuit in an operating state, the opening of any contact will stop the output.

The OR function, symbolized by Fig. 13·1*b*, is equivalent to a circuit in which two or more (three in the sketch shown) *open contacts are connected*

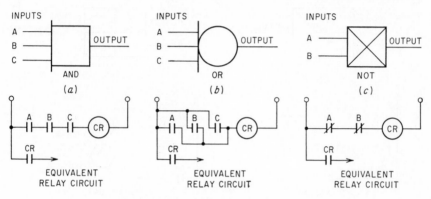

Fig. 13·1   Basic logic functions.

in parallel so that the closing of any one of them will result in an output. Thus, with signal A OR B OR C (or any combination of them) present, the circuit will be closed.

The third of the fundamental logic functions is the NOT (Fig. 13·1*c*); it is in effect a normally *closed contact* which makes an output possible when there is *no* input signal. In the case of a two- (or three-) input NOT, A and B (and C), for example, no output will be delivered if either one of the signals A or B (or C) is present; conversely, the output circuit will be closed if all input signals are removed.

Two kinds of MEMORY functions, the OFF-RETURN MEMORY and the RETENTIVE MEMORY, are encountered in static switching work; they are represented in their standard forms in Fig. 13·2. In the first of these, similar to the familiar sealed-in relay circuit, a momentary signal to the A section will turn the unit ON and the output will continue indefinitely until a B input signal is applied to the OFF section. Also, in the event of a power failure, the OFF-RETURN MEMORY will return the output to the OFF state in exactly the same way as in the conventional relay circuit with a sealing contact across the START button. The RETENTIVE MEMORY, as the name implies, always remembers the state which last existed after the restoration of power following a power failure. It is, indeed, turned on and off exactly as is the first type of unit, but as indicated, it provides the additional functions as well. Specifically, when power returns after an outage, the circuit will be restored to the on state if that was its original condition or the off state if the circuit was originally deenergized. In relay circuitry, RETENTIVE MEMORY is accomplished by a device that has two coils, one of them designed to pick up the relay (equivalent to the A input in the logic function), after which a mechanical latch

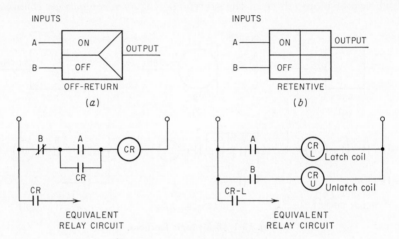

Fig. 13·2   Memory logic functions.

holds it in the energized position following the deenergization of the operating coil, and the second coil (corresponding to the B input in the logic function) arranged to unlatch the unit when it is desired to open the relay.

DELAY components perform precisely the same functions as timers in conventional relay circuits, and as in the latter, there are two general types of them. In the on-delay device, Fig. 13·3a, an output is produced following a definite intentional time delay after the input is *energized;* the off-delay device, Fig. 13·3b, on the other hand, removes an output following a definite intentional time delay after its input is *deenergized.*

It is particularly important to note that, unlike relays which are generally energized by a single input to give one or more outputs (one exciting and one or a multiplicity of contacts), the static device produces a single output when one or more input signals are applied. The latter arrangements are clearly illustrated by the diagrams of Fig. 13·1, where each component is provided with two or more inputs and a single output.

In addition to the foregoing five kinds of logic functions, two other components are generally necessary to make static-switching circuits operative. These are (1) the *signal converter*, Fig. 13·4a, which, as a voltage divider, acts to lower the pilot-device emf to a suitably lower voltage for the logic units, and (2) the *amplifier*, Fig. 13·4b, whose function

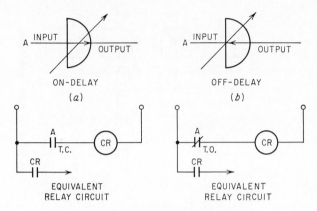

ON-DELAY
(a)

OFF-DELAY
(b)

EQUIVALENT
RELAY CIRCUIT

EQUIVALENT
RELAY CIRCUIT

Fig. 13·3   Delay logic functions.

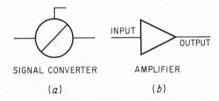

SIGNAL CONVERTER   AMPLIFIER
(a)   (b)

Fig. 13·4   Other static-switching components.

it is to enlarge the extremely low signal power output from the logic device (in the milliwatt range) to a reasonably high operating power level.

Static-switching applications involving logic components. Neglecting for the present the internal construction and wiring of the various logic devices of the foregoing article, it will now be desirable to illustrate how they are used in actual circuits. This will be done first for a simple arrangement of control relays, limit switches, and a START-STOP station that is designed to actuate a pair of solenoid valves $SV1$ and $SV2$. Note particularly in the relay circuit diagram of Fig. 13·5a that each of two control relays $CR2$ and $CR3$ have associated N.O. and N.C. contacts and that control relay $CR1$ is connected as an OFF-RETURN MEMORY device. Analyzing the circuit further, it should be clear that: (1) relay $CR1$ must be picked up before $SV1$ or $SV2$ can be energized, since the operation of relays $CR2$ or $CR3$ depends primarily upon the closing of a

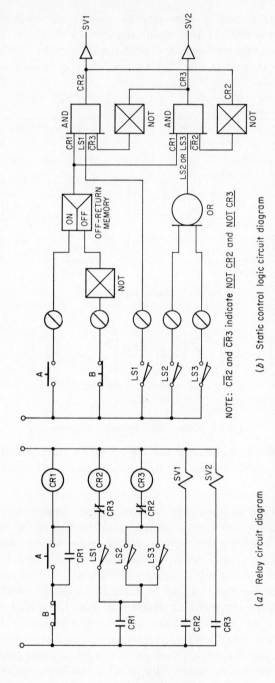

(a) Relay circuit diagram

(b) Static control logic circuit diagram

NOTE: $\overline{CR2}$ and $\overline{CR3}$ indicate NOT CR2 and NOT CR3

Fig. 13-5   Diagrams illustrating equivalent relay and logic circuits for a solenoid application.

$CR1$ contact; (2) when pilot device A (a START button) is operated, relay $CR1$ is picked up and is sealed in by its interlock $CR1$; (3) relay $CR1$ can be made to drop out by operating pilot device B (a STOP button), i.e., B must be associated with a NOT device in static switching logic; (4) since relay $CR3$ is dropped out when relay $CR2$ is picked up, and vice versa, only one solenoid can be energized at a time; (5) relay $CR3$ can be energized by the operating of $LS2$ or $LS3$.

The relay circuit diagram is translated into static switching logic in Fig. 13·5b. Observe first that each of the pilot devices (A, B, $LS1$, $LS2$, and $LS3$) is connected to a logic unit through a signal converter and that the solenoid valves are now energized by amplifier outputs. Comparing the circuit with its relay counterpart note that: (1) signals $CR2$ and $CR3$ replace, respectively, N.O. contacts $CR2$ and $CR3$, and these signals are eventually responsible for the operation of solenoid valves $SV1$ and $SV2$; (2) the $CR2$ signal can result only when signals $CR1$ and $LS1$ *are* present AND signal $CR3$ is *not* present; this combination of actions, therefore, calls for a three-input AND as indicated, with the NOT $CR3$ signal (indicated by $\overline{CR}3$) supplied from the output of a second AND through a NOT unit; (3) the $CR3$ signal can result only when $CR1$ AND either $LS2$ or $LS3$ *are* present AND signal $CR2$ is *not* present; this combination of actions, therefore, requires a second three-input AND as shown, with the $\overline{CR}2$ signal supplied from the output of the first AND through a NOT unit; (4) the two-input OR is employed to supply an output signal to the second AND when either $LS2$ or $LS3$ are closed; (5) a NOT is used in the OFF section of the MEMORY to turn off the system when the STOP button is pressed; this merely removes the $CR1$ signal from *both* ANDS.

A second circuit, Fig. 13·6, represents a simple machine-tool application in which a small motor can be started if either of two limit switches $LS1$ OR $LS2$ is closed AND a START button is pressed. Moreover, after the motor is started, the foregoing limit switches are to be ineffective, and the opening of an N.C. main limit switch LSM OR the pressing of an emergency STOP button will halt the machine. Figure 13·6a shows how this is accomplished with conventional relay circuitry, and its counterpart is illustrated with three two-input AND logic units in Fig. 13·6b. Referring to the latter diagram note that: (1) with the motor at rest, AND 1 will deliver a CR signal when AND 2 and AND 3 both have outputs; the latter obtains when AND 2 receives signals from the START button and the closing of $LS1$ or $LS2$, and when AND 3 receives inputs through LSM and emergency STOP button B; (2) with the motor running, the output signal of AND 2 can be removed because a feedback signal CR replaces the former and maintains the output to the motor; (3) if either one of the inputs to AND 3 is removed (by opening LSM or B), one of the inputs to AND 1 will disappear; AND 1 will, therefore, have no output CR and the motor will stop.

Although circuits like Figs. 13·5b and 13·6b illustrate the use of logic with static-switching devices, they are far too simple for practical application. This is because these control systems do not offer the advantages of economy, improved reliability and service, and reduced space requirements until the application becomes rather complex and involves large

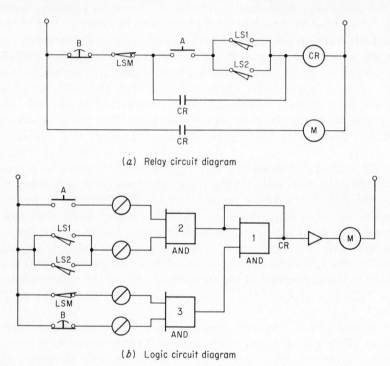

(*a*) Relay circuit diagram

(*b*) Logic circuit diagram

Fig. 13·6   Diagrams illustrating equivalent relay and logic circuits for a small machine-tool application.

numbers of components. To be sure, conventional relay circuits are still used, in the main, for most power installations but are gradually being replaced by the newer developments as the installations, particularly the large machine tools and processing lines, make more exacting demands on control equipment. It should be understood, therefore, that the foregoing examples were given for illustrative purposes only and are intended to indicate how logic devices are employed. With this in mind, it will now be desirable and appropriate to discuss a unique kind of mathematical analysis that is frequently used to advantage in the design of complex static-switching circuits.

Boolean algebra. This brief discussion will be devoted to an extremely useful and convenient notation for expressing logical networks. Named for George Boole who first developed the system of logic, *Boolean algebra* utilizes a special notation that permits manipulation and rearrangement of two-terminal series-parallel networks into a variety of equivalent and often simpler forms. It can indeed yield solutions that eliminate redundant contacts (not always apparent) to give circuits with a minimum number of components.

Contact Notation. In Boolean algebra, the following notation is used for relays, their contacts, and the manner in which the latter are connected in AND (series) and OR (parallel) circuits.

1. All relays are designated by particular letters, and their associated contacts are represented by corresponding letters.
2. N.O. contacts are indicated by simple capital letters like A, B, C, etc., while N.C. contacts are shown with *bars* above them, e.g., $\bar{A}$, $\bar{B}$, $\bar{C}$, etc., read A bar, B bar, C bar.
3. When contacts are connected in *series* (the AND circuit), the arrangement is denoted by using the multiplication symbol ($\cdot$), although the latter is generally omitted if there is no ambiguity. Thus, a series or AND combination of three contacts will be shown as $A \cdot B \cdot C$ or ABC and will be read A *and* B *and* C.
4. When contacts are connected in *parallel* (the OR circuit), the arrangement is denoted by using the plus symbol ($+$). Thus, a parallel or OR combination of three contacts will be shown as $A + B + C$ and read A *or* B *or* C.
5. Regarding an N.O. contact as standard, an N.C. contact is, therefore, *not* standard. Thus, $\bar{A}$ (A bar), $\bar{B}$ (B bar), and $\bar{C}$ (C bar) would be NOT elements, i.e., NOT A, NOT B, and NOT C.
6. Since relays and contacts can be in either of two states, i.e., open or closed (they perform digital functions), they are arbitrarily given either of two values. With the contact *closed*, transmission *can* occur; a "1" is used to indicate this condition, which could be regarded as a circuit with infinite admittance. With the contact open, transmission *cannot* occur; a "0" is used to indicate this condition, which could be regarded as a circuit with zero admittance.

Postulates. Considering the foregoing statements, the following identities should be clear:

1. For two N.O. contacts in series, Fig. 13·7a, $0 \cdot 0 = 0$, $0 \cdot 1 = 0$ $1 \cdot 1 = 1$, $A \cdot 0 = 0$, $A \cdot 1 = A$, $A \cdot A = A$.
2. For two N.O. contacts in parallel, Fig. 13·7b, $0 + 0 = 0$, $1 + 0 = 1$, $1 + 1 = 1$, $A + 0 = A$, $A + 1 = 1$, $A + A = A$.

3. For N.C. contacts in series, Fig. 13·7c, $\bar{A} \cdot \bar{A} \cdot \bar{A} \cdots = \bar{A}$.
4. For N.C. contacts in parallel, Fig. 13·7d, $\bar{A} + \bar{A} + \bar{A} + \cdots = \bar{A}$.
5. For N.O. and N.C. contacts in series, Fig. 13·7e, $A \cdot \bar{A} = 0$.
6. For N.O. and N.C. contacts in parallel, Fig. 13·7f, $A + \bar{A} = 1$.

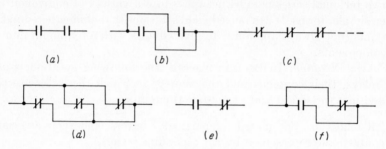

Fig. 13·7   Simple contact arrangements.

Theorems. The following equivalents are given for a number of contact arrangements:

1. Fig. 13·8a $(A + B)(A + C) = A + BC$
2.   b $(\bar{A} + \bar{B})(\bar{A} + \bar{C}) = \bar{A} + \bar{B}\bar{C}$
3.   c $A + AB = A$
4.   d $A(A + B) = A$
5.   e $A + \bar{A}B = A + B$
6.   f $A(\bar{A} + B) = AB$
7.   g $\bar{A}(A + \bar{B}) = \bar{A}\bar{B}$

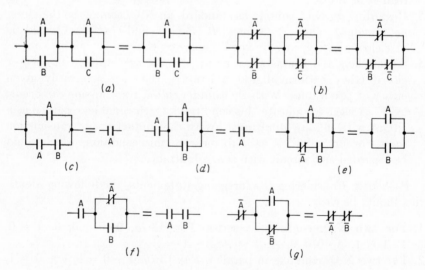

Fig. 13·8   Several contact arrangements and their equivalents.

The following proofs are given for equalities 1 to 7:

1. $(A + B)(A + C) = AA + AC + AB + BC$
$$= A(1 + C + B) + BC = A + BC$$
3. $A + AB = A(1 + B) = A$
4. $A(A + B) = AA + AB = A + AB = A(1 + B) = A$
5. $A + B = A + B(A + \bar{A}) = A + AB + \bar{A}B$
$$= A(1 + B) + \bar{A}B = A + \bar{A}B$$

Other Useful Equivalents

8.   Fig. 13·9a $AB + \bar{A}\bar{B} = (A + \bar{B})(\bar{A} + B)$
9.   b $\bar{A}B + A\bar{B} = (\bar{A} + \bar{B})(A + B)$
10.   c $A\bar{B} + \bar{A}B = (A + B)(\bar{A} + \bar{B})$

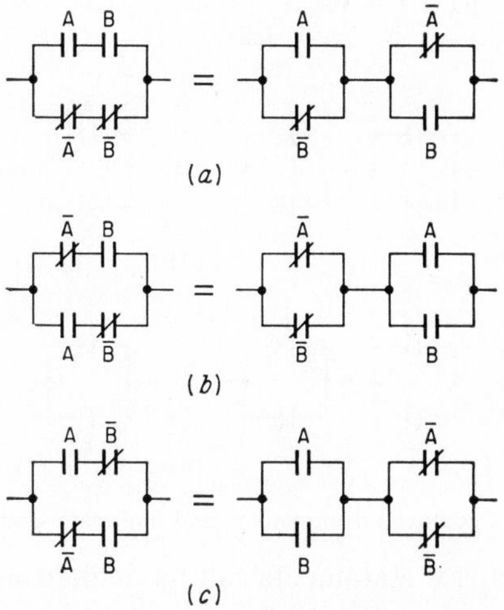

Fig. 13·9   Parallel-series and equivalent series-parallel contact arrangements.

11. Fig. 13·10a $AB + BC + \bar{A}C = AB + \bar{A}C$
12.   b $(\bar{A} + \bar{B})(A + \bar{C})(\bar{B} + \bar{C}) = (\bar{A} + \bar{B})(A + \bar{C})$
13.   c $(A + B)(\bar{A} + C)(B + C) = (A + B)(\bar{A} + C)$

The following proofs are given for equalities 8 to 13:

8. $(A + \bar{B})(\bar{A} + B) = A\bar{A} + AB + \bar{A}\bar{B} + B\bar{B} = AB + \bar{A}\bar{B}$
9. $(\bar{A} + \bar{B})(A + B) = A\bar{A} + \bar{A}B + A\bar{B} + B\bar{B} = \bar{A}B + A\bar{B}$
10. $(A + B)(\bar{A} + \bar{B}) = A\bar{A} + A\bar{B} + \bar{A}B + B\bar{B} = A\bar{B} + \bar{A}B$
11. $AB + BC(A + \bar{A}) + \bar{A}C = AB + ABC + \bar{A}BC + \bar{A}C$
$$= AB(1 + C) + \bar{A}C(B + 1) = AB + \bar{A}C$$

12. $(\bar{A} + \bar{B})(A + \bar{C})(\bar{B} + \bar{C}) = (\bar{A}\bar{C} + A\bar{B} + \bar{B}\bar{C})(\bar{B} + \bar{C})$
$$= \bar{A}\bar{B}\bar{C} + \bar{A}\bar{C} + A\bar{B} + A\bar{B}\bar{C} + \bar{B}\bar{C}$$
$$= \bar{B}\bar{C}(\bar{A} + A + 1) + A\bar{B} + \bar{A}\bar{C}$$
$$= \bar{B}\bar{C} + A\bar{B} + \bar{A}\bar{C}$$
$$= A\bar{B} + \bar{A}\bar{C} = (A + \bar{C})(\bar{A} + \bar{B})$$

13. $(A + B)(B + C)(\bar{A} + C) = (AC + \bar{A}B + BC)(B + C)$
$$= ABC + AC + \bar{A}B + \bar{A}BC + BC$$
$$= BC(A + \bar{A} + 1) + AC + \bar{A}B$$
$$= BC + AC + \bar{A}B$$
$$= AC + \bar{A}B = (A + B)(\bar{A} + C)$$

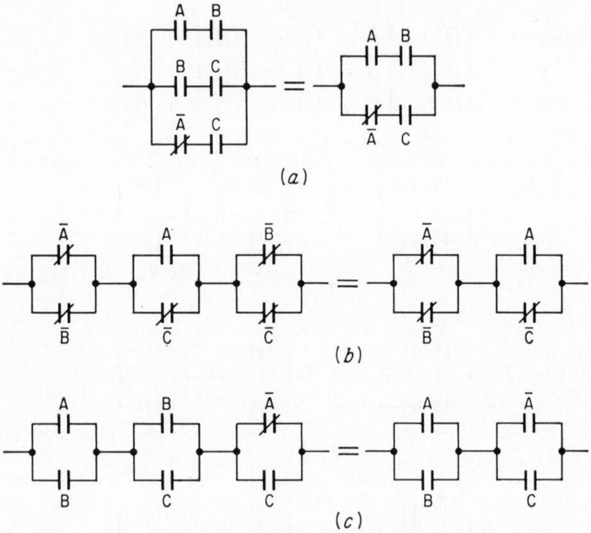

Fig. 13·10 Six-contact arrangements and their four-contact equivalents.

The following general statements can be made concerning certain contact combinations:

a. An N.O. contact in *series* with a similarly labeled N.C. contact is equivalent to an open circuit; e.g., $A\bar{A} = B\bar{B} = C\bar{C} = 0$.

b. An N.O. contact in *parallel* with a similarly labeled N.C. contact is equivalent to a closed circuit; e.g., $(A + \bar{A}) = (B + \bar{B}) = (C + \bar{C}) = 1$.

c. A contact X in series with any combination of parallel-connected contacts one of which has a value of 1 is equivalent to the contact X; e.g., $X(1 + A + B + C + \cdots) = X$.

d. When a function is represented by the sum of all possible series-paired combinations of two variables X and Y, the entire function (or circuit) is equivalent to a 1; e.g.,

$$(XY + X\bar{Y} + \bar{X}Y + \bar{X}\bar{Y}) = X(Y + \bar{Y}) + \bar{X}(Y + \bar{Y}) = (X + \bar{X}) = 1$$

Logical Switching-circuit Applications. When the foregoing principles of Boolean algebra are applied to the analysis of complex switching functions, it is frequently possible to simplify greatly an original circuit by rearranging contacts in more suitable forms or, what is often extremely important, eliminate unnecessary contacts and reduce a given circuit to its most elementary form. Since such changes and modifications are, as indicated, accomplished by logical mathematical procedures rather than those based on judgment (or perhaps errors of judgment), the resulting circuit will generally be a best solution for the given operating conditions. Several illustrations will now be given to represent the method.

Figure 13·11 shows what appears to be a complex combination of contacts that are actuated by four relays A, B, C, and D. Remembering

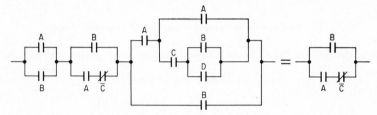

Fig. 13·11   Example illustrating conversion of a series-parallel circuit.

that the operation of a given relay always closes all associated N.O. contacts and simultaneously opens its N.C. contacts, it should be observed by inspection that transmission will occur (1) if the B relay is operated to close the B contacts, OR (2) if the A relay closes the A contacts AND the C relay is NOT operated; moreover, the D relay has no function and the C relay by itself cannot produce transmission. The final solution should therefore have the equation $B + A\bar{C}$. Proceeding by Boolean algebra to see if this is the case, the complete equation is first written as

$$(A + B)(B + A\bar{C})\{B + A[A + C(B + D)]\}$$

Reducing the first two sets of terms,

$$(A + B)(B + A\bar{C}) = AB + A\bar{C} + B + AB\bar{C} = B(A + 1 + A\bar{C}) + A\bar{C}$$
$$= B + A\bar{C}$$

Reducing the remaining sets of terms,

$$B + A[A + C(B + D)] = B + A + ABC + ACD$$
$$= B + A(1 + BC + CD) = B + A$$

Combining the two derived results,

$$(B + A\bar{C})(B + A) = B + AB + AB\bar{C} + A\bar{C} = B(1 + A + A\bar{C}) + A\bar{C}$$
$$= B + A\bar{C}$$

The final result is precisely the one predicted above and is represented by the equivalent sketch of Fig. 13·11.

A second example, Fig. 13·12, concerns a combination of contacts that form a familiar bridge. Here, the solenoid valve SV is energized

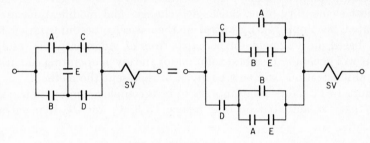

Fig. 13·12   Example illustrating conversion of a bridge circuit.

when transmission takes place through one of four paths, namely,

$$AC \text{ or } BD \text{ or } AED \text{ or } BEC$$

Thus
$$SV = AC + BD + AED + BEC$$
$$= C(A + BE) + D(B + AE)$$

Note that the rearranged (equivalent) circuit includes three additional contacts (A, B, and E) but it does eliminate the shunt contact E in the original arrangement. It may, in this respect, offer a possible advantage.

Figure 13·13 shows a parallel-series circuit with a combination of 12 N.O. and N.C. contacts which, after conversion, yields a much simpler

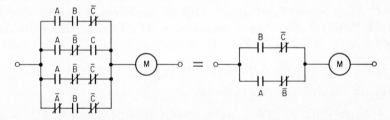

Fig. 13·13   Example illustrating conversion of a parallel-series circuit.

equivalent arrangement with only 4 contacts. Here again, as in Fig. 13·11, a large number of contacts are shown to be redundant and are, therefore, eliminated by logical analysis. Referring to the original circuit,

$$M = AB\bar{C} + A\bar{B}C + A\bar{B}\bar{C} + \bar{A}B\bar{C}$$

Grouping and manipulating terms *one* and *four* and terms *two* and *three*,

$$M = B\bar{C}(A + \bar{A}) + A\bar{B}(C + \bar{C}) = B\bar{C} + A\bar{B}$$

In analyzing *both* configurations it is well to note that transmission to the M component will occur (1) if the A relay operates and the B relay does not, OR (2) if the B relay operates and the C relay does not, OR (3) if the A and B relays operate and the C relay does not, OR (4) if the A and C relays operate and the B relay does not.

Logical design of control circuit with static-switching components. Boolean algebra can be used to advantage in the design of complex static-switching circuits. Although this can be done in several ways depending upon the available information and/or existing circuitry, it will be desirable for our purposes to apply the principles discussed in the foregoing article to the conversion of an actual relay problem and its solution to static-switching operation. This will be done for the circuit diagram Fig. 12·21 of Prob. 21, page 435, which concerns the operation of a three-speed two-winding 3-phase induction motor. The control circuit is reproduced here for convenience, Fig. 13·14; its performance, as previously stated, calls for starting the motor at any of the three speeds L, M, or H, but a shift from a higher to a lower speed must first be preceded by the pressing of the STOP button.

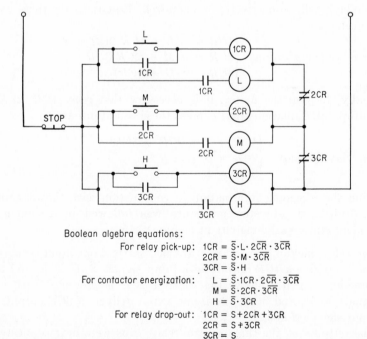

Boolean algebra equations:

For relay pick-up:
$$1CR = \bar{S} \cdot L \cdot 2\overline{CR} \cdot 3\overline{CR}$$
$$2CR = \bar{S} \cdot M \cdot 3\overline{CR}$$
$$3CR = \bar{S} \cdot H$$

For contactor energization:
$$L = \bar{S} \cdot 1CR \cdot 2\overline{CR} \cdot 3\overline{CR}$$
$$M = \bar{S} \cdot 2CR \cdot 3\overline{CR}$$
$$H = \bar{S} \cdot 3CR$$

For relay drop-out:
$$1CR = S + 2CR + 3CR$$
$$2CR = S + 3CR$$
$$3CR = S$$

Fig. 13·14 Conventional relay circuit for three-speed motor. Boolean logic equations are given for conversion to static-switching circuit, Fig. 13·15.

Referring to Fig. 13·14, note that there is an N.C. STOP button $\bar{S}$; three N.O. START buttons L, M, and H, two N.C. contacts $2\overline{CR}$ and $3\overline{CR}$, and six N.O. contacts, two each for those marked $1CR$, $2CR$, and $3CR$. Corresponding relay coils are labeled $1CR$, $2CR$, and $3CR$, and the contactor coils are designated L, M, and H.

To start the motor in *low*, *medium*, or *high*, it is, of course, necessary that the $1CR$, $2CR$, or $3CR$ relays, respectively, pick up. Consulting the diagram, the following Boolean algebra equations can be written:

For relay pick-up $\begin{cases} 1CR = \bar{S}·L·2\overline{CR}·3\overline{CR} \\ 2CR = \bar{S}·M·3\overline{CR} \\ 3CR = \bar{S}·H \end{cases}$

After the motor is running in *low*, *medium*, or *high*, the L, M, or H contactors will, respectively, be energized. Equations for these conditions are:

For contactor energization $\begin{cases} L = \bar{S}·1CR·2\overline{CR}·3\overline{CR} \\ M = \bar{S}·2CR·3\overline{CR} \\ H = \bar{S}·3CR \end{cases}$

Also, while the motor is operating in *low*, *medium*, or *high*, one of the three relays will, respectively, be energized. Equations for these conditions are:

For relay energization $\begin{cases} 1CR = \bar{S}(L + 1CR)2\overline{CR}·3\overline{CR} \\ 2CR = \bar{S}(M + 2CR)3\overline{CR} \\ 3CR = \bar{S}(H + 3CR) \end{cases}$

Finally, to stop the motor, it is necessary that relay $1CR$, $2CR$, or $3CR$ drop out. Equations that express these conditions are:

For relay dropout $\begin{cases} 1CR = S + 2CR + 3CR \\ 2CR = S + 3CR \\ 3CR = S \end{cases}$

Using the foregoing information in conjunction with the diagram of Fig. 13·14, a step-by-step procedure was followed in designing the equivalent static-switching circuit of Fig. 13·15.

1. For relay pickup, three ANDS are necessary. Four-input AND 1 was drawn to give a $1CR$ output with input signals $\bar{S}$, L, $2\overline{CR}$, and $3\overline{CR}$; three-input AND 2 shows that output $2CR$ is produced with input signals $\bar{S}$, M, and $3\overline{CR}$; two-input AND 3 delivers a $3CR$ output with input signals $\bar{S}$ and H.

2. The outputs of the three control-relay ANDS were next connected to ON sections of appropriate OFF-RETURN MEMORYS. Thus, outputs $1CR$, $2CR$, and $3CR$ were connected, respectively, to MEMORYS 1, 2, and 3.

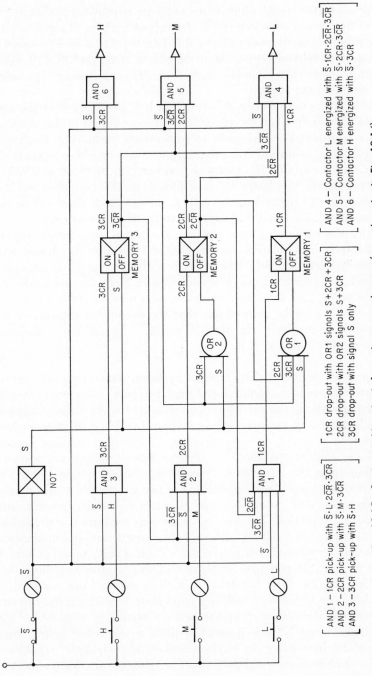

Fig. 13-15 Static-switching circuit for a three-speed motor (see relay circuit, Fig. 13-14).

$$\left[\begin{array}{l} \text{AND 1 – 1CR pick-up with } \overline{S} \cdot L \cdot 2\overline{CR} \cdot 3\overline{CR} \\ \text{AND 2 – 2CR pick-up with } \overline{S} \cdot M \cdot 3\overline{CR} \\ \text{AND 3 – 3CR pick-up with } \overline{S} \cdot H \end{array}\right]$$

$$\left[\begin{array}{l} \text{1CR drop-out with OR1 signals } S + 2CR + 3CR \\ \text{2CR drop-out with OR2 signals } S + 3CR \\ \text{3CR drop-out with signal } S \text{ only} \end{array}\right]$$

$$\left[\begin{array}{l} \text{AND 4 – Contactor L energized with } \overline{S} \cdot 1CR \cdot 2\overline{CR} \cdot 3\overline{CR} \\ \text{AND 5 – Contactor M energized with } \overline{S} \cdot 2CR \cdot 3\overline{CR} \\ \text{AND 6 – Contactor H energized with } \overline{S} \cdot 3CR \end{array}\right]$$

3. For contactor energization, three ANDS are required. Four-input AND 4 was drawn to give a low-speed (L) output with input signals $\bar{S}$, $1CR$, $2\overline{CR}$, and $3\overline{CR}$; three-input AND 5 illustrates that a *medium-speed* (M) output is delivered with input signals $\bar{S}$, $2CR$, and $3\overline{CR}$; two-input AND 6 shows that there is a *high-speed* (H) output when input signals $\bar{S}$ and $3CR$ are applied.

4. To stop the motor after it is running, it is necessary that the appropriate relay drop out by applying the proper signal to the OFF section of a MEMORY. For *high-speed* (H) operation, MEMORY 3, this requires an S signal, obtained from the output of a NOT whose input is an $\bar{S}$; for *medium-speed* (M) operation, MEMORY 2, an S or a $3CR$ signal is necessary, the latter being supplied to the two-input OR 2; for *low-speed* (L) operation, MEMORY 1, an S or a $2CR$ or a $3CR$ signal is required, the last being supplied to the three-input OR.

5. After the various logic components are drawn (with proper regard to their positions relative to one another) and the input and output terminals are properly labeled, it is a simple matter to draw in the various connecting lines. The pilot devices $\bar{S}$, H, M, and L, the signal generators, and the amplifiers must, or course, be included as shown. It should also be pointed out that the outputs of MEMORYS 2 and 3 furnish, respectively, signals $2\overline{CR}$ and $3\overline{CR}$ in addition to the operating signals $2CR$ and $3CR$.

Types of static-switching control systems. Several manufacturers of control equipment have developed logic systems that exercise switching functions without the need for moving contacts. As already mentioned and illustrated by examples, these employ five kinds of static-switching device (AND, OR, NOT, MEMORY, and DELAY) which, performing essentially the same control operations as electromagnetic relays, are completely different from the latter in both construction and action. All systems, however, make use of one or the other of two general kinds of component, namely, magnetic amplifiers and transistors, which possess unique switching properties. The first of these was explained and applied to control circuits in Chap. 9, although its construction and application as a switch differ somewhat from those of a control unit. As will be explained subsequently, the magnetic amplifier displays suitable switching characteristics because it is constructed with a special core material. The transistor, on the other hand, is a solid-state (semiconductor) device of more recent development. It has become extremely popular for static-switching service because it occupies little space, is fast-acting, rugged, completely insensitive to environment, and extremely reliable, and requires practically no maintenance. The elements are, for the most part, generally combined with such other parts as resistors, capacitors, and

diodes and, as packaged and encapsulated logic components, are manufactured under various trade names.

Magnetic-amplifier-type logic devices. Except for certain constructional details and the use of special so-called "square-loop" core material,
magnetic amplifiers as switching devices are similar to and follow the
same principles as those used for control circuits (Chap. 9). Consisting
essentially of a closed iron core around which are placed a *gate* (output),
feedback, bias, and suitable control windings and equipped with one or
more rectifiers to give it self-saturating properties, it displays the general *amplifier-transfer characteristic* shown in Fig. 13·16b. Note particularly that the curve of *output* (load current) *versus input* (signal current)

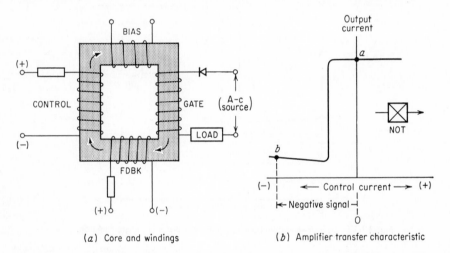

(*a*) Core and windings   (*b*) Amplifier transfer characteristic

Fig. 13·16 Magnetic amplifier as a static switch. Device is a NOT when *no bias* is used.

is digital in character, since the device has only two states; it is either
ON (when there is no signal) or OFF (when a negative signal is applied),
and with this particular arrangement no bias is used. Moreover, to obtain
the digital response indicated, it is necessary to have a proper value of
positive feedback. The device as shown is obviously a NOT, since an output results, point *a*, when there is no input signal, and the application
of a *negative* signal causes the output to disappear, point *b*.

When a negative bias is permanently applied to the device, the entire
characteristic is shifted to the right (or the Y axis is moved to the left)
and, with no input signal, the output is removed. Then, if two or more
signal windings are employed, the presence of any one or any combination of positive input signals will turn the unit ON. This is illustrated in
Fig. 13·17, which, in effect, is an OR.

The two-input AND logic function is obtained by connecting two units together as illustrated by Fig. 13·18a. With the two *gate* windings joined in series in a common load circuit and with individual energized bias windings on the two cores, an output will be produced when *both* input windings receive signals. Under this condition, *both* cores will be saturated, an essential requirement if the device is to be turned ON; an output will *not* result if only one signal is applied because its own core, not the other, will be saturated. The transfer characteristic of the device, which is, in effect, a combination of two OR units, is given in Fig. 13·18b. Note

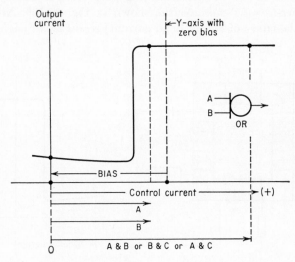

Fig. 13·17   Transfer characteristic for a two-input OR logic function.

that signals A and B must be present if an output is to be delivered; A or B by itself cannot do so.

It should be understood that the digital characteristic curves of Figs. 13·16 to 13·18 were obtained with a properly proportioned number of turns in the feedback windings; with *fewer* than the desired turns, the output current would have a positive slope, Fig. 13·19b, and, being analogue in character, would not be suitable for static-switching service. Moreover, if *more* than the ideal feedback turns are used, the characteristic is altered further, although, as Fig. 18·19c shows, digital current changes would result upon both the application and removal of signal input. The action of the device then resembles the operation of a conventional relay where the latter does not pick up until a certain critical voltage is reached, under which condition it snaps closed; the static device also behaves like a relay that drops out at a voltage that is *lower* than the pickup value.

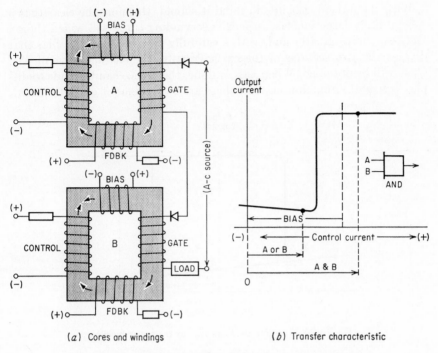

(a) Cores and windings   (b) Transfer characteristic

Fig. 13·18   Magnetic amplifier designed as a two-input AND.

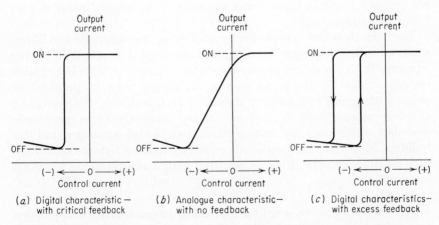

(a) Digital characteristic —   (b) Analogue characteristic—   (c) Digital characteristics—
with critical feedback   with no feedback   with excess feedback

Fig. 13·19   Transfer characteristics with different degrees of feedback-winding turns.

With the foregoing points in mind it should, therefore, be clear that a substantially large increase in feedback-winding turns will widen the ON and OFF curves greatly and, with a carefully adjusted negative bias that divides the two sections of the curve equally, the characteristic of Fig. 13·20 will be obtained. When this is done, the device can be made to display a MEMORY function because there are two values of output current,

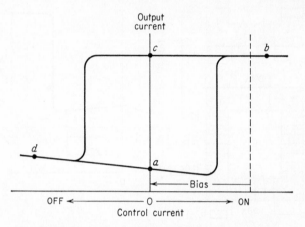

Fig. 13·20   Transfer characteristic of a MEMORY function.

high and low, for a wide range of signal inputs (between the two vertical parts of the characteristic). Thus, starting at point *a*, the application of a proper *positive* input will drive the output to point *b* and turn the device ON. Removal of the positive input shifts the operating point to *c* with no change in output current. Next, the application of a proper *negative* input drives the output to point *d*, which is OFF. Finally, removing the OFF negative input shifts the unit to its original point *a*, also in the OFF state.

The delay function is obtained by employing two magnetic amplifiers, each one with its own gate winding, that are coupled by a mutual linking winding. With the device in the OFF state both cores will be in negative saturation, and to obtain an ON output, signals must be provided to saturate the cores in a positive direction. In this device, the saturating effect is accomplished in increments, not suddenly as in previously described components, by making the mutual linking winding retard flux buildup. Moreover, the magnitude of the increments, which control the turn-on time, i.e., the delay, is adjusted with a resistor in the coupling-coil circuit.

Pulse-type magnetic-amplifier switching components. An interesting type of static-switching device that uses *pulse power* techniques to energize the gate circuits of magnetic amplifiers was developed by the General Electric Company. The components are constructed with square-loop core material and have the usual complement of windings previously referred to, i.e., gate, bias, control, and feedback windings.

In its elementary form (Fig. 13·21), gate voltage is applied in *pulses* by a special pulsing transformer, and negative bias and control signals

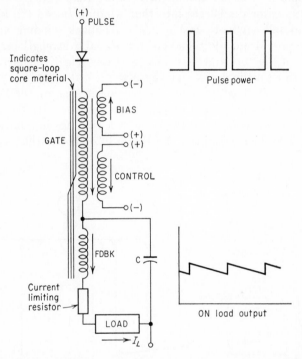

Fig. 13·21   Basic pulse-operated static-switching magnetic amplifier.

are derived from available d-c sources. Assuming the core to be in negative saturation and with negative potential applied to the bias winding, the first gate-voltage pulse drives the core toward positive saturation. However, since a large opposing emf is induced in the gate winding, a comparatively low value of magnetizing current passes through the gate and into the load by way of the feedback winding; the pulse also charges the capacitor slightly, which promptly discharges through the load, since the blocking rectifier prevents a reversed gate current. Before the second pulse is applied, the negative bias drives the core to negative saturation, and this *resets* it. Subsequent pulses then repeat the actions described with the result that a *very small* direct current (with sawtooth variations) passes into the load; the device is, therefore *not* in the ON state.

Next, if a positive signal is applied to the control winding, its mmf neutralizes the effect of the bias winding and thus prevents the latter from *resetting* the core. The second and succeeding pulses then proceed to saturate the core completely, so that the device, now turned ON, permits a rather large current to pass into the load. The fact that a feedback winding is used implies, as in previously described components, that a digital response is achieved and furthermore that a succession of superimposed capacitor discharges into that winding causes the load current to rise and fall sawtooth fashion. The feedback winding acts, in this respect, to provide an additional positive signal to strengthen the positive saturation of the core and to require a strong negative signal when it is necessary to turn the device OFF sharply.

The basic unit illustrated by the schematic diagram of Fig. 13·21 is used to construct the several types of logic components. The three-input OR would, for example, have three control windings in addition to gate, feedback, and bias windings. As Fig. 13·22 shows, the negative bias

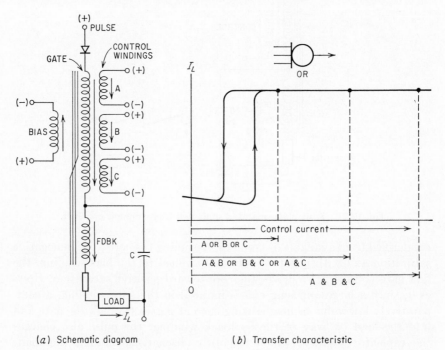

(a) Schematic diagram   (b) Transfer characteristic

Fig. 13·22   Pulse-operated OR switching component.

locates the zero control reference to the left of the vertical portions of the transfer characteristic and an overstrengthened feedback winding separates the ON and OFF digital responses for the purposes previously mentioned. Note that inputs A or B or C or any combination of these signals will turn the device ON. Moreover, the output will not be affected appreciably if more than one control winding is energized.

To obtain a two-input NOT or a two-input AND a slightly different constructional arrangement is necessary. For either of these logic functions, *two cores must be used with common gate and feedback windings*, and each core is surrounded by an independent control winding that is energized separately. Since a bias signal is not required for the NOT, the bias winding is omitted (or remains unenergized) in this device. Bias must, however, be used in the AND, and this signal is supplied by a common bias winding that surrounds both cores. Figure 13·23 represents a

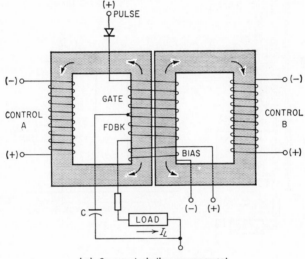

(*a*) Core and winding arrangement

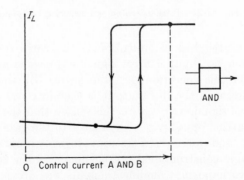

(*b*) Transfer characteristic

Fig. 13·23 Pulse-operated AND component.

two-input AND and clearly shows how the various windings are arranged. Note particularly that an output will be produced only when the cores within the common gate winding are saturated, and this condition is realized when *both* control signals are present. The energization of one control winding can saturate one core only no matter how much the signal current.

The schematic diagram of Fig. 13·24 illustrates a two-input NOT and should be interpreted as being similar in construction to Fig. 13·23 with

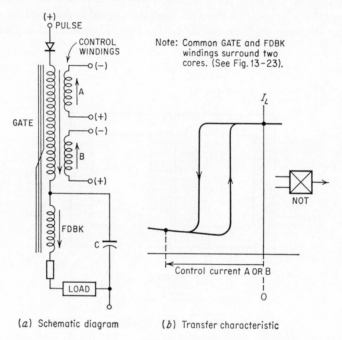

(*a*) Schematic diagram   (*b*) Transfer characteristic

Fig. 13·24   Pulse-operated NOT switching component.

the omission of the bias winding. Since the cores remain in positive saturation when no input control signal is present, an output *will be* produced under this condition; i.e., the device will thus be turned ON. To put the device in the OFF state it is therefore necessary to apply a negative control signal, and it is for this reason that the control windings are shown polarized oppositely with respect to the gate winding.

OFF-RETURN and RETENTIVE-MEMORY logic components (Fig. 13·2) employ the same constructional arrangements as the basic design (Fig. 13·21) with two important modifications. As was previously discussed in connection with Fig. 13·20, these involve the use of a feedback winding with many turns to widen the ON-OFF loop and a carefully adjusted shown ...ase is connected to three input terminals A, B, and C through resist... R_B, and R_C and to a positive (reverse) bias through the resist... to prevent collector-base leakage current and avoid thermal runa... i.e., destructive temperature rise. When no *negative* input sign... resent at A NOR at B NOR at C, there is no emitter-base current and ... transistor is, in effect, an open circuit between collector and emit... under this condition, resistor $R2$ and the load are in series acro... ne potential $-V$ and ground, and a voltage will appear across a lo... (output) resistance. However, if a negative input is applied to A ... or C (or any combination of them), the base becomes negative wi... respect to the emitter and current *does* pass from emitter to base th... gh the proper resistor (or resistors); this action thus grounds the co... tor with the result that the load resistor is short-circuited and there is ... output. It should be pointed out that the ohmic value of the biasing re... stor $R1$ is made rather high with respect to input resistors R_A, R_B, a... R_C to minimize its effect on the input current and that the line re... stance $R2$ is low enough to give an output that is almost equal to th... supply voltage.

The NOT, OR, *and* AND *Element.* Since the fundamental NOR circuit of Fig. 13·27 delivers an output when there is no input signal at A NOR at B NOR at C, it should be clear that it would become a simple NOT logic element, Fig. 13·28a, if only one input were used. Also, as Fig. 13·28b shows, an OR element can be formed by cascading a two-input unit with one that has a single input. The two-input AND, on the other hand, requires three NORs as illustrated by Fig. 13·28c. Note particularly in the case of the latter two elements, i.e., the OR and the AND, that the output (final) NOR merely performs an *inverter* or NOT function and must be used to give an output when signals are present at the input terminals. This aspect of the NOR system of static switching, therefore, adds to the number of required units as compared with the so-called English method of control, but it does have the advantage of simplicity and low cost. However, in complex control circuits where many components are required, it is often possible to reduce the number of elements by eliminating double conversions; this will be illustrated subsequently by practical examples.

The OFF-RETURN MEMORY *Element.* This logic function is shown in Fig. 13·29 where two NORs are interconnected (1) to give an output when there is an input signal A and (2) to remove an output when a B signal is applied or if a power interruption occurs. With neither of the two input terminals A or B energized, transistor $T1$ will deliver an output which will be impressed on the base of transistor $T2$; the latter will therefore be short-circuited (between emitter and collector), and no output will be produced. Also, the base of $T1$ will receive a zero feedback

---

negative bias that centers the Y axis between the vertical portions of the transfer characteristic.

When the device is used as an OFF-RETURN MEMORY, a sequencing procedure in the power supply always applies the bias signal *before* pulse power energizes the gate and feedback windings. This action always drives the unit to the OFF state, point d in Fig. 13·25b, regardless of

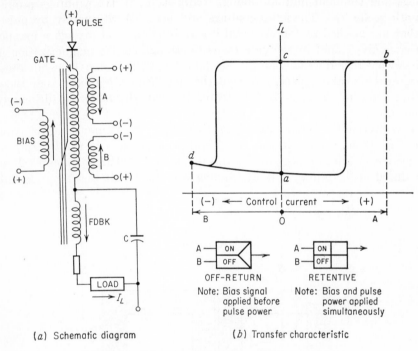

(a) Schematic diagram   (b) Transfer characteristic

Fig. 13·25 Pulse-operated MEMORY **switching component.**

its previous output condition. Thus, after service is restored following a power failure, the device will always be turned OFF and a new input signal is necessary to drive it ON to point b.

In the RETENTIVE MEMORY, where the device resumes its last output condition after the removal and restoration of power, bias signal and pulse power are not sequenced (as in the OFF-RETURN MEMORY) but are applied at approximately the same time. Under this mode of operation, the output condition, ON or OFF, will be determined by the magnetic state of the core at the time power fails. This means, of course, that, while the device is delivering an output, the operating point will shift from b to c during a service interruption; thus, when power is reestablished, the MEMORY is automatically turned ON because operation is resumed at point c.

The DELAY logic function (Fig. 13·3) provides a preset time delay between the application of an input signal and the delivery of an output. In the ON-DELAY unit, the output comes on after energization, while the OFF-DELAY component removes the output after deenergization. The actions indicated are accomplished in the pulse-operated device by using a two-core construction with a gate winding around each of the outside legs and common mutual-linking, feedback, and bias windings around both inside legs. The gate windings are energized alternately by pulses that are applied at 180-electrical-degree intervals and in such a manner that a given mmf in one gate tends to strengthen the magnetization in its own core and weaken flux strength in the other. Moreover, while these changes are taking place, the strengthening effects are greater than those that tend to demagnetize the core, so that the latter is ultimately saturated.

Neglecting the action of the bias winding for the moment and referring to Fig. 13·26, a pulse in *gate No.* 1 will drive the flux in the *left* core to positive saturation; also, during this period of flux change, an emf will be induced in the mutual-linking winding with the result that the *right* core

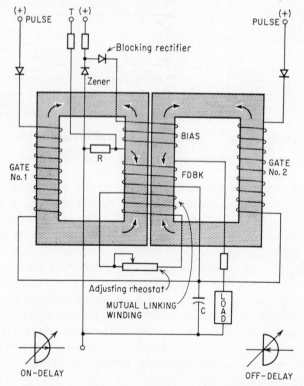

Fig. 13·26   Pulse-operated DELAY switching component.

will be driven to negative saturation. Moreover, if there we same flux changes, except for direction, would occur in bo a 180° interval a pulse will energize *gate No.* 2 to drive to positive saturation and *reset* the *left* core. Since each st of pulses repeats the magnetizing effects of the preceding p it should be clear that each pulse drives its own core to posi tion and resets the other core. However, if a resistance is co the *mutual-linking* winding circuit where a voltage drop will mmf that resets the core will be reduced and partial core rese result. The core, therefore, reaches saturation by increments, time the element turns ON.

In order to turn the device OFF and hold it OFF when there is input, a *bias* winding is placed around the center legs and is energi a *permanent* input. The bias then ensures total core reset in the ab of an ON input. To remove the bias, a control input is applied to termin this produces a voltage drop across resistor *R* and forces the bias cu to pass through the Zener diode, thus permitting the component to turn

Transistor-type switching components. One of the more recent se conductor developments that possesses ideal properties for static-switc ing logic systems is the *transistor*. Used primarily in common emitt circuits with the simple addition of resistors, it is capable of providin both voltage and current gains that will drive several similar component without supplementary amplification. Like the magnetic-amplifier types of component previously described, transistors are designed to furnish the basic logic functions NOT, OR, AND, MEMORY, and DELAY. A number of circuit configurations are presently in use to accomplish this, but the one that appears to offer important advantages and is, therefore, ex- tremely popular employs the PNP transistor as its switching element in a so-called NOR circuit. It will be discussed in the following articles.

The basic NOR element is represented by Fig. 13·27. In the sketch

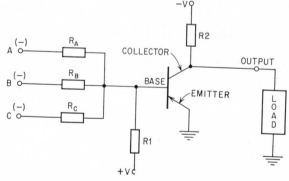

Fig. 13·27   The basic NOR logic element.

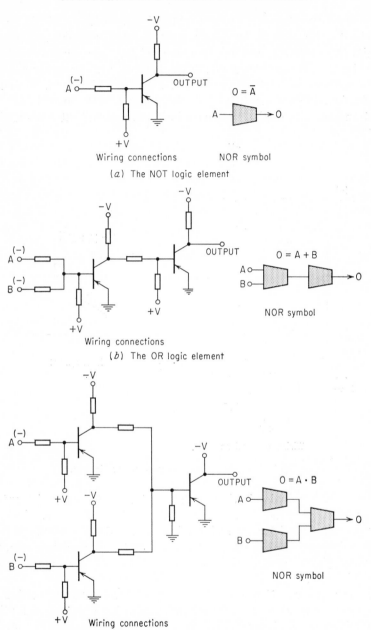

$O = \overline{A}$

Wiring connections   NOR symbol

(*a*) The NOT logic element

$O = A + B$

NOR symbol

Wiring connections

(*b*) The OR logic element

$O = A \cdot B$

NOR symbol

Wiring connections

(*c*) The two-input AND logic element

Fig. 13·28   **Three fundamental NOR types of switching device.**

signal from the output and through the lower input terminal. To turn the element ON, a negative signal is applied to terminal A. This short-circuits transistor $T1$, removes its output signal, and causes a zero feedback signal to be impressed across the base of $T2$. The latter collector-to-emitter short circuit is thus removed, and the element is turned ON. In addition, the output delivers a negative feedback signal to the base of $T1$ and permits the latter to sustain its zero output after the A signal

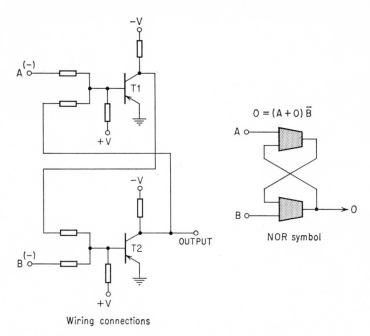

Wiring connections

Fig. 13·29   NOR-type OFF-RETURN-MEMORY element.

is removed; the device is thereby locked in. To turn the element OFF it is merely necessary to apply a negative signal to terminal B; this short-circuits transistor $T2$ and the load resistor. Also, should there be a power interruption, the element will revert to the OFF state and, with the restoration of power, will deliver a signal to the base of $T2$ to keep the device OFF.

    The TIME DELAY *Element—After Energization.* It will be recalled that the ON-DELAY timer produces an output following a definite intentional time delay *after* its input is *energized.* Moreover, when the output does appear after the adjusted time period has elapsed, an N.O. contact will close (represented by the symbol N.O.T.C.). Figure 13·30 illustrates how this is accomplished with a NOR, a transistorized timer, and NOR MEMORY (Fig. 13·29). Note that the timer comprises, in the main, a transistor $T2$,

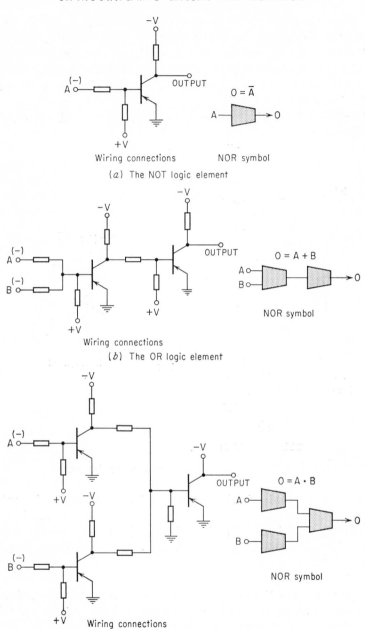

$O = \bar{A}$

(a) The NOT logic element

$O = A + B$

(b) The OR logic element

$O = A \cdot B$

(c) The two-input AND logic element

Fig. 13·28   Three fundamental NOR types of switching device.

signal from the output and through the lower input terminal. To turn the element on, a negative signal is applied to terminal A. This short-circuits transistor $T1$, removes its output signal, and causes a zero feedback signal to be impressed across the base of $T2$. The latter collector-to-emitter short circuit is thus removed, and the element is turned on. In addition, the output delivers a negative feedback signal to the base of $T1$ and permits the latter to sustain its zero output after the A signal

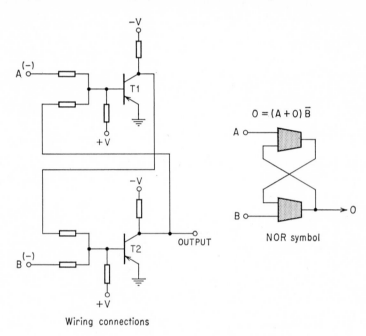

$$0 = (A + 0)\,\bar{B}$$

NOR symbol

Wiring connections

Fig. 13·29   NOR-type OFF-RETURN-MEMORY element.

is removed; the device is thereby locked in. To turn the element off it is merely necessary to apply a negative signal to terminal B; this short-circuits transistor $T2$ and the load resistor. Also, should there be a power interruption, the element will revert to the off state and, with the restoration of power, will deliver a signal to the base of $T2$ to keep the device off.

   The TIME DELAY *Element—After Energization.* It will be recalled that the on-delay timer produces an output following a definite intentional time delay *after* its input is *energized.* Moreover, when the output does appear after the adjusted time period has elapsed, an N.O. contact will close (represented by the symbol N.O.T.C.). Figure 13·30 illustrates how this is accomplished with a NOR, a transistorized timer, and NOR MEMORY (Fig. 13·29). Note that the timer comprises, in the main, a transistor $T2$, negative bias that centers the Y axis between the vertical portions of the transfer characteristic.

When the device is used as an OFF-RETURN MEMORY, a sequencing procedure in the power supply always applies the bias signal *before* pulse power energizes the gate and feedback windings. This action always drives the unit to the OFF state, point d in Fig. 13·25b, regardless of

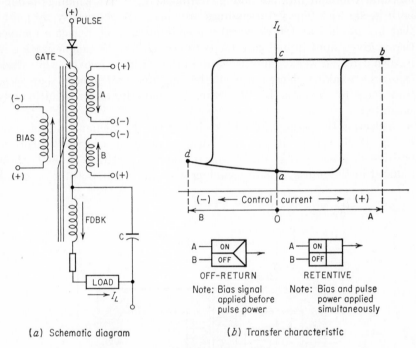

(a) Schematic diagram   (b) Transfer characteristic

Fig. 13·25   Pulse-operated MEMORY switching component.

its previous output condition. Thus, after service is restored following a power failure, the device will always be turned OFF and a new input signal is necessary to drive it ON to point b.

In the RETENTIVE MEMORY, where the device resumes its last output condition after the removal and restoration of power, bias signal and pulse power are not sequenced (as in the OFF-RETURN MEMORY) but are applied at approximately the same time. Under this mode of operation, the output condition, ON or OFF, will be determined by the magnetic state of the core at the time power fails. This means, of course, that, while the device is delivering an output, the operating point will shift from b to c during a service interruption; thus, when power is reestablished, the MEMORY is automatically turned ON because operation is resumed at point c.

The DELAY logic function (Fig. 13·3) provides a preset time delay between the application of an input signal and the delivery of an output. In the ON-DELAY unit, the output comes on after energization, while the OFF-DELAY component removes the output after deenergization. The actions indicated are accomplished in the pulse-operated device by using a two-core construction with a gate winding around each of the outside legs and common mutual-linking, feedback, and bias windings around both inside legs. The gate windings are energized alternately by pulses that are applied at 180-electrical-degree intervals and in such a manner that a given mmf in one gate tends to strengthen the magnetization in its own core and weaken flux strength in the other. Moreover, while these changes are taking place, the strengthening effects are greater than those that tend to demagnetize the core, so that the latter is ultimately saturated.

Neglecting the action of the bias winding for the moment and referring to Fig. 13·26, a pulse in *gate No.* 1 will drive the flux in the *left* core to positive saturation; also, during this period of flux change, an emf will be induced in the mutual-linking winding with the result that the *right* core

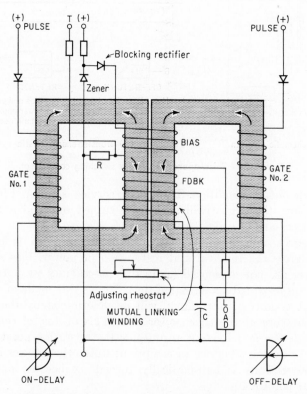

Fig. 13·26 Pulse-operated DELAY switching component.

will be driven to negative saturation. Moreover, if there were no losses, the same flux changes, except for direction, would occur in both cores. After a 180° interval a pulse will energize *gate No.* 2 to drive the *right* core to positive saturation and *reset* the *left* core. Since each successive pair of pulses repeats the magnetizing effects of the preceding pair of pulses, it should be clear that each pulse drives its own core to positive saturation and resets the other core. However, if a resistance is connected into the *mutual-linking* winding circuit where a voltage drop will occur, the mmf that resets the core will be reduced and partial core resetting will result. The core, therefore, reaches saturation by increments, at which time the element turns ON.

In order to turn the device OFF and hold it OFF when there is no ON input, a *bias* winding is placed around the center legs and is energized by a *permanent* input. The bias then ensures total core reset in the absence of an ON input. To remove the bias, a control input is applied to terminal T; this produces a voltage drop across resistor R and forces the bias current to pass through the Zener diode, thus permitting the component to turn ON.

Transistor-type switching components. One of the more recent semi-conductor developments that possesses ideal properties for static-switching logic systems is the *transistor*. Used primarily in common emitter circuits with the simple addition of resistors, it is capable of providing both voltage and current gains that will drive several similar components without supplementary amplification. Like the magnetic-amplifier types of component previously described, transistors are designed to furnish the basic logic functions NOT, OR, AND, MEMORY, and DELAY. A number of circuit configurations are presently in use to accomplish this, but the one that appears to offer important advantages and is, therefore, extremely popular employs the PNP transistor as its switching element in a so-called NOR circuit. It will be discussed in the following articles.

The basic NOR element is represented by Fig. 13·27. In the sketch

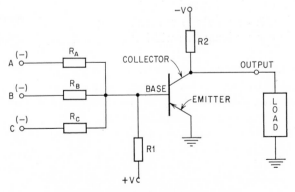

Fig. 13·27   The basic NOR logic element.

shown, the base is connected to three input terminals A, B, and C through resistors R_A, R_B, and R_C and to a positive (reverse) bias through the resistor $R1$ to prevent collector-base leakage current and avoid thermal runaway, i.e., destructive temperature rise. When no *negative* input signal is present at A NOR at B NOR at C, there is no emitter-base current and the transistor is, in effect, an open circuit between collector and emitter; under this condition, resistor $R2$ and the load are in series across line potential $-V$ and ground, and a voltage will appear across a load (output) resistance. However, if a negative input is applied to A or B or C (or any combination of them), the base becomes negative with respect to the emitter and current *does* pass from emitter to base through the proper resistor (or resistors); this action thus grounds the collector with the result that the load resistor is short-circuited and there is no output. It should be pointed out that the ohmic value of the biasing resistor $R1$ is made rather high with respect to input resistors R_A, R_B, and R_C to minimize its effect on the input current and that the line resistance $R2$ is low enough to give an output that is almost equal to the supply voltage.

The NOT, OR, *and* AND *Element.* Since the fundamental NOR circuit of Fig. 13·27 delivers an output when there is no input signal at A NOR at B NOR at C, it should be clear that it would become a simple NOT logic element, Fig. 13·28a, if only one input were used. Also, as Fig. 13·28b shows, an OR element can be formed by cascading a two-input unit with one that has a single input. The two-input AND, on the other hand, requires three NORs as illustrated by Fig. 13·28c. Note particularly in the case of the latter two elements, i.e., the OR and the AND, that the output (final) NOR merely performs an *inverter* or NOT function and must be used to give an output when signals are present at the input terminals. This aspect of the NOR system of static switching, therefore, adds to the number of required units as compared with the so-called English method of control, but it does have the advantage of simplicity and low cost. However, in complex control circuits where many components are required, it is often possible to reduce the number of elements by eliminating double conversions; this will be illustrated subsequently by practical examples.

The OFF-RETURN MEMORY *Element.* This logic function is shown in Fig. 13·29 where two NORs are interconnected (1) to give an output when there is an input signal A and (2) to remove an output when a B signal is applied or if a power interruption occurs. With neither of the two input terminals A or B energized, transistor $T1$ will deliver an output which will be impressed on the base of transistor $T2$; the latter will therefore be short-circuited (between emitter and collector), and no output will be produced. Also, the base of $T1$ will receive a zero feedback a Zener diode, a neon tube (that breaks down at about 73 volts), and an *RC* circuit whose time constant can be adjusted with the rheostat. A discussion of the operation follows.

With switch *S open*, transistor *T*1 delivers an output signal to point *m* and in turn to the lower NOR element of the MEMORY. The latter is thus switched to the OFF state because it furnishes no signal to output terminal *X*. Since output signal *m* is also applied to the base of transistor *T*2, that component is short-circuited (between emitter and collector)

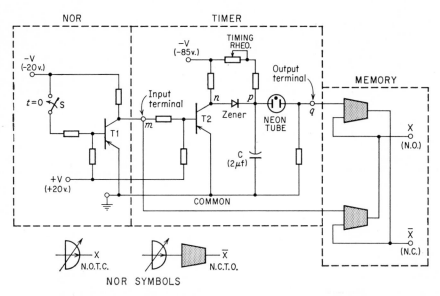

Fig. 13·30 NOR-type TIME DELAY element after energization (ON-DELAY).

and connects points *n* and *p* to ground potential. The timer therefore produces no output at point *q*, and the upper NOR element of the MEMORY, now switched ON, furnishes a signal to the lower NOR to keep it locked OFF.

To start the timing operation the signal to the input terminal *m* of the timer must be removed by closing switch *S*. This causes transistor *T*1 to conduct (between emitter and collector) and removes the signal to the base of *T*2. Since no signal now appears at point *p*, the capacitor charges from the 85-volt source through the fixed resistor and the timing rheostat. Then, after the capacitor emf builds up to the breakdown potential of the neon tube (about 73 volts), an output signal is delivered to the upper NOR of the MEMORY, i.e., terminal *q*. With no output furnished by the upper NOR to the lower NOR, the MEMORY is switched ON at terminal *X*. During this switching process the $\bar{X}$ output changes from *X* (open) to $\bar{X}$ (closed).

The TIME DELAY *Element—After Deenergization.* In this OFF-DELAY timer, Fig. 13·31, an output is removed following a definite intentional

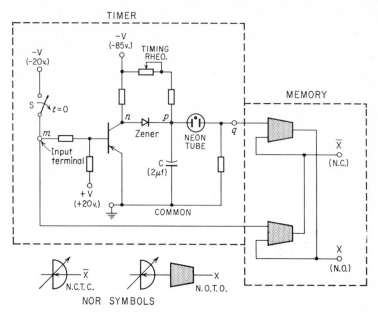

Fig. 13·31 NOR-type TIME DELAY element after deenergization (OFF-DELAY).

time delay *after* its input is *deenergized.* Moreover, when the output is disconnected, after the adjusted time period has elapsed, an N.O. contact, closed during energization, will open (designated by the symbol N.O.T.O.). Referring to the diagram, note that the device is made up of the same components as in Fig. 13·30 without the input NOR element. A discussion of its operation follows.

Before the switch S is *opened* at $t = 0$, a signal is delivered to the base of the transistor, and with that component short-circuited (between emitter and collector), points n and p are connected to ground potential. This means that no output exists at terminal q of the timer and at the input terminal of the upper NOR. The latter element is, therefore, switched ON and provides an output to terminal X. A negative signal is also applied to the lower NOR (from terminal m) with the result that its zero output is fed into the upper NOR to lock that element ON.

To start the timing operation the signal to the input terminal m of the timer must be removed by opening switch S. This deenergizes the circuit, switches the transistor to the OFF state, and causes the capacitor to charge through the fixed resistor and the timing rheostat from a source of -85 volts. Then, after the capacitor emf builds up to the breakdown potential of the neon tube (about 73 volts), an output signal is delivered to the upper NOR. Since the output of the latter element is now zero, terminal X is deenergized. Also, with no signal furnished to the lower NOR, the output signal is delivered to the upper NOR to lock it ON. During the switching process the $\bar{X}$ output changes from $\bar{X}$ (closed) to X (open).

The transistor relay. Since the output of a transistorized switching device is comparatively low, about 5 to 15 ma, it is customary, as previously indicated, to raise the current level by connecting an amplifier into the circuit between the logic system and the operating unit. This practice can, however, be modified in small power applications by employing a rather simple transistor-type relay which, with other standard inexpensive components, functions directly from the output. A diagram illustrating how this can be accomplished is given in Fig. 13·32.

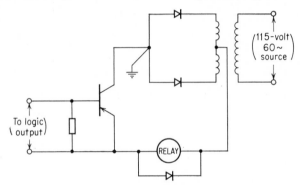

Fig. 13·32   The transistor relay.

The "package," as shown, consists of a relay (about 12 volts) in combination with a power transistor, a small transformer, several diodes, and a fixed resistor. With no signal delivered by the logic device, a small pulsating leakage current passes through the transistor (from emitter to base to collector) and the relay coil; the latter therefore remains dropped out. When the transistor is switched ON by the logic (NOR) element, full-wave output is delivered by the transformer and its diodes; the relay then picks up as a relatively large current (about 300 ma) energizes the operating coil.

NOR circuit applications. Several examples will now be given and discussed to illustrate how NOR elements are employed in practical installations.

1. The first application concerns the circuit for a three-speed motor, for which a conventional relay diagram and a standard static-switching

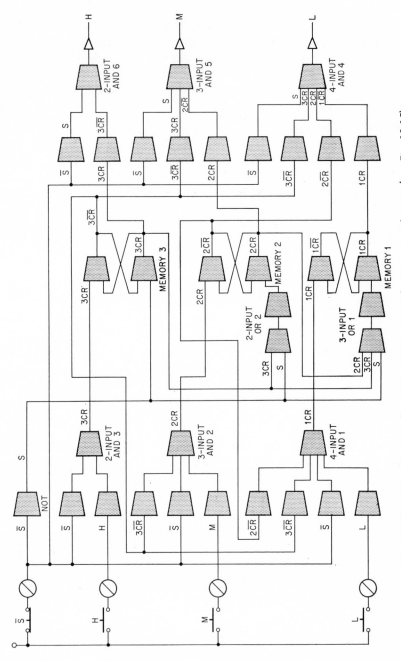

Fig. 13·33 NOR equivalent of a static-switching circuit for three-speed motor (see Fig. 13·15).

equivalent were presented, respectively, in Figs. 13·14 and 13·15. This will be done in two steps (a) by first replacing each of the standard logic elements (NOT, ANDS, ORS, and MEMORYS) of Fig. 13·15 with an equivalent NOR component as in Fig. 13·33 and then (b) by simplifying the converted circuit to eliminate several NORS that perform double-conversion functions only and obtaining certain feedback signals by making direct connections to available terminals. To understand how this was accomplished, it will be desirable to compare Figs. 13·15 and 13·33 carefully and note that each of the conventional symbols in the first diagram is replaced by an equivalent NOR symbol in the second. The changes indicated can be followed readily, since the six ANDS, the two ORS, and the three MEMORYS are similarly numbered in both diagrams. Another extremely important point that must be recognized is that the inputs to NORS 4, 5, and 6 in Fig. 13·34 are the *converse* of the inputs to ANDS 4, 5, and 6 in Fig.

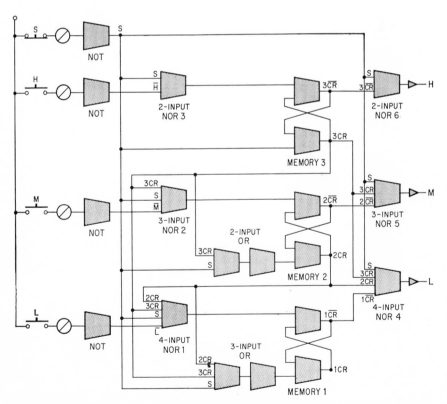

Fig. 13·34 Simplified NOR equivalent of a static-switching circuit for a three-speed motor (see Fig. 13·33).

13·15; this difference is fundamental to all conventional- and NOR-logic switching systems because a signal conversion does *not* take place in the first set of components and *does* result in the second set. For example, AND 4 in Fig. 13·15 yields an L output with inputs $\overline{S}$, $3\overline{CR}$, $2\overline{CR}$, and $1CR$, while NOR 4, its equivalent, produces an L output with inputs S, $3CR$, $2CR$, and $1\overline{CR}$. It is also significant that ANDs 1, 2, and 3 in Fig. 13·33 are replaced by the simpler NOT-NOR combinations in Fig. 13·34. Thus, the NOT-NOR 1 pair of elements in Fig. 13·34 performs the same function as the five-unit AND 1 of Fig. 13·33.

2. The second example is represented by the solenoid application of Fig. 13·5, which is shown converted to an equivalent static-switching system in Fig. 13·35. In this control circuit, it will be recalled, only one solenoid valve will be operated at a time because $LS1$ is open when

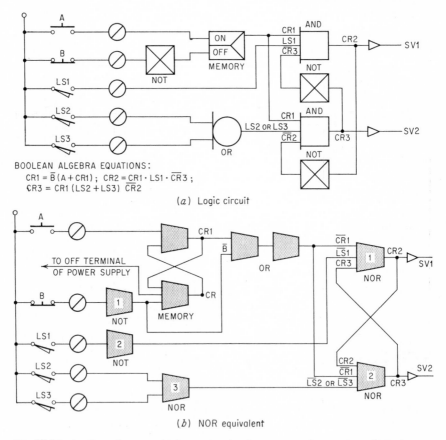

BOOLEAN ALGEBRA EQUATIONS:
CR1 = $\overline{B}$(A + CR1) ; CR2 = $\underline{CR1}$ · LS1 · $\overline{CR}$3 ;
CR3 = CR1 (LS2 + LS3) $\overline{CR}$2

(*a*) Logic circuit

(*b*) NOR equivalent

Fig. 13·35 Logic and NOR equivalent static-switching circuits for a solenoid application (see Fig. 13·5).

$LS2$ *or* $LS3$ are closed and $LS1$ is closed when $LS2$ and $LS3$ are open (see the relay diagram, Fig. 13·5a). With the power supply disconnected, all NORS are deenergized and the system is, of course, in an OFF state.

When the main switch is closed, an N.C. relay contact in a special power-supply assembly remains closed until the emf, as it builds up, reaches about 22 volts. Up to this potential and while the contact is still closed, a pulse is delivered to the lower unit of the MEMORY. Since the zero output of the latter is connected to the input of the upper MEMORY unit, a signal appears at the output terminal ($\overline{CR}1$) of this element. Then, when the relay contact opens (above 22 volts), the output of the lower MEMORY unit is still zero and the same conditions prevail. Following through from terminal $\overline{CR}1$ it will thus be found that NOR 1 does *not* produce an output $CR2$ to actuate solenoid valve $SV1$.

To get $SV1$ to operate, NOR 1 must receive three zero-input signals, and these conditions are fulfilled when button A is pressed and $LS1$ closes, assuming that $LS2$ and $LS3$ remain open. The pressing of the A button and the closing of $LS1$ establish zero inputs at the upper two terminals of NOR 1, and the zero output of NOR 2, i.e., $CR3$, gives the third zero input to NOR 1.

To turn off $SV1$ it is merely necessary to press the B button. This causes NOT 1 to deliver a signal to the lower MEMORY unit which, in turn, through a feedback, produces one input signal at the OR and another that is fed from the output of NOT 1.

Assuming next that $LS1$ opens and $LS2$ closes, the pressing of the A button will initiate a series of actions, similar to those described for the operation of $SV1$, that will energize $SV2$.

The OR unit shown is incorporated to bypass the MEMORY should the operator inadvertently press buttons A and B simultaneously. Under this condition, a signal will be delivered by the OR unit to the $\overline{CR}1$ terminal with the result that NORS 1 and 2 will have no outputs. Without such an OR unit, the pressing of buttons A and B would mean that the closed contact at the A button would override the open contact at the B button and energize the $SV1$ solenoid valve; this is contrary to the function performed when the two buttons are pressed in the relay circuit.

3. Machine tools that perform a repetitive series of operations once they are started are excellent candidates for static-switching systems. The latter generally control reversing motors whose speeds may or may not be changed during each operating cycle. Good examples of such applications are planing, milling, and slotting machines, gear shapers, key setters, planer-millers, and many others. The control circuit for one of these, a reciprocating planer, will now be discussed.

In this application, a line-start reversing motor moves a table in one direction, FORWARD, as a metal cutter machines the workpiece. After the table reaches its position limit, the motor is stopped momentarily to permit an air jet to remove metal chips; the motor then reverses and moves the table back to the initial position where another motor reversal takes place for the next cutting operation. This reciprocating motion continues automatically until the STOP button is pressed.

However, before the operations described are permitted to take place, a number of conditions must be fulfilled and several protective devices must be set to indicate that the equipment will function properly. These are: (a) a lubrication pump must provide a suitable, above-minimum pressure, and a coolant motor must be in operation; (b) FORWARD and REVERSE buttons must be included to permit the operator to start the table, while at rest, in a desired direction; (c) a JOG button must be provided so that the operator may have the option of running the table at his discretion for testing and observation purposes; (d) emergency limit switches must be incorporated so that the machine will be stopped automatically at either end should the table move beyond the safe limits of travel; (e) overload protection must be included.

A NOR static-switching circuit that will satisfy the requirements listed is given in Fig. 13·36. Assuming that the *coolant pump* (CP) and *lubrication pressure* (LP) contacts are closed and that overload contacts are closed, signals are delivered to the lower two-input elements of AND 3, i.e., points a and b. Then, after the START button is pressed, a third signal is impressed on the upper unit of AND 3, point c, and permits the latter component to have an output which, in turn, energizes the lower two elements of ANDS 1 and 2 at terminals d and e. The foregoing preliminary actions thus have the effect of instructing the upper portion of the control system that normal operation may begin. Assuming that the table is at the far left, the *left limit switch LLS* will be closed, a signal will be delivered to the upper element of AND 1, point f, and the forward relay F will be actuated. A cutting operation will then be performed as the table moves *forward*, i.e., to the *right*. Note also, that a zero signal exists simultaneously at point g so that AND 2 is automatically turned OFF.

When the table reaches its *right* limit of travel, the *right limit switch RLS* is actuated. The upper MEMORY, therefore, deenergizes the FORWARD relay F, and after a time delay which permits the air jet to blow away metal chips, the REVERSE relay is actuated; the table thus returns to its original position where *LLS* will again close momentarily to energize the F relay for a second cutting operation. The reciprocating action described continues until the STOP is pressed.

Should the table be at rest at some intermediate position between the two limits of travel, the operator can start the motion in either direction. This can be done by pressing the FOR. button for motion to the *right* or the REV. button for motion to the *left;* normal operation will then continue as described above.

Again with the table at rest, a jogging operation can be performed by first pressing the FOR. (or REV.) button and then depressing the JOG button for short intervals or as long as motion is desired.

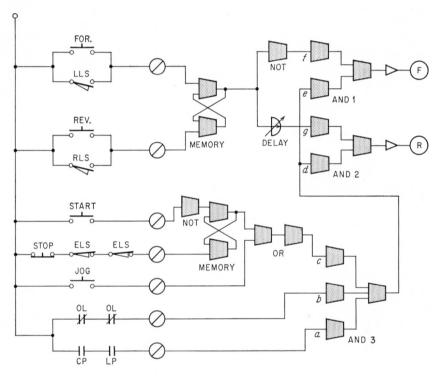

Fig. 13·36   NOR static-switching circuit for an automatically reversing planing machine.

Finally, with the machine operating normally, an automatic stop will take place if an overload trips or should contacts *CP, LP,* or *ELS* open. The opening of any one of these contacts has the same effect as the pressing of the STOP button, i.e., it removes the output from AND 3 and therefore the signals to terminals *d* and *e* of ANDS 1 and 2.

QUESTIONS AND PROBLEMS

1. Name five basic types of static-switching device and indicate what each one is equivalent to in electromagnetic systems of control.

2. Distinguish between OFF-RETURN and RETENTIVE MEMORY logic functions. How does each correspond to the familiar relay-type circuits?

3. How do ON-DELAY and OFF-DELAY logic devices differ from each other?

4. Why are signal converters and amplifiers needed in static-switching circuits?

5. What important difference exists between relays and static-switching devices with regard to the number of inputs and outputs?

6. After describing the operation of the relay control circuit of Fig. 13·5a for the operation of two solenoid valves, explain carefully how the static-switching circuit of Fig. 13·5b performs exactly the same operating functions.

7. Modify circuit diagram Fig. 13·5 so that a third solenoid valve $SV3$ will be energized when $SV1$ or $SV2$ is operated.

8. Modify circuit diagram Fig. 13·6 so that a second START button C bypasses START button A and limit switches $LS1$ and $LS2$. Thus, when START button C is pressed, the motor will start even though $LS1$ or $LS2$ are open.

9. Three relays A, B, and C have N.O. and N.C. contacts arranged in four parallel paths given by the Boolean equation

$$ABC + \bar{A}\bar{B}\bar{C} + AB\bar{C} + A\bar{B}\bar{C}$$

Show that this equation can be reduced to the equivalent simplified form

$$AB + \bar{B}\bar{C}$$

Draw diagrams to illustrate both equations.

10. Draw the circuit diagram represented by the Boolean equation

$$A\bar{D}(CE + \bar{B}\bar{E}) + A\bar{B} + C[B + A(B + \bar{D}\bar{E})]$$

and show that this equation can be simplified to

$$A\bar{B} + BC$$

11. The Boolean equation $(A + B)(B + C)(C + A)$ represents a *series-parallel* combination of contacts that connects in series three paired sets of parallel-connected contacts A, B, and C. Show that this circuit is equivalent to a parallel-series arrangement of contacts represented by the equation

$$(AB + BC + CA)$$

12. Design a static-switching control circuit for the two-speed motor illustrated by Fig. 7·20 following essentially the procedure outlined on pages 235 to 238 for the diagrams shown in Figs. 13·14 and 13·15. To do this, first write the Boolean equations for (a) the pickup of relays $1CR$, $2CR$, and $3CR$; (b) the energization of contactors HF, LF, and HR; (c) the energization of relays $1CR$, $2CR$, and $3CR$; and (d) the dropout of relays $1CR$, $2CR$, and $3CR$.

13. Applying the same technique as indicated in Prob. 12, design a static-switching circuit for the two-speed reversing motor illustrated by the conventional relay-contactor diagram of Fig. 7·21.

14. What are the basic parts of a magnetic amplifier that is used as a static switch? Indicate by transfer characteristics how such a static switch can be made to exhibit NOT, OR, and AND properties.

15. Explain the action of a magnetic amplifier for different degrees of feedback.

16. How can a magnetic amplifier be made to operate as a static-switching MEMORY function?

17. Referring to Fig. 13·21, explain the behavior of the basic pulse-operated static-switching magnetic amplifier.

18. How is the pulse-operated magnetic amplifier made to act like (a) an OR component? (b) an AND component? (c) a NOT component? (d) a MEMORY component?

19. In the pulse-operated MEMORY switching component illustrated by Fig. 13·25, explain why the time of application of the bias signal determines whether the component will display OFF-RETURN or RETENTIVE MEMORY characteristics.

20. Explain the action of the pulse-operated DELAY switching component illustrated by Fig. 13·26.

21. Referring to the basic NOR logic element illustrated by Fig. 13·27: (a) Why is there an output when *no* input signals are applied to A, B, or C? (b) Why is no output present when negative signals *are* applied to A or B or C? (c) Why must the biasing resistance $R1$ be high with respect to the input resistances R_A, R_B, and R_C?

22. Show that one or more basic NOR static-switching elements can be made to function as a NOT, an OR, or an AND. Refer to Fig. 13·28.

23. Explain the operation of the NOR OFF-RETURN MEMORY element (Fig. 13·29) (a) with zero signals at A and B, (b) for *turn-on* with a signal at A, (c) for *turn-off* with a signal at B, (d) during a power interruption while the element is ON.

24. Compare the NOR equivalent static-switching circuit of Fig. 13·33 with the logic circuit of Fig. 13·15 noting particularly their similarities and differences. Also explain how the simplified circuit of Fig. 13·34 is obtained from the first design of Fig. 13·33.

25. Carefully describe the operation of the NOR static switching circuit of Fig. 13·36 designed for an automatically controlled reversing planing machine.

Answers to Numerical Problems

Chapter 2
3. $R_{1-2} = 1.56$ ohms; $R_{2-3} = 1.69$ ohms
4. 575 amp
9. (a) 33 amp, (b) 58 amp, (c) 42 amp
17. 0.166 ohms, PF $= 0.91$
18. $X = 0.0693$ ohm; $L = 0.184$ mh; PF $= 0.247$
19. (a) $X = 0.144$ ohm; (b) $I_{\text{inrush}} = 5.13 I_{FL}$
20. (a) $I_M = 199$ amp, $I_L = 157$ amp; (b) $I_M = 318$ amp, $I_L = 282$ amp
21. (a) $T_{50\%} = 163.5$ lb-ft; (b) $T_{80\%} = 417$ lb-ft

Chapter 3
20. 0.95 ohm,
22. $R_1 = 0.56$ ohm; $R_2 = 0.55$ ohm
26. $R_P = 0.74$ ohm
27. $I_P = 221$ amp
28. 8.35 lb

Chapter 4
7. (a) 1,685 rpm, (b) 1,585 rpm, (c) 425 rpm
13. (a) $e_{\max} = 2,050$ volts; $P = 133$ watts
    (b) $R_d = 295$ ohms; $P = 179$ watts
14. (a) 2.53 amp, (b) 5 sec, (c) 1.47 amp, (d) 1 sec, (e) 460 volts
15. (a) $P_f = 1,470$ watts; (b) $P_f = 258$ watts; $P_r = 357$ watts

Chapter 6
7. $R_e = 3.12$ ohms; $R_e = 2.03$ ohms
10. I_{inrush} (at 440 volts) $= 484$ amp; T_{ST} (at 440 volts) $= 1,047$ lb-ft
    (a) I (at 65% E) $= 314$ amp; T (at 65% E) $= 442$ lb-ft
    (b) $E = 356$ volts; I_{inrush} (at 356 volts) $= 392$ amp
27. $I_{\text{inrush}} = 37.8$ amp; $T_{ST} = 521$ lb-ft

34.

rpm	260	480	720	1,000
f	47	36	24	10
E	212	162	108	45

41. (a) 30,240, (b) 30,240, (c) 15,120, (d) 20,160

Chapter 7
24. 6 poles, 10 poles; opposite; 40 cps

Chapter 10
1. (a) 5°, (b) 10°, (c) 350°, (d) 360°
8. 25 hp, 40 hp, 60 hp, 75 hp
9. 26 lb per sq in.
10. (a) 1.59 sec, (b) 7.62 revolutions
11. 0.88 sec

Index